AF333208

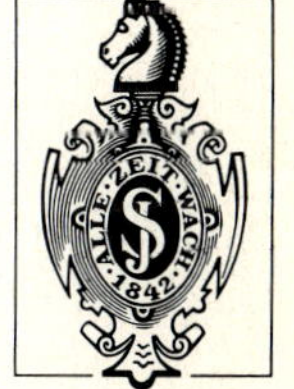

Springer Series in Synergetics   Editor: Hermann Haken

Synergetics, an interdisciplinary field of research, is concerned with the cooperation of individual parts of a system that produces macroscopic spatial, temporal or functional structures. It deals with deterministic as well as stochastic processes.

Dynamical Problems in Soliton Systems

Proceedings of the
Seventh Kyoto Summer Institute, Kyoto, Japan
August 27–31, 1984

Editor: S. Takeno

With 102 Figures

Springer-Verlag Berlin Heidelberg New York Tokyo

Professor Dr. Shozo Takeno
Department of Physics, Faculty of Engineering and Design,
Kyoto Institute of Technology, Matsugasaki,
Kyoto 606, Japan

Series Editor:

Professor Dr. Dr. h. c. Hermann Haken
Institut für Theoretische Physik der Universität Stuttgart, Pfaffenwaldring 57/IV,
D-7000 Stuttgart 80, Fed. Rep. of Germany

ISBN 3-540-15372-1 Springer-Verlag Berlin Heidelberg New York Tokyo
ISBN 0-387-15372-1 Springer-Verlag New York Heidelberg Berlin Tokyo

Preface

This volume contains most of the papers presented in the oral session of the 7th Kyoto Summer Institute (KSI) on *Dynamical Problems in Soliton Systems*, held in Kyoto from August 27 to 31, 1984. Furthermore, it contains contributions of R.K. Bullough, H.H. Chen, A.S. Davydov, and N. Sanchez, who unfortunately could not attend. Thirty-six papers were presented in the oral session and 17 papers in the poster session. The meeting brought together 109 physicists and mathematicians, of which 22 were from abroad (see group photograph).

The KSI is an international meeting organized by the Research Institute for Fundamental Physics (RIFP), Kyoto University to discuss various current problems of fundamental importance in theoretical physics. The 7th KSI was the first international meeting on solitons in Japan. Early in 1983, it was felt in the RIFP that the time was ripe for a conference dealing with problems concerning solitons. The RIFP asked us to organize the conference. The Organizing Committee consisted of:

R. Hirota (Hiroshima)	*T. Taniuti* (Nagoya)
Y.H. Ichikawa (Nagoya)	*M. Toda* (Tokyo)
Z. Maki (Kyoto)	*M. Wadati* (Tokyo)
S. Takeno (Kyoto)	*N. Yajima* (Fukuoka)

Since its discovery, the study of the soliton as a stable particle-like state of nonlinear systems has caught the imagination of physicists and mathematicians. Elaborate numerical experiments and satisfactory results of rigorous soliton theory developed during the last two decades have further motivated physicists' appreciation, leading to a dramatic rise in research activity related to the soliton concept and its application. As everywhere in science, two main tendencies are visible here, namely, the achievement of more and more detailed and sophisticated results on the one hand, and the development of new unifying ideas on the other hand. The 7th KSI sought the latter as its main goal with the observation that nonlinear ideas of solitons have been a unifying influence on various fields in physics ranging from fluid dynamics, plasma physics and solid state physics to statistical mechanics, field theory, general relativity and biophysics. We have even noticed that several other areas in natural sciences such as biochemistry, geophysics, materials science, etc., have become influenced by solitons. The soliton problem can now genuinely claim to be one of the few interdisciplinary subjects of modern physics.

In many soliton meetings, efforts were made to achieve a balance of physicists and mathematicians so that each would benefit from the other's expertise and outlook. The 7th KSI was intended to be broadly based with emphasis placed on interdisciplinary aspects of the soliton problem. All of the investigative methods were presented there, theoretical, computational and experimental. The conference subject was divided into six parts:

1) Mathematical Theory of Solitons such as Inverse Scattering Methods, the Hirota Theory, and the Painlevé Analysis
2) Field Theory and Statistical Mechanics of Solitons
3) Solitons in Hydrodynamics and Plasma Physics
4) Solitons in Condensed Matter Physics
5) Solitons in Biological Systems
6) Solitons and Chaos

Like the city of Kyoto, where the old and the new in Japan coexist and the new is even affected by the old, a harmony was noted in the 7th KSI between the classical aspects of the soliton theory and its novel features, such as solitons in biological molecules and coherent structures and pattern selection in perturbed or near-soliton systems.

The present proceedings aim at giving a coherent survey of current activity in the various fields of soliton physics. I hope that the reader of this volume will feel the same enthusiasm the participants of this meeting felt when listening to the talks, looking at the posters and finding common features of solitons treated in different areas. This is the first volume on solitons in the Springer Series in Synergetics, which has endeavored to attain new unifying ideas in diverse areas in natural science. I also hope that this survey of the soliton problem will serve to highlight some of the more pressing and universal problems in nonlinear physics and also be accepted as a tribute to synergetics.

I could not include the works presented in the poster session in this volume due to the limitation of the overall size of it. I would like to thank all contributors for their kind cooperation. Various forms of interesting posters and nearly endless discussions there made this meeting even more successful.

I am grateful to our colleagues who have contributed to this volume for their skill, knowledge and continuing patience.

This meeting on solitons was made possible by generous support from various science foundations in Japan. I would like to take this opportunity to thank the Japan Society for the Promotion of Science, the Nishina Memorial Foundation, the Yamada Science Foundation, the Shimazu Science Foundation, the Yoshida Foundation for Science and Technology, and the Yukawa Foundation for their financial support. My thanks also go to the Faculty of Science, Kyoto University for providing us with a fine lecture room for the meeting. I am also grateful to Dr. H. Lotsch of Springer-Verlag for his

kind advice in publishing the proceedings of the 7th KSI in the Springer
Series in Synergetics. Last but not least, I wish to express my thanks to the
secretaries in the RIFP, Dr. M. Toya, Mrs. H. Hioki, Mrs. K. Honda, Mrs.
S. Matsumoto, Miss T. Sumide, and Miss K. Suzuki for their invaluable
assistance in organizing the 7th KSI and editing these proceedings.

On behalf of the Organizing Committee   *S. Takeno*
Kyoto, October 1984

Opening Address

Ladies and gentlemen,

It is a great pleasure for me to present the opening address at this 7th Kyoto Summer Institute on *Dynamical Problems in Soliton Systems*. On behalf of the Research Institute for Fundamental Physics, Kyoto University and the Organizing Committee, I am delighted to welcome all of you to this international meeting. I would like to thank all participants, especially those who have traveled long distances to contribute to the meeting. I wish everyone a pleasant stay in Kyoto.

The Kyoto Summer Institute, called KSI for short, is one of the most important activities of the Institute in recent years. The KSI, which is something between a conference and a summer school, is a small-size international meeting dedicated to various problems of fundamental importance in theoretical physics, and has been held every year since 1978. The topics of the KSI in the past have covered such fields as: (1) Particle Physics (1978), (2) Physics of Low Dimensional Systems (1979), (3) Fundamental Physics of Amorphous Semiconductors (1980), (4) Grand Unified Theories and Related Topics (1981), (5) Microscopic Theories of Nuclear Collective Motions (1982), (6) Chaos and Statistical Mechanics (1983).

This year the soliton problem was taken up. I believe that the selection of this subject is timely and particularly suited to the spirit of the KSI. As you know, the soliton has been a unifying influence on the natural sciences in recent years, and its universal appreciation in the physics community is very much a current development. Scholars have noticed the existence of common mathematical structures in various fields in physics, ranging from fluid dynamics, mathematical physics, plasma physics to solid state physics, statistical physics, field theory, general relativity, and biophysics. Due to its richness of structures and ubiquity in applications, the soliton has become one of the corner-stones of modern physical science. In view of current developments in soliton problems, the 7th KSI aims at being an interdisciplinary meeting of researchers interested in solitons in various fields in physics.

I would like to mention something more about the soliton and our Institute which is the host of the KSI. Our Institute has its own annual research project supported by physicists in our country. More than twenty years ago, "Lattice Dynamics and Related Problems" were taken up as one of the projects, and it lasted about ten years. In this project most people were interested in exact tractable problems in solid state physics, and they often

treated models encountered in lattice dynamics. It is not too much to say
that several outstanding contributions by Japanese physicists to the soliton
problem have their own roots in this research project.

I hope this meeting will lead to further development of soliton physics
through lively discussion among scientists from different parts of the world
and from different disciplines.

Before closing my address, I should like to thank the Japan Society for
the Promotion of Science, the Nishina Memorial Foundation, the Yamada
Science Foundation, the Shimazu Science Foundation, the Yoshida Foundation for Science and Technology, and the Yukawa Foundation for their
financial support, and the Faculty of Science of Kyoto University for kindly
providing us with a lecture room. Thank you.

Ziro Maki
Director of the Research Institute
for Fundamental Physics

Contents

Part III Solitons in Plasma Physics and Hydrodynamics

Part IV Solitons in Condensed Matter Physics

Part V Solitons in Biological Systems

Part VI Solitons and Chaos

Introduction

Miki Wadati

Institute of Physics, College of Arts and Sciences, University of Tokyo, Komaba, Tokyo 153, Japan

It is now a well-known story that John Scott-Russell on his horse found and chased a solitary wave propagating in the canal near Edinburgh, Scotland. It was some evening in August, 1834. In a sense, the 7th Kyoto Summer Institute on "Dynamical Problems in Soliton Systems" is the 150-year anniversary of the first observation of a soliton.

In 1965, using high-speed computers, Zabusky and Kruskal studied the Korteweg-de Vries equation as a model for the Fermi-Pasta-Ulam problem. They reconfirmed the recurrence phenomena which Fermi, Pasta and Ulam had found in the numerical analysis of a one-dimensional nonlinear chain ten years before. At the same time, Zabusky and Kruskal discovered the novel and surprising nature of the solitary waves in the Korteweg-de Vries equation. Its solitary waves behave like particles. They run, collide and interact nonlinearly, but preserve their identities against mutual collisions. The solitary wave with particle properties was named "soliton" by Zabusky and Kruskal.

Solitons are essentially nonlinear phenomena. They are the elementary excitations in nonlinear dispersive systems. Physicists overlooked them for a long time. After the discovery by Zabusky and Kruskal, a remarkable property of solitons has aroused a surge of interest in physics society. Nowadays, the soliton concept is successfully applied to almost every field of physics; hydrodynamics, plasma physics, nonlinear optics, low temperature physics, solid state physics, elementary particle physics, astrophysics, biophysics, etc. We may say that soliton physics has started.

Japan is one of the most active countries in the study of solitons. The 7th Kyoto Summer Institute was the first international conference on soliton physics held in Japan. During the 5-day meeting, we reviewed the present status and discussed the future problems in soliton physics.

Developments in the mathematical physics of nonlinear evolution equations made a great contribution to the establishment of the soliton concept. Analytical methods, such as the inverse scattering method, Bäcklund transformations and Hirota's direct method, enable us to solve nonlinear evolution equations exactly. It should be emphasized that the solvable models, such as the sine-Gordon equation and the Toda lattice, played an important role in the development of the soliton theory. Also it is to be noted that methods, such as the averaging method and the reductive perturbation method, were invented in good time to present canonical soliton equations. The discoveries of new analytical methods and new solvable models have given not only a firm foundation to the soliton theory but have also made a big impact on mathematicians. Many works are being done to find criteria for solvability. Mathematical physics of solitons will continue to be very exciting.

In 1811, the Fourier series was introduced to treat the diffusion equation. For over 150 years, the superposition principle and the Fourier series even governed our way of thinking. The inverse scattering method is an extension of the Fourier transform when we regard scattering data as the generalization of momentum space. It has been a unique method whereby the initial value problem of nonlinear evolution equations can be solved. Furthermore, by using the inverse scattering method, we can prove that the soliton system is a completely integrable Hamiltonian system. Soliton perturbation theory based on the inverse scattering method is extremely important. In practical problems the system may not be completely integrable due to dissipation and/or an external force. Therefore, perturbation theory is useful and has wide applications. More recently, the inverse scattering method has been extended to quantum theory. We call it the quantum inverse scattering method. The quantum inverse scattering method is a powerful method for analyzing quantum completely integrable systems. Moreover, through its relations to the Bethe ansatz method and the transfer integral method, it provides us with a unified viewpoint on solvable models in field theory and statistical mechanics.

Physics in low-dimensional materials has made remarkable progress. We now have many experimental results done with reliable samples and much theoretical work based on analytically tractable models. In particular, the soliton picture has been applied to one-dimensional organic materials, for instance, polyacetylene. In plasma physics and hydrodynamics various soliton phenomena have been predicted and observed. Even if systems are complicated, a soliton there is a macroscopic object. We can create, control and detect solitons. On the other hand, condensed matter physics deals with a soliton which is in most cases microscopic. Therefore, the identification of a soliton mode sometimes becomes controversial. A statistical description of solitons is often necessary. This remark also applies to the solitons in biological systems. The soliton concept in biophysics is fascinating. The dynamics of Davydov solitons and solitons associated with open states in the DNA molecule have been studied theoretically. To examine these theories, it is desirable to have experimental techniques which select the soliton mode from others. The development of soliton spectroscopy will give a firmer basis to soliton physics. We emphasize here that the motion of magnetic fluxes in Josephson junctions has been observed. The agreement between the experiments and the theories is very good.

We have some examples where the soliton system has more than two independent variables. It is a challenging problem to ask whether in higher dimensional spaces the soliton picture should be altered or not. Research in this direction is still in its infancy. The soliton resonance found, for instance, in the Kadomtsev-Petviashvili equation (two-dimensional Korteweg-de Vries equation) is quite interesting. Due to the resonant interaction between solitons, there appears a spatial structure where 3-solitons branch away from a point in space. The dynamics of soliton resonance and its physical significance must be studied further. More generally, the study of the dynamics of solitons in higher dimensional space will enrich soliton physics. Another problem related to the extensions of the soliton concept is the influences of external force (or noise) and/or dissipation on solitons. In some cases soliton perturbation theory gives sufficiently accurate results. It will be interesting to see how the soliton picture and the chaos picture compete or compromise in those systems.

A brief outline of the developments in soliton theory has been given. The study of solitons is the first systematic research on nonlinear phenomena with a consistent concept. We now have a guiding principle and new analytical methods. In the following excellent contributions, we will find detailed discussions on the many interesting issues. In conclusion, we summarize the Introduction by emphasizing that interdisciplinary research among physicists, mathematicians, chemists, biologists and engineers is crucial, as we have experienced in the past two decades.

Part I

Mathematical Theory of Solitons

Some Aspects of Soliton Dynamics

M. Toda

5-29-8-108 Yoyogi, Shibuya-ku, Tokyo 151, Japan

1. Integrable Systems and Solitons

It is about 30 years since the recent study of nonlinear waves started
with the work by FERMI et al. [1]. They numerically integrated the
equations of motion of certain one-dimensional nonlinear lattices, and
found recurrence to initial state at least for small nonlinearity and
smooth initial excitation. The idea that for sufficiently smooth waves,
a lattice can be approximated by a continuum led ZABUSKY and KRUSKAL [2]
to the numerical study of the Korteweg-de Vries equation

$$\partial u/\partial t + u\partial u/\partial x + \delta^2 \partial^3 u/\partial x^3 = 0 \ . \tag{1}$$

According to their results published in 1965, a smooth initial wave
turns into a series of several profound pulses, each of which is a
solitary wave solution of the KdV equation. They proceed freely, collide mutually, interact nonlinearly, and then recover their profiles.
Zabusky and Kruskal coined the name "solitons" for these stable pulses.

The fact that the numerical solutions to the nonlinear wave equation
were composed of solitons was so exciting that the analytic method of
integration was intensively investigated. Thus in 1967, the so-called
inverse scattering method was found [3]. The method was elucidated
by Lax, and further extended by ABLOWITZ et al. [4] so as to include
many nonlinear evolution equations, e.g., modified KdV equation, and
sine-Gordon equation. The method of integration was also extended to
cover periodic boundary conditions. In short, the method of inverse
spectral transform thus established gives the solution at arbitrary
time in terms of a set of initial data.

On the other hand, the one-dimensional lattice with exponential
interaction potential

$$\phi(r) = \frac{a}{b} \{e^{-b(r - \sigma)} - 1\} + a(r - \sigma) \tag{2}$$

between adjacent particles was presented in 1967, and was shown to
have soliton solutions as well as periodic solutions [5]. The inverse
scattering method for the exponential lattice was developed by FLASCHKA
[6] in 1974, and for periodic case by DATE and TANAKA [7], and independently by KAC and MOERBEKE [8].

A simple transformation connects the exponential lattice equation,
through integrable regime, to the KdV equation [9]. Similar transformations can be found between certain discrete systems and their continuum counterparts.

Integrability of nonlinear evolution equations including certain
generalized exponential lattices is discussed by referring to the
Painlevé equation [10], the Lie algebras [11], and other mathematical
theories [12]. The ingenious method of bilinear transform invented

by HIROTA [13], and the method of τ-functions by SATO et al. [14] provide not only powerful methods of solution, but also a very wide viewpoint over interrelations between so many nonlinear equations.

Now, the KdV equation, originally an equation for shallow water waves, was found applicable to plasma waves [15]. There are other equations for surface waves and waves of interface between two fluids. For example we have the Benjamin-Ono equation, and more generally equations of the form [16]

$$\partial u/\partial t + u\partial u/\partial x + \int K(x - y) \; dy \; \partial^2 u(y)/\partial y^2 = 0 \; . \tag{3}$$

In this equation, the kernel $K(x - y)$ implies some effect from a distance. The same kind of equations can be considered in the biological population problem, ecological problems, and so forth, where effect or information at a distance is important. Soliton solutions of such equations are also well-known. In general, solitons are moving entities of localized energy, information, stress, and so forth.

There are many other important subjects to be investigated. Among them, the study of the role of solitons in conformational change in biological molecules seems to be promising. Scattering and destruction of solitons by impurities [17] are related to the problem of thermal conductivity of solids. Life time of solitons in almost integrable systems, onset of chaotic behavior, and ergodicity are fundamental problems awaiting further investigations.

In the following, I would like to give remarks on some problems I have been interested in, and which are related to soliton dynamics.

2. Chopping Phenomena

As we have seen in the Zabusky-Kruskal's figure, sharp solitons emerge even if the initial wave is very smooth. In other words, initially disperse, widespread stress may yield profound localized pulses of stress. Numerical calculations clearly show that such localization of stress can give rise to the breakdown of materials into pieces [18]. Such phenomena can take place only when the material is nonlinear with respect to the relation between stress and strain. It will be serious if great earthquakes or strong winds cause such nonlinear damages to constructions such as chimneys, tall buildings, bridges, highways, and like.

3. Pressure and Expansion due to Solitons

The thermal expansion of ordinary solids is generally explained as a result of nonlinearity of the force between constituent atoms. Usually the attraction force gets weaker for larger separation, and thermal expansion is clear from the asymmetry of the interaction potential between atoms. For an exponential lattice with the potential given by (2), thermal expansion is positive when the constant b is positive. We shall limit ourselves to this case. The equation of motion for the lattice is

$$m \, \frac{d^2 y_n}{dt^2} = f_n - f_{n+1} \tag{4}$$

where $f_n = -d\phi(r_n)/dr_n$ with $r_n = y_n - y_{n-1}$, or

$$f_n = a[\exp\{-b(r_n - \sigma)\} - 1] \tag{5}$$

is the force between adjacent particles. The soliton solution can be written as

$$f_n = \frac{m}{b} \beta^2 \operatorname{sech}^2 (\kappa n - \beta t) \ , \qquad \beta = \sqrt{\frac{ab}{m}} \sinh \kappa \ . \tag{6}$$

We see that a soliton is a compressed pulse ($r_n < 0$).

Now, motion in the lattice is described as a collection of solitons and ripples. However, since solitons generally carry almost all the energy, and as we shall see in the following, the problem is also pertinent to motion without ripples, we may disregard ripples in the present argument. Then the facts that the lattice expands when the energy is raised and that solitons give rise to contraction of the lattice look contradictory.

This paradox is solved when we consider the pressure exerted by solitons. We note that a soliton has an excess mass due to contraction, which amounts to

$$M = -m \sum_{n=-\infty}^{\infty} (r_n - \sigma) = \frac{2m}{b} \kappa \ . \tag{7}$$

The velocity of the soliton is

$$v = \frac{\beta}{\kappa} \tag{8}$$

and the linear momentum associated with the soliton turns out to be

$$p = \sum_{n=-\infty}^{\infty} m\dot{y}_n = \frac{2m}{b} \beta = Mv \ . \tag{9}$$

Suppose that a soliton comes to a fixed end and is reflected. The momentum changes from p to $-p$, and so it will exert an impulse $2p$ on the boundary. To confirm this assertion we describe this process in terms of a two-soliton solution. We think of an infinite lattice with two solitons of the same strength, but running in opposite directions to meet at $n = 0$. As they go through each other, the particle $n = 0$ remains still, and so it plays the role of a fixed end. The stress during the process is given as

$$f_n = \frac{m}{b} \frac{d^2}{dt^2} \log \psi_n \qquad \text{with} \tag{10}$$

$$\psi_n = \cosh 2\kappa (n - \tfrac{1}{2}) + B\cosh(2\beta t) \ , \qquad B = \cosh \kappa \ . \tag{11}$$

Thus $f_0 = f_1$ as it should be for the fixed end $n = 0$. The impulse exerted by the soliton turns out to be

$$I_0 = \int_{-\infty}^{\infty} f_0 \, dt = \frac{m}{b} \frac{d}{dt} \log \psi_0 \Big|_{t=-\infty}^{\infty}$$

$$= \frac{4m}{b} \beta = 2p \tag{12}$$

as we have expected. It is also clear that

$$I_n = \int_{-\infty}^{\infty} f_n \, dt = I_0 \tag{13}$$

which implies that the momentum flow is responsible for the impulse. Further, if there are solitons confined between two ends, they will give outward impulse to these end particles, and give rise to expan-

sion of the lattice when there is no external pressure present. Again,
we see that a periodic lattice or an infinite lattice will expand by
the pressure due to solitons if no external pressure is applied.

A particular solution, called a cnoidal wave, provides an example
of the expansion due to a periodic wave. This can be written as

$$f_n = \frac{m}{b} (2K\nu)^2 [\mathrm{dn}^2 \{2(\frac{n}{\lambda} - \nu t)K\} - \frac{E}{K}] \tag{14}$$

where the frequency ν is a certain function of the wave length λ and
the modulus of the elliptic function dn, and K and E are complete elliptic integrals of the first and the second kind with the same modulus. If we take the average over a period, we have

$$\overline{f_n} = 0 . \tag{15}$$

Thus the above cnoidal wave is a solution with no external pressure.
The cnoidal wave (14) can be written in an alternative form as

$$f_n = \frac{m}{b} [\sum_{l=-\infty}^{\infty} \beta^2 \mathrm{sech}^2 \{\kappa(n - \lambda l) - \beta t\} - 2\beta\nu] \tag{16}$$

where β is a function of κ, which is related to the modulus mentioned
above. The first term on the r.h.s. of (16) expresses a succession
of solitons, each having the form $\mathrm{sech}^2(\kappa n)$, and the speed modified
by the interaction among themselves. The second term on the r.h.s.
of (16) is the contraction force which comes from expansion of the
lattice. Now, since

$$\int_{-\infty}^{\infty} \mathrm{sech}^2(\beta t) \, dt = \frac{2}{\beta} ,$$

and ν solitons pass in unit time, the time average of the stress due
to solitons is

$$\overline{f_n}(\text{solitons}) = \frac{m}{b} \overline{\sum_{l=-\infty}^{\infty} \beta^2 \mathrm{sech}^2 \{\kappa(n - \lambda l) - \beta t\}}$$

$$= \frac{m}{b} 2\beta\nu \tag{17}$$

which is just cancelled by the second term of (16), and thus $\overline{f_n} = 0$;
that is, there is no pressure on the average.

4. Thermodynamic Consideration

As we have seen, the pressure P of the lattice is the time average of
the force (5):

$$P = \overline{a\{e^{-b(r - \sigma)} - 1\}} . \tag{18}$$

We note that (2) and (18) yield

$$\overline{\phi(r)} = \frac{P}{b} + a(\overline{r} - \sigma) \tag{19}$$

and that the internal energy of the lattice is

$$E = \sum_{n=1}^{N} \frac{m}{2} \dot{y}_n^2 + \sum_{n=1}^{N} \overline{\phi(r_n)} . \tag{20}$$

In classical case, the law of equipartition of energy gives

$$\frac{m}{2}\,\overline{\dot{y}_n{}^2} = \frac{kT}{2} \ .$$

Thus we have an interesting equation for the internal energy of the lattice:

$$E = \frac{NkT}{2} + \frac{NP}{b} + a(L - N\sigma) \tag{21}$$

where $L = N\overline{r}$ is the length of the lattice. That is, the internal energy E of the exponential lattice is a linear function of T, P, and L.

We use the thermodynamic relation $dE = TdS - PdL$, and require that $S(T,L)$ is a total differential. Then we have

$$\frac{\partial P}{\partial T} - \frac{N}{bT}\frac{\partial P}{\partial L} = \frac{a + P}{T} \ . \tag{22}$$

Introducing $X = (a + P)/T$, we eliminate the r.h.s. of (22) to have

$$T\,\frac{\partial X}{\partial T} - \frac{N}{b}\frac{\partial X}{\partial L} = 0 \ . \tag{23}$$

Therefore

$$X = F(L - N\sigma + \frac{N}{b}\ln T + \text{const.}) \tag{24}$$

where F is an arbitrary function, which cannot be obtained in the frame-work of thermodynamics. When we denote the inverse function of F by $(N/b)\psi$, then we may write the equation of state of the lattice as

$$L = N\sigma - \frac{N}{b}\left\{\ln\left(\frac{bkT}{a}\right) + \psi\left(\frac{a + P}{bkT}\right)\right\} \ . \tag{25}$$

By applying statistical mechanics [19], we see that the function ψ is identified to the digamma function

$$\psi(z) = \frac{d\ \ln\Gamma(z)}{dz} \ .$$

References

1. E. Fermi, J. Pasta and S. Ulam, _Collected Papers of E. Fermi_, Univ. of Chicago Press, 1965
2. N. Zabusky and M.D. Kruskal, Phys. Rev. Lett. 15 240 (1965)
3. C.S. Gardner, J.M. Greene, M.D. Kruskal and R. Miura, Phys. Rev. Lett. 19 1095 (1967)
4. M.J. Ablowitz, D.J. Kaup, A.C. Newell and H. Segur, Studies Appl. Math. 53 249 (1974)
5. M. Toda, _Theory of Nonlinear Lattices_, Springer Series of Solid-State Sciences vol.20 (1981)
6. H. Flaschka, Phys. Rev. B9 1924 (1974)
7. E. Date and S. Tanaka, Prog. Theor. Phys. 55 457 (1976); Prog. Theor. Phys. Suppl. 59 107 (1976)
8. M. Kac and P. van Moerbeke, Proc. Nat. Acad. Sci. USA 72 1627, 2879 (1975)
9. N. Saitoh, J. Phys. Soc. Japan 49 409 (1980)
10. A. Ramani, B. Dorizzi and B. Grammaticos, Phys. Rev. Lett. 49 1539 (1982)
11. M.A. Olshanetsky and A.M. Perelomov, Phys. Rep. 71 313 (1981)

12. H. Yoshida, Celes. Math. _31_ 363, 381 (1983), _Nonlinear Integrable Systems_ (ed. by M. Jimbo and T. Miwa) p.273, _World Scientific_ 1983
13. R. Hirota and J. Satsuma, Prog. Theor. Phys. Suppl. _59_ 64 (1976)
14. see for example, E. Date, M. Kashiwara, M. Jimbo and T. Miwa in _Nonlinear Integrable Systems_ loc. cit., p.39
15. T. Taniuti and C.C. Wei, J. Phys. Soc. Japan _24_ 941 (1968)
16. H. Ono, J. Phys. Soc. Japan _39_ 1082 (1975); J. Satsuma in _Nonlinear Integrable Systems_ loc. cit., p.183
17. M. Toda, R. Hirota and J. Satsuma, Prog. Theor. Phys. Suppl. _59_ 148 (1976)
18. S. Watanabe and M. Toda, J. Phys. Soc. Japan _50_ 3443 (1981)
19. M. Toda and N. Saitoh, J. Phys. Soc. Japan _52_ 3703 (1983)

Approximations for the Inverse Scattering Transform

D.J. Kaup

Department of Physics, Clarkson University, Potsdam, NY 13676, USA

I. Introduction

What I would like to present here is a discussion of some of the various techniques and approximations which have been found to be useful in the IST (inverse scattering transform) [1]. Although the IST is an exact method, its complexity does require that one be able to make use of some approximations. And while I shall be discussing approximations, let me note that I also shall be emphasizing the methods and not the mathematical details.

My reasons for emphasizing approximation methods for the IST are very simple and practical. Right now, I would say that if one had to model some simple physical system, one would find that there is actually an 'excess' of integrable systems to choose from. The number of known tabulated integrable systems has just simply grown to be that large. In fact, we have gotten to be so sophisticated that, for example, if you give Yuji Kodama [2] a physical system close to some basic simple integrable system, then he could tailor-make a more complicated integrable system for you which would be as close to the original physical system as you would wish to specify. So once upon a time there may have been a dearth of integrable systems. However, that is no longer the case and now we can even tailor-make integrable systems to our specifications.

But at the same time one should note that there is certainly no surplus in the application of what we do have. Although we do have this wealth of integrable systems to choose from, very little has been done in the way of applying what we do have to the physical universe. Most of the effort has been devoted to the theory and the mathematical aspects. It is true that this effort was once both necessary and needed. However, now I believe that it is time for us to switch the emphasis. I say this because given any integrable system, there is no doubt in my mind that with sufficient time, then someone will be able to come up with some scheme for solving the initial value problem and for obtaining the soliton solutions. We now have had enough experience to know that such is true. And we have indeed come a long way. Particularly when one can solve such systems as the Benjamin-Ono equation [3] and even perform an inversion of the heat equation [4].

But what is the purpose of all this that we have done? I would like to think that we do want to make use of our results, or at least to see to it that others are able to make use of them. Unfortunately that is not what is occuring, except in some isolated cases. In part this is due to the difficulty in designing a simple experiment to demonstrate solitons. But also it is due in part to the novelty of the IST and the elegant mathematical magic which it contains. One can easily become beguiled by the complexity and also the simplicity which is contained within it. However be that, there are always others who are more practical-minded. They ask about how to use this IST. And when we can only show them how to treat what they consider to be very, very simple models, and when our repertoire of approximation techniques is very limited, well it is then not surprising that some others would consider the IST to be only a mathematical toy. Briefly put, the IST will be judged to be just about as useful as we can demonstrate that it can be useful.

As an example, everyone is familiar with the trains of internal wave solitons observed in the seas of Southeast Asia [5]. Experimental data on these trains have been collected. Well, what does one need in order to compare theory with the experimental data? This data was collected for the intermediate-depth case

"

[5]. That means that the integrable equation describing these trains of solitons is the one originally investigated by Joseph [6] and later by Kubota et al. [7]. N-soliton solutions for this intermediate-depth equation were found by Chen and Lee [8], while Kodama et al. [9] have studied the direct- and the inverse-scattering problems. So one would expect that one should have to use this Kodama et al. IST (which is a rather complicated IST) in order to properly analyze the data. However such was not the case. Instead Osborne [10] found that all he needed to explain the experimental data was the very simple IST for the KdV equation, which is the well-known Schroedinger eigenvalue problem. He numerically evaluated the scattering data of the Schroedinger equation by using experimental data for the potential. Then he numerically solved the Gel'fand-Levitan equation in order to obtain the time evolution. And the agreement that he found, to within the experimental error of the data, was quite good. In other words, the experimental data just was not able to distinguish the shallow-water KdV case from the more complex intermediate-depth case. So it then became quite appropriate to analyze the data with the more simple KdV model.

As a result of this, one could easily say that, in some sense, our theories now have well outstripped the needs of the experimentalist. In other words, one could say that we have overshot the target, and have developed a structure far too complex to be useful to others in the intermediate future. What is really needed is more applications of the IST in real-life situations just as Osborne has done, and the development of the appropriate mathematical tools for doing such.

What I shall do here is to describe three such mathematical tools which I have helped develop for handling certain classes of problems involving the IST. First I shall describe the singular perturbation theory for integrable systems and I shall use a very recent example to demonstrate how one may use this theory to predict the birth of solitons in a strongly damped system involving SRS (stimulated Raman scattering). Second, I shall discuss what I call 'forced integrable systems' which is really just another name for boundary-value problems for integrable systems. Here there is no doubt in my mind that these forced integrable systems are exactly solvable. But the complexity involved in obtaining such an exact solution is just too much to be useful at the present. So, I shall instead discuss how one may glean information from these systems even if an exact method of solution is not known. Lastly, I shall consider immersing a soliton system into a temperature bath, and then describe some of the considerations involved in determining the temperature effects on the soliton motion.

II. Singular Perturbation Expansions

One of the most important features of the IST is that it maps an integrable field into a set of action-angle variables (the scattering data) in what we shall call 'scattering space'. And in scattering space, the equations of motion for the scattering data become very simple. In fact the spectrum becomes invariant and the phases execute a very simple rotation about the spectrum [11]. From the early AKNS [1] work, we were able to obtain exact expressions for how the scattering data evolved in time for any system, integrable or not. All that one had to do was to specify how q and r evolved in time. Of course, only for the integrable systems were these equations tractable. For a nonintegrable system, one just could not obtain a close form for the integrals involved. From this it was a short step to develop a perturbation theory for scattering data. The idea was very simple. For an integrable system the spectrum was invariant. Now let the integrable system be slightly perturbed. We then expect the spectrum to also be only slightly perturbed. And the evolution of the scattering data away from its initial spectrum would follow from the appropriate AKNS equations [1].

The details of these equations were first worked out by myself in 1976 [12] for the Zakharov-Shabat IST [13]. Shortly after that Keener and McLaughlin [14] did the same from the point of view of a Green's function theory, by introducing a two-time procedure. And Karpman and Maslov [15] developed a perturbation theory similar to mine, but for the KdV equation instead. Then Newell and I

[16] summarized what had been done up to that time, generalized the techniques
as much as we could, and then discussed several different examples in order to
illustrate the possible applications of the method. Since that time, there has
been an expanding number of examples studied and published, which I shall not
attempt to delineate. Instead I shall proceed directly to a new example of an
application of this singular perturbation expansion.

This example arises from some recent experimental studies involving SRS
(stimulated Raman scattering) with intense laser beams [17,18] wherein an
anomalous reversal of the pump depletion was observed. The experimental
arrangement involved the simultaneous passage of both a Stokes pulse and a pump
pulse through a Raman-active medium. When both the Stokes pulse and the pump
pulse are simultaneously present in the Raman-active medium, the depletion of
the pump is tremendously enhanced due to the presence of this Stokes pulse. The
Stokes pulse does not have to be large (typically about 10μ J) to cause this,
whereas the pump pulse is usually quite large (typically about 10 J) and the
widths are all in the order of 100 nsec. During this process, the population
difference between the excited molecules and the ground-state molecules remained
constant to within a few percent [17]. Thus in this experimental setup, one
could assume an infinite source of ground-state molecules. In this case, one
does not need the full set of SRS equations [19,20] to describe the theory, but
rather one may use the simpler equations which were first given by Chu and
Scott [21]. These equations, with damping, are

$$\partial X/\partial\tau = -\,\varepsilon X + A_1 A_2^{*} \tag{1a}$$

$$\partial A_1/\partial\xi = -\,XA_2 \tag{1b}$$

$$\partial A_2/\partial\xi = X^{*}A_1 \tag{1c}$$

where ε is the damping constant, $A_1(A_2)$ is the slowly-varying envelope of the
pump wave (Stokes wave), and X is the off-diagonal density-matrix element for
the Raman transition. The coordinates τ and ξ are related to the standard
space-time coordinates, t and z, by

$$\tau = t - z/c \tag{2a}$$
$$\xi = z \; . \tag{2b}$$

When the damping constant ε is zero, these equations are integrable [21].
However for this system, the damping is not weak, but instead is quite strong
and is of order one. In fact, as we shall see, it very significantly affects
the evolution of the system and its solitons.

Even though the damping is large, one can still use the IST to obtain results
and make predictions about this system. This may seem to be surprising at first,
since with the strong damping the system is undoubtedly not integrable. But let
me emphasize what I have stated above and which has been known since the days of
AKNS. And that is that scattering data can be defined for <u>any</u> system,
integrable or not. And that that scattering data will evolve in some fashion,
which AKNS tells us how to calculate. <u>If</u> we can do the integrals. And in this
case we can indeed do the integrals, at least initially and that is sufficient.

Let me demonstrate this with a minimum of mathematics. As shown by Chu and
Scott [21], the Lax pair [11] for this system <u>when the damping constant is zero</u>
is

$$\partial u_1/\partial\xi - i\lambda u_1 = Xu_2 \tag{3a}$$

$$\partial u_2/\partial\xi + i\lambda u_2 = -X^{*}u_1 \tag{3b}$$

$$\partial u_1/\partial\tau + iS_3 u_1/\lambda = S_{+}u_2/\lambda \tag{4a}$$

$$\partial u_2/\partial\tau - iS_3 u_2/\lambda = -\,S_{-}u_1/\lambda \qquad \text{where} \tag{4b}$$

$$S_3 = (\,A_1^{*}A_1 - A_2^{*}A_2)/4 \tag{5a}$$

14

$$S_+ = iA_2^* A_1/2 \tag{5b}$$

$$S_- = S_+^* . \tag{5c}$$

Now the initial value problem for this system is where one is given $X(\tau=-z/c,\ \xi=z)$ and $A_i(\tau=t,\ \xi=0)$ for $i = 1$ or 2. Thus to solve the initial value

problem, one then must use (4) to define the initial scattering data. This is because you are not given sufficient information to use (3) to define scattering data, although it is the well-known Zakharov-Shabat (ZS) [13] eigenvalue problem. But you are given sufficient information for solving (4) initially. Although (4) do not look like a familiar eigenvalue problem, they can indeed be transformed [20] into the Zakharov-Shabat form. Thus all the AKNS results will apply to the eigenvalue problem (4). The transformed form of (4) is [22]

$$\partial v_1/\partial T + i\rho v_1 = q v_2 \tag{6a}$$

$$\partial v_2/\partial T - i\rho v_2 = -q^* v_1 \qquad \text{where} \tag{6b}$$

$$T = \int_{-\infty}^{\tau} A\ d\tau \tag{7}$$

$$q = i\{\ \partial[\ e^{i\gamma}\sin\beta\ ]/\partial T\ \}/[\ 2\cos\beta\ ] . \tag{8}$$

In (6), ρ is the eigenvalue. The quantities γ and β are phases which are defined from the S's as follows.

$$\partial\gamma/\partial\tau = \cos\beta\ \partial\sigma/\partial\tau \tag{9}$$

$$S_3 = A\ \cos\beta \tag{10a}$$

$$S_+ = A\ e^{i\sigma}\sin\beta \tag{10b}$$

where A is the normalized total intensity and is defined by

$$A = (\ A_1^* A_1 + A_2^* A_2)/4 . \tag{10c}$$

The main purpose of giving the above definitions is to point out that the initial scattering data for this system is almost trivial. One notes for small β that q is also small. And as AKNS has shown, when q becomes small, then the scattering data becomes simply the linear Fourier transform. This is indeed the initial state that we do have because the initial Stokes pulse(A_2) is approximately one-millionth of the pump(A_1). (These laser pulses also have relatively stable phases. So we may consider σ to be small also.)

Well if q is indeed small initially, then there can be no solitons present initially and the spectrum must consist of solely the continuous spectrum, or 'radiation'. But let us look at what happens as the system evolves in ξ.

When the phase of the Stokes pulse relative to the pump remains fairly constant, we see normal pump depletion as shown in Fig. 1. The top curve is the incident pump pulse and the bottom curve is the depleted pump, after the pump has passed through the medium. As the pump travels through the medium, it is simply eaten away more and more. (In SRS, the distance traversed through the medium scales as the inverse of the pump power. So a higher incident power is equivalent to a thicker medium.) This is just ordinary pump depletion. But when the phase of the Stokes pulse relative to the pump would happen to reverse itself about midway along the Stokes pulse, we see the phenomenon shown in Fig. 2. Here, a very distinctive peak (a soliton) has been formed which corresponds to a reversal of the normal pump depletion. And as the pump propagates further into the medium, the soliton becomes narrower and narrower.

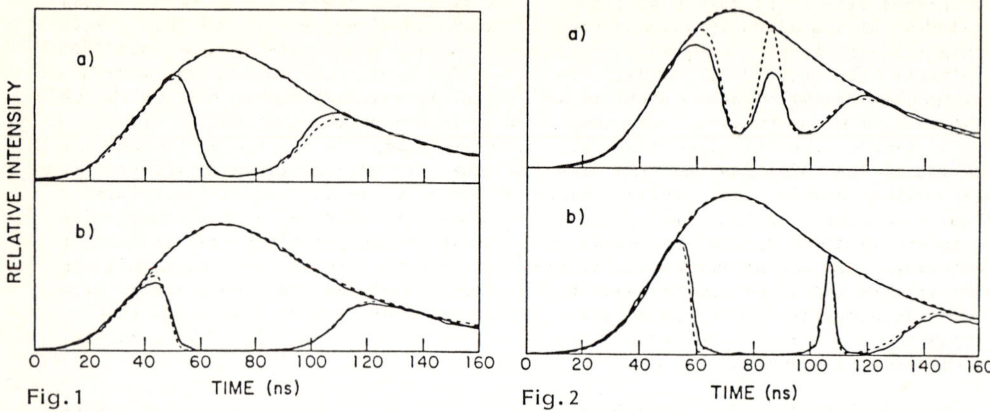

Fig.1   TIME (ns)   Fig.2   TIME (ns)

Fig. 1 Comparisons of experimental and theoretical pulse shapes of the incident (upper curve) and depleted (lower curve) pump in 'normal' pump depletion. The incident pump power in (b) is 38% larger than that of (a), which is equivalent to a 38% longer length. The solid lines are the experimental shapes while the dashed lines are the theoretical fits. (From Ref. 17 and courtesy of K. Druhl and the Phys. Rev. Letters)

Fig. 2 Comparisons of experimental and theoretical pulse shapes of the incident (upper curve) and depleted (lower curve) pump showing the anomalous reversal of pump depletion. The incident pump power in (b) is approximately 40% larger than that of (a). The solid lines are the experimental shapes while the dashed lines are the theoretical fits. (From Ref. 17 and courtesy of K. Druhl and the Phys. Rev. Letters)

    The dashed curves in Figs. 1 and 2 are theoretical fits of the solutions of (1) to the experimental data. As one can see, the fit is quite remarkably good.

    Now, what would happen if we turned off the damping? We cannot do this experimentally, but we can do it with the theoretical numerical solutions. When this is done, we see results as shown in Fig. 3. The reversal of the pump depletion still tries to occur, but it just is not able to fully form, and at

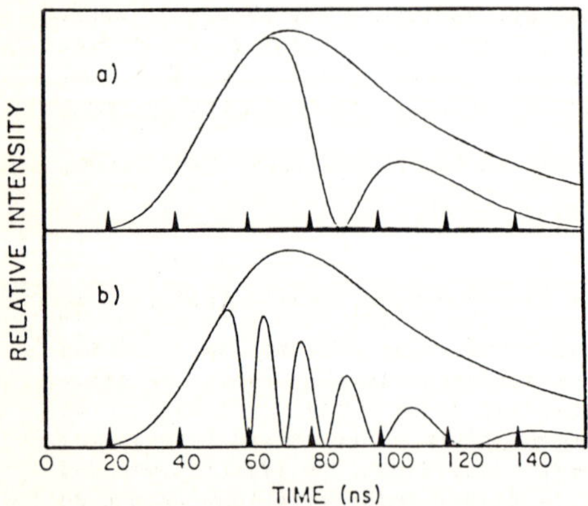

TIME (ns)

Fig. 3 The theoretical depleted pulse shape (lower curve) for the given incident (upper curve) pulse shape in the absence of damping. Note the oscillations which just pack up into the front of the pulse as the gain is increased. The incident power in (b) is 4 times that in (a). (Courtesy of K. Druhl)

most, one simply has some oscillations forming at the front of the pulse. This
failure of a soliton to form is exactly as one would expect because when $\varepsilon = 0$,
then (1) are exactly integrable. And since we have already determined that no
solitons can be present initially, then as long as the equations remain
integrable, no solitons can ever appear. Thus one concludes that the birth of
the soliton very critically depends on the presence of the damping.

I would like to conclude this section by briefly describing how one can
determine the conditions for the creation of one or more of these solitons. As
you will recall, the initial state of this system is in effect the 'linear
limit' because q is initially so small. And in this limit we may determine the
scattering coefficients by expanding them in powers of the potential, q. In
particular, we may determine the transmission coefficient, $\underline{a}$, and look and see
if it has any zeros in the upper-half ρ-plane. And the appearance of any such
zero would correspond to the creation of a soliton. From the perturbation
expansion and upon taking the linear solutions to be valid initially, we find
the following expansion for $\underline{a}$.

$$a = 1 + \Delta_+\Delta_-\exp[\,-4i\xi/(2\rho - i\varepsilon/A_+)] \tag{11}$$

where

$$\Delta_{\pm} = [\varepsilon/(2A_+)]\int_{-\infty}^{\infty} S_{\pm}(\xi=0,\tau)\,\exp[-(\underline{+})\varepsilon\tau]\,d\tau\ . \tag{12}$$

Several approximations have been made to obtain the above simplified result. We
have approximated the incident pump pulse profile to be a square pulse of
amplitude A_+, we have assumed that the duration of the pump pulse is much longer
that the damping period, and we have assumed that the value of ρ is near the
singularity at $i\varepsilon/(2A_+)$. If ρ is far from this singularity, then $\underline{a}$ never does
become significantly different from its initial value of approximately unity.
Thus we only expect to find zeros of $\underline{a}$ (eigenvalues) near this value.

In the experimental situation, it is believed that the phase of the envelope
of the Stokes pulse remains constant in τ, except whenever it flips its sign
[23]. If this is true, then the phase of S_+ is constant and of no consequence.
We may then ignore it and even assume $S_+(\xi=0,\tau)$ to be real. In this case, (12)
shows that Δ_+ will then be real. Now when S_+ as a function of τ never changes
its sign, by (11), $\underline{a}$ can never have a zero in the upper-half ρ-plane. So in
order for a zero to occur, there must be a change of sign. Let us now suppose
that $S_+(\xi=0,\tau)$ does flip its sign somewhere near the middle of the Stokes
pulse. Then from (12) we may expect Δ_+ and Δ_- to be of opposite signs. This is
because, according to (12), Δ_+ is determined by the sign of S_+ when τ is small
while Δ_- is determined by the sign of $S_-(=S_+$ if S_+ is real) when τ is large.
Thus, if S_+ ever flips its sign between the front and the back of the pulse, the
product of $\Delta_+\Delta_-$ could become negative, particularly if the sign-flip were

located near the midpoint. And if this ever occurred, then $\underline{a}$ would have a zero
at

$$\rho = i\varepsilon/(2A_+) + 2i\xi/\{\ln[-\Delta_+\Delta_-]\}\ . \tag{13}$$

As shown by (13), the zero begins at $i\varepsilon/(2A_+)$ when $\xi = 0$ and then drifts down
toward the real ρ-axis at a constant rate. Thus, we expect to see a soliton
form which as ξ increases has an increasing width due to this decay in the value
of the zero of $\underline{a}$. This is what is observed numerically [24]. It remains to
check the numerical value of this rate of decay against (13).

The solution given above is expected to remain valid until a significant pump
depletion has occurred. Up until that value of ξ where pump depletion begins to
be significant, the assumption that one may use the linear solutions remains
valid. But as the pump starts to deplete, the soliton takes on more and more
its characteristic shape, and one may now better approximate the evolution by
considering only a single localized soliton. When this is done, then one finds

that the increasing width has ceased and has reversed into a decreasing width, with the pulse becoming narrower and narrower [22,24] as we have seen above in Fig. 2.

III. Forced Integrable Systems

There is a very important class of problems which as of yet still remain unsolved. These are the boundary-value problems for nonlinear integrable systems, which I called 'forced integrable systems'. In some elementary cases such as simple harmonic generation [25], self-induced transparency (SIT) [26,27,28], and SRS [19,20], such boundary-value problems can indeed be solved. However, in a sense, these are trivial examples, and the solvability of these cases depends very crucially on the fact that the eigenfunctions of each of the two Lax operators (the x-eigenvalue problem and the t-eigenvalue problem) have exactly the same analytical properties. In general this is not true. In general the eigenfunctions of one of the Lax operators will have analytical properties which are different from that of the eigenfunctions of the other operator. And that then creates a much more complicated problem, which is in need of a solution.

As an example of such a forced integrable system, let us consider the generation of Langmuir waves by an intense rf-wave which was being directed directly straight up into the ionosphere [29]. The excitation of Langmuir waves would occur because the rf-wave would penetrate up until it reaches its WKB turning-point, at which point the rf-wave energy would accumulate due to the WKB swelling. This intense collection of rf-energy could then be expected to excite Langmuir waves in the direction of the electric-field polarization. These Langmuir waves would then propagate away from this source and their propagation would be governed by the nonlinear Schroedinger equation (NLS). So the important questions which one would like to answer are what type of Langmuir waves may we expect to see propagating away from the source? solitons? continuous spectrum? And how would this all be dependent on the rf-source?

Well at the moment this system cannot be solved. And it is of interest to see why it cannot be solved and where the difficulty lies. Let us take the NLS in the form

$$i\partial_t q = -\partial_x^2 q + 2 (rq) q \tag{14a}$$

$$i\partial_t r = \partial_x^2 r - 2 (rq) r \tag{14b}$$

where

$$r = \pm q^* . \tag{15}$$

The eigenvalue problem is the ZS eigenvalue problem

$$\partial_x v_1 + i \rho v_1 = q v_2 \tag{16a}$$

$$\partial_x v_2 - i \rho v_2 = r v_1 \tag{16b}$$

while the time-evolution operator is given by the ABC equations [1]

$$i\partial_t v_1 = A v_1 + B v_2 \tag{17a}$$

$$i\partial_t v_2 = C v_1 - A v_1 \tag{17b}$$

where for the NLS

$$A = 2 \rho^2 + rq \tag{18a}$$

$$B = 2 i \rho q - \partial_x q \tag{18b}$$

$$C = 2 i \rho r + \partial_x r . \tag{18c}$$

Let us specify that the forced NLS has the boundary conditions

$$S(x) = q(x,t=0) \tag{19a}$$
$$Q(t) = q(x=0,t) \tag{19b}$$
$$P(t) = \partial_x q(x=0,t) \ . \tag{19c}$$

From the linear problem one may show that $P(t)$ and $Q(t)$ are <u>not</u> linearly independent and that only one may be specified or else the problem would be overdetermined. Presumably the same is true for the nonlinear problem also.

In these forced integrable systems, one uses the usual eigenvalue problem to define the scattering data, which in this case is the ZS eigenvalue problem, (16). In particular, we may use it at $t = 0$ to determine the initial scattering data, which will only require the value of $S(x)$.

Once we have the initial scattering data, then we would like to know how it evolves in time. To do this, it is necessary to integrate the ABC equations at $x = 0$. If we rewrite these ABC equations, we can put them into the form

$$\partial_t v_1 + i(\ 2\rho^2 + rq\ )v_1 = (\ 2\rho q + i\partial_x q\ )v_2, \tag{20a}$$

$$\partial_t v_2 - i(\ 2\rho^2 + rq\ )v_2 = (\ 2\rho r - i\partial_x r\ )v_1 \ . \tag{20b}$$

This is a familiar set of equations. In fact it is another well-known eigenvalue problem. This eigenvalue problem has been used to study the Thirring model [30,31] and the derivative-NLS [32,33]. [Equations (20) do differ from those eigenvalue problems by the term 'rq' which has been added to the $2\rho^2$ term. But this term could easily be eliminated by simply rotating the phases of v_1 and v_2 in opposite directions.] Thus we understand quite well the analytical properties of its eigenfunctions [33,34]. But first let us note that we are not able to integrate (20) even at $x = 0$ because we do not know what $P(t) = \partial_x q(x=0,t)$ is. In the solution of the linear problem, $P(t)$ can be determined from $S(x)$ and $Q(t)$. Presumably the same is true for the nonlinear problem. But how is one to determine what $P(t)$ is to be in the nonlinear case? If we knew this, then we would have a solution to this problem.

The problem of having to determine a second or dependent potential will occur in almost all forced integrable systems. The only exceptions are those trivial cases mentioned above. Whether or not one can solve this problem remains to be seen. But even if it is possible to 'crack' this one, I would anticipate that the exact solution may be so involved that it may well be worthwhile for one to at least hesitate and ponder whether or not there might be an easier way to obtain some information on these systems. And it was in this spirit that I started this investigation of forced integrable systems. I have known for several years that these systems were an important set of problems. And I finally decided to just simply force my way through one of the cases and find some method to glean some information from it. My first result was with the forced Toda lattice which has been used as a model in molecular dynamics studies [35,36]. The same problem existed there as in the forced NLS, namely one had two potentials appearing in the eigenvalue problem for the time-evolution of the scattering data and only one potential could be specified without overdetermining the problem. The way I then bypassed that was to approximate the unknown potential, and then determine the time-evolution of the scattering data by using that approximation [37]. It turned out that in this case, this approximation worked quite well and that one could use it to predict fairly accurately the soliton birth-rate in the molecular dynamics model [38]. I would expect that similar approaches could also work in other systems.

IV. <u>Stochastic Effects on Integrable Systems</u>

One of the first treatments of an integrable system with some type of a stochastic property was that by John Elgin and myself [39]. The question that we addressed was how would the stochastic nature of a realistic laser profile

affect the number and amplitude of solitons which could arise from a given
initial profile. The difficulty being addressed here was the fact that
realistic laser profiles are not simple. When a laser system operates well
above its threshold, although its amplitude may be fairly constant, its phase
will vary randomly. And any phase variation across an initial profile will
usually reduce the number of solitons present in the profile [28].

To model this stochastic behavior of the phase we used the 'phase diffusion
model' [40]. From this we were then able to calculate an exact result for the
statistical averages of the scattering coefficients, $\underline{a}$ and $\underline{b}$, and were able to
give a very simple prescription for determining how many solitons and of what
amplitude would arise from such a stochastic initial profile.

Wadati [41] has studied the stochastic KdV equation

$$\partial_t u - 6u\partial_x u + \partial_x^3 u = \sigma(t) \tag{21}$$

where σ is a stochastic forcing term. This equation is integrable because one
can use Galilean invariance [41] to reduce (21) to the standard simple KdV
equation with $\sigma = 0$. However even so, it is still not a trivial case. By using
the integrability, Wadati was able to construct the full solution, and in
particular to evaluate the average of the soliton solution. Taking the soliton
solution to initially be

$$u(x,t) = -2\eta^2 \mathrm{sech}^2\theta \quad \text{where} \tag{22}$$

$$\theta = \eta(x - x_o - 4\eta^2 t) \tag{23}$$

Wadati found the average value of $u(x,t)$ in the limit of t approaching $+\infty$ to be

$$\langle u(x,t) \rangle = - \{\eta/[3\pi\varepsilon t^3]^{1/2}\}\exp[-\theta^2/(48\eta^2 t^3)] \tag{24}$$

where ε indicates the strength of the stochastic variable. This form clearly
indicates the diffusion in the average value of the soliton. One can interpret
this diffusion as occuring because the soliton is being randomly kicked
back-and-forth due to the stochastic term σ. At one time we might find the
soliton on the right. Then at a later time we might find it on the left. Where
we find it tends to be random. The stochastic term is causing the soliton to
execute a random walk about its 'classical path' (the path it would follow if σ
was zero). And as time moves along, the soliton just tends to wander farther
and farther from its classical path. Whence in the limit of t approaching $+\infty$,
the averaging over the field $u(x,t)$ can only see the average of this random walk
and is independent of the structure of the soliton. Note that by this argument,
we would expect to see the same type of result for MKdV solitons, sine-Gordon
solitons, etc. The average over the field for time approaching $+\infty$ would
approach that of a diffusion about the classical path, as in (24).

About the same time, I took a look at a different type of a stochastic
problem. In the above example, one was interested in the average value of a
field variable. There are other cases where one might be interested in the
average value of something else, e.g. the mobility of a soliton or some detail
of its structure. In these cases, one must use a different viewpoint from the
above in order to obtain a meaningful result. And that viewpoint is that one
has to 'track' with the soliton while one is doing the averaging. As a simple
example, let us consider the change in the shape of a soliton due to thermal
fluctuations. The first thing that I want to make clear is what I do mean by
the change in the shape. I do not mean the above diffusion obtained by Wadati.
This diffusion is what one would see if one were to remain stationary and then
to average the field that he observed. What I mean by the change in the shape
is the following. Follow the soliton as it is being bounced back-and-forth by
the stochastic driving term. This requires one to define the 'center' of the
soliton so that it is clear how one is to 'follow' it. This can be done, and in
fact, singular perturbation theory tells one how it should be done [42,43]. The
answer is that the center of the soliton should be considered to be an ordinary
point particle and that this center is to follow the classical equations of

motion for a point particle. Then by knowing what the stochastic force is across the width of a soliton, one can calculate the average stochastic force at the soliton's center, and thereby determine the equation of motion of the soliton's center. (If we would now average the motion of the soliton's center, we would obtain Wadati's result.) We now follow this center, moving always with it and then averaging the field that we observe. And how this average differs from the shape of a stationary soliton is what I mean by the change of shape.

Let's consider this for the overdamped sine–Gordon equation

$$\partial_t \theta = \partial_x^2 \theta - \sin\theta + \xi(x,t) \tag{25}$$

where ξ is the stochastic driving term. In lowest order, this stochastic term causes the averaged shape of the soliton to become [43]

$$\langle\theta\rangle = 4\tan^{-1}[\exp(Z)] \tag{26}$$

where

$$Z = x(1 - \varepsilon) + (\varepsilon/4) \tanh(x) [1 + (\text{sech}^2 x) / 3] \tag{27}$$

where ε indicates the strength of the stochastic term. Obviously the soliton has flattened out from what it would have been in the absence of the stochastic term.

V. Acknowledgements

This research has been supported in part by the Air Force Office of Scientific Research and the National Science Foundation.

REFERENCES

1. M.J. Ablowitz, D.J.Kaup, A.C. Newell, and H. Segur, Stud. Appl. Math. 53, 249 (1974).
2. Yugi Kodama, 'Nearly Integrable Systems', Preprint–Nagoya Univ. # DPNU–11–83, (1983).
3. A.S. Fokas and M.J. Ablowitz, Stud. Appl. Math. 68, 1 (1983).
4. M.J. Ablowitz, D. Bar Yaacov, and A.S. Fokas, Stud. Appl. Math. 69, 135 (1983).
5. A.R. Osborne and T.L. Burch, Science 208, 451 (1980).
6. R.J. Joseph, J. Phys. A 10, L255 (1977).
7. T. Kubota, D.R.S. Ko, and L.D. Dobbs, J. Hydronaut. 12, 157 (1978).
8. H.H. Chen and Y.C. Lee, Phys. Rev. Letters 43, 264 (1979).
9. Y. Kodama, J. Satsuma, and M.J. Ablowitz, J. Math. Phys. 23, 564 (1980).
10. A.R. Osborne, Poster session at Solitons '82, Scott Russell Centenary Conference, Edinburgh, Scotland (23 Aug. 1982).
11. P.D. Lax, Comm. Pure Appl. Math. 21, 467 (1968).
12. D.J. Kaup, Siam J. Appl. Math. 31, 121 (1976).
13. V.E. Zakharov and A.B Shabat, Zh. Eksp. Teor. Fiz. 61, 118 (1971) [Soviet Phys. JETP 34, 62 (1972)].
14. J.P. Keener and D.W. McLaughlin, Phys. Rev. A 16, 777 (1977).
15. V.I. Karpman and E.M. Maslov, Phys. Letters 60A, 307 (1977).
16. D.J. Kaup and A.C. Newell, Proc. Roy. Soc. London A361, 413 (1978).
17. K. Druhl, R.G. Wenzel, and J.L. Carlsten, Phys. Rev. Lett. 51, 1171 (1983).
18. J.L. Carlsten, R.G. Wenzel, and K. Druhl, SPIE 380, 201 (1983).
19. H. Steudel, Physica 6D, 155 (1983).
20. D.J. Kaup, Physica 6D, 143 (1983).
21. F.Y.F. Chu and A.C. Scott, Phys. Rev. A12, 2060 (1975).
22. D.J. Kaup, 'Using Dissipation for Creating Solitons', (Preprint in Preparation).
23. R.G. Wenzel (private communication).
24. K. Druhl (private communication).
25. D.J. Kaup, Stud. Appl. Math. 59, 25 (1978).
26. G.L. Lamb, Jr., Phys. Rev. Lett. 31, 196 (1973).
27. M.J. Ablowitz, D.J. Kaup, and A.C. Newell, J. Math. Phys. 15, 1852 (1974).

28. D.J. Kaup, Phys. Rev. $\underline{A16}$, 704 (1977).
29. D.R. Nicholson (private communication).
30. A.V. Mikhailov, Pis'ma Zh. Eksp. Teor. Fiz. $\underline{23}$, 356 (1976) [Sov. Phys.-JETP
    Lett. $\underline{23}$, 320 (1976)].
31. D.J. Kaup and A.C. Newell, Lett. al Nuovo Cimento $\underline{20}$, 325 (1977).
32. D.J. Kaup and A.C. Newell, J. Math. Phys. $\underline{19}$, 798 (1978).
33. T. Kawata and H. Inoue, J. Phys. Soc. Japan $\underline{44}$, 1968 (1978).
34. V.S. Gerdzhikov, V.S. Ivanov, and P.P. Kulish, Teor. Math. Fiz. (USSR) $\underline{44}$,
    342 (1980) [Theor. Math. Phys. (USA) $\underline{44}$, 784 (1980)].
35. B.L. Holian and G.K. Straub, Phys. Rev. B $\underline{18}$, 1593 (1978).
36. T.G. Hill and L. Knopoff, J. Geophys. Res. $\underline{85}$, 7025 (1980).
37. D.J. Kaup, J. Math. Phys. $\underline{25}$, 277 (1984).
38. D.J. Kaup and D.H. Neuberger, J. Math. Phys. $\underline{25}$, 282 (1984).
39. J.N. Elgin and D.J. Kaup, Optics Comm. $\underline{43}$, 233 (1982).
40. J. Sargent III, M.O. Scully and W.E. Lamb, Laser Physics (Addison-Wesley
    Publishing Co., 1974).
41. M. Wadati, J. Phys. Soc. Japan, $\underline{52}$, 2642 (1983).
42. D.J. Kaup, Phys. Rev. B$\underline{29}$, 1072 (1984).
43. D.J. Kaup, Phys. Rev. B$\underline{27}$, 6787 (1983).

Hamiltonian Structures of Soliton Equations via Constrained Variational Calculus

Tu Gui-zhang

Computing Center of Chinese Academy of Sciences, Beijing, PR China

1. MAIN RESULT: A TRACE IDENTITY

We prove the following trace identity

$$(\delta/\delta u_\alpha)\,\mathrm{Tr}(K\,\partial U/\partial\eta) = (\partial/\partial\eta)\,\mathrm{Tr}(K\,\partial U/\partial u_\alpha), \tag{1}$$

where K satisfies the equation $K_x=[U,K]$ and $U=U(u_\alpha,\eta)$. We show that a number of known formulae, obtained previously by complicated calculation, are the special cases of this remarkable identity.

2. PRINCIPAL SPECTRAL PROBLEM AND ADJOINT REPRESENTATION

In the soliton literature the word 'integrable' (or 'completely integrable') is used quite frequently, however there is as yet no clear and authorized definition for it. We shall take the following working definition: A linear or nonlinear equation will be called integrable if it can be written as the integrability condition

$$U_t - V_x + [U , V] = 0 \tag{2}$$

of a couple of linear equations $\psi_x = U\psi$, $\psi_t = V\psi$, where $U=U(u,\eta)$ and $V=V(u,\eta)$ which are $N \times N$ matrices depending on a potential $u=(u_\alpha)$ and a spectral parameter η, and $[U , V] = UV - VU$.   We call

$$\psi_x = U \psi \tag{3}$$

the principal spectral problem and call

$$K_x = [U , K] \tag{4}$$

its adjoint representation.

3. CONSTRAINED VARIATIONAL CALCULUS

Suppose that we are given a scalar or matrix equation $R(U,W)=0$ which defines implicitly the function $W=W(U)$, and we want to calculate the variational derivative $\delta f(W)/\delta U$, where $f(W)$ is a scalar function of W, and $\delta/\delta U=(\delta/\delta U_j^{\ j})$, $\delta/\delta v= \Sigma(-D)^i \partial/\partial (D^i v)$ with $D=d/dx$. Viewing $R(U,W)=0$ as a constraint imposed upon U and W we introduce first a Lagrangian multiplier $\Lambda =(\Lambda_i^{\ j})$ which is of the same dimension as R, and make the sum $\phi = f(W)+\mathrm{Tr}(\Lambda^T R)$, then we set to zero the variational derivative $\delta\phi/\delta W$

$$0 = \hat\delta\phi/\delta W = \hat\delta f(W)/\delta W + (\hat\delta/\delta W)\mathrm{Tr}(\Lambda^T R), \tag{5}$$

where $\hat{}$ indicates that W, U and Λ are treated as independent in calculating $\hat{\delta}/\delta W$. After doing that we have finally

$$\delta f(W)/\delta U = (\ \hat{\delta}/\delta U) Tr(\Lambda^T R).$$ (6)

For the proof of the above procedure see /1/.

In most cases (6) can be solved for Λ in terms of $K= \delta f(W)/\delta U$. Substituting this expression into (5) we shall obtain an equation for K which is usually what we need. As an example of application of the constrained variational calculus we have given in /1/ a simple proof of the following theorem:

THEOREM 1. Let $\psi_x = U\psi$ be a linear differential equation, where $\psi=(\psi_1,\cdots,\psi_N)$, $U = (U_i^{\ j})$, and set $Y=(Y_i^{\ j})$, $Y_i^{\ j} = \psi_i/\psi_j$, $H=(UY)_D = diag(h_1,\cdots,h_N)$, the diagonal part of UY, and $K=(\delta h_s/\delta U)^T$ for a fixed s, then we have $TrK= 1$ and

$$K_x = [\ U\ ,\ K\ ]\ ,$$ (7)

$$K_i^{\ j} = Y_i^{\ m} K_m^{\ j}\ ,\quad \text{(for any m, } 1\leq m \leq N\ ).$$ (8)

The equation (7) was found previously by us in /2/. As we showed there, for many known spectral problems this basic equation provides us with the needed Hamiltonian structures.

To give the second example we need to introduce the operator $V_u(f)=\Sigma(\partial f/\partial(D^i u))D^i$, where $f=f(u,u_x,u_{xx},\cdots)$. The formal conjugation of $V_u(f)$ is defined in the usual way: $V_u^{\#}(f)= \Sigma(-D)^i(\partial f/\partial(D^i u)\cdot)$. The operator V_u is closely related to the variational derivative $\delta/\delta u$, we have developed in /3/ a series of formulae dealing with V_u and $\delta/\delta u$. One fact among others is that $(\delta/\delta u)fg=V_u^{\#}(f)g+V_u^{\#}(g)f$, which can be verified easily by the definition. If in particular g is independent of u in the above formula, then we have

$$(\delta/\delta u)fg = V_u^{\#}(f)g.$$ (9)

Now we consider the spectral problem /4,5,6/

$$Ly = (\ \overset{N}{\underset{k=0}{\Sigma}}\ u_k D^k)y = \eta y,\qquad (u_N=1,\quad u_{N-1}=0).$$ (10)

Setting $\chi = y_x/y$ we find

$$D^k y = P_k(\chi)y$$ (11)

where $P_0(\chi)=1, P_1(\chi)= \chi,\cdots, P_k(\chi)= (D+\chi)^k 1$. From (10) and (11) we obtain $\Sigma\ u_k P_k(\chi) - \eta = 0$. To calculate $\delta\chi/\delta u_k$, we introduce $\phi= \chi + \lambda(\ \Sigma\ u_k P_k(\chi)-\eta)$, and set according to the constrained variational calculus that $0=\hat{\delta}\phi/\delta\chi=1+\Sigma(\hat{\delta}/\delta\chi)(P_k(\chi)u_k\lambda)$, which, by (9), is the same as $1+ \Sigma\ V_\chi^{\#}(P_k)(\lambda u_k)=0$. Furthermore we have $\delta_\chi/\delta u_k = (\hat{\delta}/\delta u_k)\Sigma\ (P_k(\chi)u_k\lambda)=P_k\lambda$, especially $S \equiv -\delta\chi/\delta u_0 = -\lambda$. Thus we are led to the equation

$$\delta\chi/\delta u_k = -P_k S,\quad k=0,1,\cdots,\ N-2,$$ (12)

and

$$\sum_{\chi} V_{\chi}^{\#}(P_k)(u_k S) = 1. \tag{13}$$

The equations (12) and (13) were obtained first by Guerrero in /6/ with a much more complicated argument.

By the way we note that (12) can be derived also from (8). Since the spectral problem (10) is equivalent to (3) with $\psi = (\psi_1, \cdots, \psi_N)$, $\psi_{k+1} = D^k y$ and $U = \sum_{i=1}^{N-1} E_i^{i+1} + \eta E_N^1 - \sum_{k=0}^{N-2} u_k E_N^{k+1}$, where $(E_i^j)_k{}^s = \delta_{ik}\delta_{js}$. We have from Theorem 1 and (11) that $Y_{k+1}^1 = \psi_{k+1}/\psi_1 = (D^k y)/y = P_k$, $h_1 = (UY)_1^1 = Y_2^1 = P_1 = \chi$, and

$$K_{k+1}^n = (\delta h_1/\delta U)_n^{k+1} = \delta h_1/\delta U_n^{k+1} = -\delta\chi/\delta u_k. \tag{14}$$

Thus the equation (12) is nothing but (8) with $i=k+1$, $j=n$ and $m=1$.

4. PROOF OF THE TRACE IDENTITY

We need the following simple formulae on trace of matrices

$$(\hat{\delta}/\delta A)\,\mathrm{Tr}(AB) = B^T, \tag{15}$$

$$(\hat{\delta}/\delta C)\,\mathrm{Tr}(A\,[B\,,\,C]) = [A\,,\,B]^T. \tag{16}$$

To prove the trace identity (1) we set $\Delta = \mathrm{Tr}(U_\eta K)$ and introduce $\phi = \Delta + \lambda(\Delta - \mathrm{Tr}(U_\eta K)) + \mathrm{Tr}(\Lambda^T(K_x - [U\,,\,K]))$ with λ and Λ being the Lagrangian multipliers. According to the constrained variational calculus we set $0 = (\hat{\delta}/\delta\Delta)\phi = 1 + \lambda$, and $0 = (\hat{\delta}/\delta K)\phi = -\lambda U_\eta^T - \Lambda_x - [\Lambda^T, U]^T$. Here we have made the use of (15) and (16). Hence we deduce that $\lambda = -1$ and

$$(\Lambda^T)_x = [U\,,\,\Lambda^T\,] + U_\eta. \tag{17}$$

Furthermore we have

$$\delta\Delta/\delta u_\alpha = (\hat{\delta}/\delta u_\alpha)(-(\Delta - \mathrm{Tr}(U_\eta K)) + \mathrm{Tr}(\Lambda^T(K_x - [U\,,\,K])))$$
$$= \mathrm{Tr}(K\,\partial U_\eta/\partial u_\alpha) + \mathrm{Tr}([\Lambda^T\,,\,K]\,\partial U/\partial u_\alpha). \tag{18}$$

Now from (17) and (4) and the Jacobi identity for matrices we have on the one hand

$$[\Lambda^T\,,\,K]_x = [\Lambda_x^T\,,\,K] + [\Lambda^T\,,\,K_x]$$
$$= [[U, \Lambda^T], K] + [\Lambda^T, [U, K]] + [U_\eta, K]$$
$$= [U, [\Lambda^T, K]] + [U_\eta, K],$$

On the other hand we have $K_{\eta x} = (\partial/\partial\eta)[U, K] = [U, K_\eta] + [U_\eta, K]$. Hence both $[\Lambda^T, K]$ and K_η satisfy the nonhomogeneous equation $G_x = [U, G] + [U_\eta, K]$. Under the assumption that the equation $G_x = [U, G]$ admits an unique solution of the form $G = \sum G_n \eta^{-n}$ for a given first nonzero G_n we can conclude that $K_\eta = [\Lambda^T, K]$, if we choose the first nonzero term in $\Lambda = \sum \Lambda_n \eta^{-n}$ to be the same as in $K_\eta = \sum (K_\eta)_n \eta^{-n}$. Substituting $K_\eta = [\Lambda^T, K]$ into (18) we find that $\delta\Delta/\delta u_\alpha = \mathrm{Tr}(K(\partial U/\partial u_\alpha)_\eta + K_\eta(\partial U/\partial u_\alpha)) = (\partial/\partial\eta)\mathrm{Tr}(K\partial U/\partial u_\alpha)$, which completes the proof.

We note that if U is of the form $U = U(\eta, u_\alpha, u_{\alpha x}, u_{\alpha xx}, \cdots)$ which is dependent not only on $u = (u_\alpha)$ but also on its x-derivatives $D^k u$, then the above trace identity (1) should be changed to, as easily seen from the above proof and the formula (9),

$$(\delta/\delta u_\alpha)\mathrm{Tr}(K\, \partial U/\partial \eta) = (\partial/\partial \eta)\mathrm{Tr}(V_{u_\alpha}^{\#}(U)K) ,$$

or

$$(\delta/\delta u_\alpha)\mathrm{Tr}(K\, \partial U/\partial \eta) = (\partial/\partial \eta)\mathrm{Tr}((\hat{\delta}/\delta u_\alpha)(UK)), \qquad (19)$$

where, as before, the hat indicates that K should be treated as being independent of u in calculating $(\hat{\delta}/\delta u_\alpha)UK$.

The formula (19) reveals a kind of commutativity of two operations $\delta/\delta u_\alpha$ and $\partial/\partial \eta$ as if K is really independent of η and u. This fact reminds us of two other intriguing commutativities: the commutativity of Backlund transformations, and the commutativity of the operators $U_u(f) \equiv \Sigma(D^i f)(\partial/\partial(D^i u))/7/$.

5. CONSEQUENCES OF THE TRACE IDENTITY

We show that various known results are the special cases of the trace identity (1).

(1) Consider the spectral problem /8/

$$(-D^2 + (c_1 + c_2 u) + \eta(d_1 + d_2 u))y = 0,$$

where c_i and d_j are constants. By setting $u_1 = c_1 + c_2 u$ and $u_2 = d_1 + d_2 u$ we can rewrite the above spectral problem in the standard form (3) with $\psi = (y, y_x)$ and the 2×2 matrix $U = E_1^2 + (u_1 + \eta u_2)E_2^1$. Then $\mathrm{Tr}(K\, \partial U/\partial u) = \mathrm{Tr}(Ku_2 E_2^1) = u_2 K_1^2$, and $\mathrm{Tr}(K\, \partial U/\partial u) = \mathrm{Tr}(K(c_2 + \eta d_2)E_2^1) = (c_2 + \eta d_2)K_1^2$. Thus the trace identity (1) gives

$$(\delta/\delta u)u_2 R = (\partial/\partial \eta)(c_2 + \eta d_2)R , \qquad (20)$$

where $R = K_1^2$. This equation was obtained by Gelfand and Dorfman /8/. In the special case when $c_1 = d_2 = 0$, $c_2 = d_1 = 1$, (20) reduces to

$$(\delta/\delta u)R = (\partial/\partial \eta)R,$$

an equation which was obtained by Gelfand and Dikii /9/, and Yusin /10/.

We note that the above quantity $R = K_1^2$ is the same as in /8/. In fact we have from Theorem 1 that $R_x = (K_1^2)_x = [U, K]_1^2 = K_2^2 - K_1^1 = 2K_1^2 Y_2^1 - 1$, hence

$$R_x = 2R\chi - 1 \qquad (21)$$

where $\chi = Y_2^1$ for which

$$\chi_x = (Y_2^1)_x = (UY)_2^1 - Y_2^1(UY)_1^1 = (u_1 + u_2\eta) - \chi^2 \qquad (22)$$

It is easily seen from (21) and (22) that

$$-2RR_{xx} + R_x^2 + 4(u_1 + \eta u_2)R^2 = 1,$$

this is just the equation (3) in /8/.

(2) We return to the spectral problem (10). Starting from the corresponding matrix U as shown in section 3 we find

$$\mathrm{Tr}(K\,\partial U/\partial \eta) = \mathrm{Tr}(KE_N^{\,1}) = K_1^{\,N} = -\delta\chi/\delta u_0 = S,$$
$$\mathrm{Tr}(K\,\partial U/\partial u_k) = \mathrm{Tr}(KE_N^{\,k+1}) = -K_{k+1}^{\,N} = \delta\chi/\delta u_k.$$

Hence the trace identity reads

$$\delta S/\delta u_k = (\partial/\partial\eta)(\delta\chi/\delta u_k). \tag{23}$$

This equation was first obtained by Gelfand and Dikii /4/ in the context of the resolvent method, and then by Guerrero /6/. Both of their derivations are quite complicated.

(3) Now we consider the well-known spectral problem (see, e.g. /11,12/)

$$\psi_x = U\psi, \qquad U = \eta A + P,$$
$$A = \mathrm{diag}(a_1,\cdots,a_N), \quad \mathrm{Tr}A = 0; \quad P = (P_i^{\,j}), \quad \mathrm{diag}\,P = 0.$$

In this case, $\mathrm{Tr}(K\,\partial U/\partial\eta) = \mathrm{Tr}(AK)$, $\mathrm{Tr}(K\,\partial U/\partial P_i^{\,j}) = \mathrm{Tr}(KE_i^{\,j}) = K_j^{\,i}$. Hence we have $(\delta/\delta P_i^{\,j})\mathrm{Tr}(AK) = (\partial/\partial\eta)K_j^{\,i}$, or equivalently

$$(\delta/\delta P^T)\mathrm{Tr}(AK) = (\partial/\partial\eta)K_F, \tag{24}$$

where K_F denotes the off-diagonal part of K.

In the special case when N=2, we have $U = \eta(E_2^{\,2} - E_1^{\,1}) + qE_1^{\,2} + rE_2^{\,1}$, and

$$K = \begin{pmatrix} 1-\Delta & S \\ R & \Delta \end{pmatrix} \tag{25}$$

with

$$R = \delta h/\delta q, \quad S = \delta h/\delta r, \quad \Delta = ZS, \quad Z = Y_2^{\,1}, \quad h = qZ.$$

and (24) reduces to

$$(\delta/\delta q)(2\Delta) = (\partial R/\partial\eta), \quad (\delta/\delta r)(2\Delta) = (\partial S/\partial\eta). \tag{26}$$

These two equations were obtained by us in /1/. Set $\Delta = \sum_{n=1}^{\infty} \Delta_n \eta^{-n}$, $Z = \sum Z_n \eta^{-n}$, the equations (26) are easily seen to be equivalent to

$$2\Delta_{n+1} \overset{D}{\sim} -nqZ_n, \tag{27}$$

where $f \overset{D}{\sim} g$ means $f - g = Dh$ for some h. The correctness of (27) can be directly verified by using the recurrence formulae of Δ_n and Z_n (see /13/).

(4) As a final example, we consider the spectral problem (3) with $U = i\eta^2(E_2^{\,2} - E_1^{\,1}) + \eta(qE_1^{\,2} + rE_2^{\,1})$, where $E_i^{\,j}$ are 2×2 matrices (see /13, 14/). In this case we have

$$\mathrm{Tr}(K\,\partial U/\partial\eta) = 2i\eta(K_2^{\,2} - K_1^{\,1}) + qK_2^{\,1} + rK_1^{\,2},$$
$$\mathrm{Tr}(K\,\partial U/\partial q) = \eta K_2^{\,1}, \quad \mathrm{Tr}(K\,\partial U/\partial r) = \eta K_1^{\,2}.$$

We have also the equation (25), or equivalently

$$K_1^{\,2} = S, \quad K_2^{\,1} = R, \quad K_1^{\,1} = 1 - ZS, \quad K_2^{\,2} = ZS.$$

Setting $\Delta = 2i\eta ZS$ as in /13/, we obtain thus from the trace identity that

$$(\delta/\delta q)(-2\Delta + qR + rS) = (\partial/\partial\eta)(\eta R),$$
$$(\delta/\delta r)(-2\Delta + qR + rS) = (\partial/\partial\eta)(\eta S).$$

These two equations are equivalent to

$$\Delta_n - (qR_n + rS_n)/2 \overset{D}{\sim} n(qZ_n), \tag{28}$$

where we have made the following expansions

$$\Delta = \sum_{k=0}^{\infty} \Delta_n \eta^{-(2n+1)}, \qquad Z = \sum Z_n \, \eta^{-(2n+1)},$$

and the similar expansions for R and S. The corrctness of (28) can be verified by using the recurrence formulae for Δ_n, Z_n, R_n and S_n (see /13/)

6. GENERATING CONSERVED DENSITY FROM ADJOINT REPRESENTATION

In /11/ Alberty, Koikawa and Sasaki proposed a method for generating the conserved density (i.e., $H=(UY)_D$, see Theorem 1) starting from the principal spectral problem (3). We point out that the trace identity (1) gives another method for generating the conserved density (i.e., the quantity $\Delta = \mathrm{Tr}(U_\eta K)$) starting from the adjoint representation (4). To examine more clearly, we consider the problem (3) with $U=\eta A +P$. In this case we have from (24) that $(\delta/\delta P^T)\mathrm{Tr}(AK)=(\delta/\delta P^T)(\partial h/\partial\eta)$, which is equivalent to

$$\mathrm{Tr}(AK) \underset{\sim}{\overset{D}{}} \partial h/\partial\eta \ . \tag{29}$$

Here we have made the fact that $K_F = \delta h/\delta P^T$, $h=(UY)_s^s$ (for any fixed s) as seen from Theorem 1. The equation (29) shows that we can choose as Hamiltonian the quantities Δ_n, where $\Delta = \mathrm{Tr}(AK) = \sum \Delta_n \eta^{-n}$. Noting that (see, e.g.,/15/) $\mathrm{Tr}(AK)=K(A,K)/(2N)$, where $K(\ , \ )$ represents the Killing–Cartan form of the Lie algebra $sl(N-1)$, we conclude that $K(A,K)$ can be chosen as Hamiltonian of the hierarchy of equations related to the spectral problem (3) with $U=\eta A+P$. This is what Wilson had done in his paper /16/.

In general it is easy to see that if $\partial U/\partial u = E_\alpha$ are constant matrices, and the number of nonzero entries in each E_α is the same, then $\mathrm{Tr}(U_\eta K)$ is a conserved density. The problem whether or not this is true in general, i.e., $\mathrm{Tr}(U_\eta K)$ is always a conserved density, remains open.

ACKNOWLEDGMENT

The author is much obliged to the organizer of the 7th Kyoto Summer Institute for the financial support. He would like also to express his hearty thanks to Tian He-zhi for the much valuable assistance during the progress of the work, and to Men Da-zhi for the helpful discussions.

REFERENCES
1. Tu G.Z., J. Eng. Math.,1(1984), 7.(in Chinese)
2. BOITI M., PEMPINELLT F. and TU G.Z., Nuovo Cimento 79B(1984), 231.
3. TU. G.Z., J.Math. Anal. Appl., 94 (1983), 348.
4. GELFAND I.M. and DIKII L.A., Funct. Anal. Appl. 10(1976), 13.(in Russian).
5. ADLER M., Invent. Math., 50(1979), 219.
6. GUERRERO G.F., J. Math. Phys., 23 (1982), 211.
7. TU G.Z., Comm. Math. Phys., 77 (1980), 289.
8. GELFAND I.M. and DORFMAN I.YA., Current problems of mathematical physics and of numerical mathematics, Moskva (1982), 102. (in Russian).

9. GELFAND I.M. and DIKII L.A., Uspekhi Nauk Matem., $\underline{30}$(1975), 67. (in Russian).
10. YUSIN B.V., ibid, $\underline{33}$(1978), 233. (in Russian).
11. ALBERTY T.M., KOIKAWA T. and SASAKI R., Physica $\underline{5D}$ (1982), 43.
12. BOITI M. and TU G.Z., Nuovo Cimento, $\underline{71B}$ (1982), 253.
13. TU G.Z., Sci. Exploration, 2 (1982), $\overline{85}$.
14. KAUP D.J. and NEWELL A.C., J.Math. Phys., $\underline{19}$ (1978), 798.
15. HELGASON S., Differential Geometry, Lie groups and symmetric spaces, Academic Press: New York, 1978.
16. WILSON G.,   Ergod. Th. Dynam. Sys., $\underline{1}$ (1981), 361.

Perturbative Studies of the Zakharov-Shabat Scattering Problem

Z.V. Lewis and J.N. Elgin

Department of Mathematics, Imperial College , London SW7 2BZ, Great Britain

K.J. Blow and N.J. Doran

British Telecom Research Laboratories, Martlesham Heath, Ipswich
Suffolk IP7 5RE, Great Britain

Optical pulse propagation through monomode fibres is described by the nonlinear Schrödinger equation (NLS)[1-3]. For an ideal system this takes the form

$$iq_t + q_{xx} + q|q|^2 = 0 \qquad\qquad (1)$$

with q a known function at t=0. In practice, equation 1 must be modified to incorporate other effects such as loss or stimulated scattering. Also, q may not be well-defined. For example, the laser source producing the optical pulse input to the fibre may not be bandwidth limited resulting in a frequency chirped input pulse. The nature of the chirp may be known or stochastic in the sense that there is a random phase variation along the pulse. It is clearly desirable to know how such effects modify the scattering data, since they will directly affect the properties of the optical solitons.

We consider here both the effect of stochastic initial data, deterministic chirp and loss. First a simple perturbative treatment is presented. In particular, for the case when a stochastic phase variation is present along the pulse, the soliton properties are known only in terms of a Gaussian distribution. Second we apply a WKB technique to study the scattering problem, equation 2, when a deterministic chirp is present on the initial pulse. Finally, we conclude with some results for the case when a linear loss term is added to the NLS.

The Zakharov-Shabat (Z-S) scattering problem is

$$L\phi(\zeta,x) = \zeta\phi(\zeta,x), \qquad\qquad L = \begin{bmatrix} i\partial_x & -iq \\ -iq^* & -i\partial_x \end{bmatrix} . \qquad\qquad (2)$$

Our notation throughout is that used in AKNS [2]. We are interested in shifts of the scattering data under the perturbation $q \to qe^{i\beta}$ or equivalently $L \to L+V$, $V = -\frac{1}{2}\beta x$. Using standard techniques of perturbation theory for non-adjoint systems, the shift in the eigenvalue $\zeta_n : \zeta_n \to \zeta_n + \Delta\zeta_n^{(1)} + \Delta\zeta_n^{(2)} + \cdots$ of the bound state $\phi_n(\zeta_n,x)$ is given by

$$\Delta\zeta_n^{(1)} = <\overline{\phi}_n|V|\phi_n> \equiv \overline{V},$$

$$\Delta\zeta_n^{(2)} = <\overline{\phi}_n|VG_0(\overline{V}-V)|\phi_n>, \qquad\qquad (3)$$

where G_0 is the propagator of the unperturbed system. Equation 3 are transformed into algebraic form using the following definitions of inner products together with the completeness relation shown

$$<x|\phi_n> \equiv \phi_n(\zeta_n,x), \qquad\qquad <\phi_n|x> \equiv -\overline{\phi}_n^{\ T}(\zeta_n^*,x)\sigma_1,$$

$$<x|\overline{\phi}_n> \equiv \overline{\phi}_n(\zeta_n^*,x), \qquad\qquad <\overline{\phi}_n|x> \equiv \phi_n^{\ T}(\zeta_n,x)\sigma_1,$$

$$<x|\phi_\xi> \equiv \phi_\xi(\xi,x), \qquad\qquad <\phi_\xi|x> \equiv -\overline{\phi}_\xi^{\ T}(\xi,x)\sigma_1,$$

$$<x|\overline{\phi}_\xi> \equiv \overline{\phi}_\xi(\xi,x), \qquad\qquad <\overline{\phi}_\xi|x> \equiv \phi_\xi^{\ T}(\xi,x)\sigma_1,$$

$$<\overline{\phi}_n|\phi_m> = <\phi_n|\overline{\phi}_m> = \delta_{nm},$$
$$<\phi_\xi|\psi_{\xi'}> = <\overline{\phi}_\xi|\psi_{\xi'}>^* = 2\pi\, a_\xi^*\, \delta(\xi - \xi'). \tag{4}$$

All other inner products are identically zero. The functions ϕ and ψ, together with their adjoints $\overline{\phi}$ and $\overline{\psi}$, are defined in AKNS [2], as is the scattering coefficient ∂_ξ. The required completeness relation is

$$\mathbf{1} = \sum_n |\phi_n><\overline{\phi}_n| + \sum_n |\overline{\phi}_n><\phi_n| + \frac{1}{2\pi} \int_{-\infty}^{\infty} \left(|\psi_\xi> \frac{1}{a_\xi} <\overline{\phi}_\xi| + |\overline{\psi}_\xi> \frac{1}{a_\xi^*} <\phi_\xi| \right) d\xi. \tag{5}$$

Let the perturbation V correspond to a phase variation (either stochastic or deterministic) along $q: q \to q e^{i\beta}$. Then, $V = -\frac{1}{2}\beta_x$. Hence

$$\Delta\zeta_n^{(1)} = -\frac{1}{2} \int \phi_n^T(\zeta_n,x)\sigma_1\, \phi_n(\zeta_n,x)\, \beta_x\, dx. \tag{6}$$

If the initial unperturbed q is real, then ζ_n is purely imaginary and the bound states $\phi_n(\zeta_n,x)$ are real quantities. The first order shift is real and corresponds to a change in the soliton's velocity. Moreover, if the unperturbed potential is an even function of x, then $\phi_n^T(\zeta_n,x)\sigma_1\phi_n(\zeta_n,x)$ is also even [3], and so only odd phase functions β contribute to the shift $\Delta\zeta_n^{(1)}$.

We are particularly interested in the case when β is a stochastic random variable [4]. The soliton properties are then no longer well-defined, but are determined by a Gaussian distribution whose width is simply related to the excess bandwidth of the input pulses. The stochastic model studied here- the phase diffusion model in the context of laser physics- has been discussed in a previous publication [4]; briefly the phase of the pulse is governed by a Langevin equation

$$\partial\beta/\partial x \equiv \beta_x = F(x), \tag{7}$$

where F describes a random Wiener process

$$<F(x)> = 0, \qquad <F(x)\, F(x')> = 2\Gamma\delta\,(x-x'). \tag{8}$$

Angular brackets indicate an ensemble averaged quantity. The quantity Γ is a measure of the correlation length of the input pulse. If $F(x)$ is a Gaussian stochastic quantity, then the n'th order correlation $<F(x_1) \cdots F(x_n)>$ can be expressed in terms of two time correlations

$$<F(x_1) \cdots F(x_n)> = 0, \qquad\qquad n\ \text{odd}$$
$$= \sum_{perm} <F(x_1)\, F(x_2)> \cdots <F(x_n)\, F(x_{n-1})>. \tag{9}$$

These considerations, together with equations 6 and 7, produce the following probability distribution for the shifts

$$p(\Delta\zeta_n^{(1)}) = \frac{1}{\sqrt{\pi\alpha\Gamma}}\, \exp\, \{-(\Delta\zeta_n^{(1)})^2/\alpha\Gamma\} \tag{10}$$

where

$$\alpha = \int_{-\infty}^{\infty} \{\phi_n^T(\zeta_n,x)\, \sigma_1\, \phi_n(\zeta_n,x)\}^2\, dx.$$

Note the importance of this result in the context of optical fibre systems: if the input pulses to the fibre are produced having a stochastic phase structure, then the soliton velocities are no longer well determined, but are known only in terms of the distribution function equation 10. Clearly, the integrity of high bit-rate transmission signals will be lost if the fibre length is too long. This point will be discussed in more detail elsewhere.

Consider now the second order shift $\Delta\zeta_n^{(2)}$. We restrict attention to the case when the unperturbed q is real. Then, using equations 3-5 in equation 2 produces

$$\Delta\zeta_n^{(2)} = i\sum_{m\neq n}^N \frac{M_{nm}^2}{\eta_m - \eta_n} + i\sum_m^N \frac{\overline{M}_{nm}^2}{\eta_m + \eta_n} \tag{11}$$

$$- \int_0^\infty \left\{\frac{M_{n\psi}M_{\phi n}}{\partial_\xi(\xi - i\eta_n)} - c.c.\right\} d\xi - \int_0^\infty \left\{\frac{M_{n\psi}M_{\phi n}}{\partial_\xi^*(\xi - i\eta_n)} - c.c.\right\} d\xi.$$

Here, all zeros of the unperturbed problem lie on the imaginary axis which we denote $\zeta_m = i\eta_m$, m=1, N. The matrix elements are defined by

$$M_{nm} = \langle\overline{\phi}_n|V|\phi_m\rangle = \langle\overline{\phi}_m|V|\phi_n\rangle,$$

$$\overline{M}_{nm} = \langle\overline{\phi}_n|V|\overline{\phi}_m\rangle = -\langle\phi_m|V|\phi_n\rangle, \tag{12}$$

$$M_{n\psi} = \langle\overline{\phi}_n|V|\psi_\xi\rangle, \qquad M_{n\psi} = \langle\overline{\phi}_\xi|V|\phi_n\rangle.$$

$$M_{n\overline{\psi}} = \langle\overline{\phi}_n|V|\overline{\psi}_\xi\rangle, \qquad M_{n\overline{\phi}} = \langle\overline{\phi}_\xi|V|\phi_n\rangle.$$

Moreover, M_{nm} and $\overline{M}_{nm}$ are easily shown to be real quantities so that the second order shift in the eigenvalue is imaginary. For the stochastic case, the average value for $\Delta\zeta^{(n)}$ then decreases by an amount proportional to Γ; more details will be published elsewhere.

Consider now the limit in which the WKB technique can be used. All the scattering data can be obtained directly by solving the Z-S scattering problem in the semiclassical limit. This is the asymptotic limit where the structure of the wavefunctions is dominated by the action function, S, which satisfies the classical limit of the system [5,6].

We introduce an ordering parameter, ε, (supposed small with respect to all other quantities) by rescaling such that $\partial/\partial x \to \varepsilon\partial/\partial x$:

$$-i\varepsilon v_{1x} + \zeta v_1 = -iqv_2, \qquad -i\varepsilon v_{2x} - \zeta v_2 = iq^*v_1. \tag{13}$$

Eliminating v_2 from 1 we get

$$\varepsilon^2 v_{1xx} - \varepsilon^2 \frac{q_x}{q} v_{1x} + \left\{-i\varepsilon\zeta\frac{q_x}{q} + (\zeta^2 + q\,q^*)\right\} v_1 = 0. \tag{14}$$

The semiclassical solutions to 14 as $\varepsilon \to 0$ is given by a WKB expansion of the form

$$v_1(x) = (A^{(0)}(x) + \varepsilon A^{(1)}(x) + \varepsilon^2 A^{(2)}(x) + \cdots)\,\exp(iS/\varepsilon)$$

$$+ (B^{(0)}(x) + \varepsilon B^{(1)}(x) + \varepsilon^2 B^{(2)}(x) + \cdots)\,\exp(-iS/\varepsilon). \tag{15}$$

Substituting 15 into 14 and equating coefficients of powers of ε we obtain a series of coupled differential equations for S, $A^{(n)}$, $B^{(n)}$. The order of approximation required depends on the complexity of the potential q. It is in principle possible to obtain the complete scattering data in analytic form, finally rescaling to eliminate ε.

In the case of the square initial pulse

$$\begin{aligned} q &= q_0, \qquad |x| < d/2 \\ &= 0, \qquad\ \ |x| > d/2 \end{aligned} \tag{16}$$

the solution 15 truncates exactly such that $A^{(n)}$, $B^{(n)}$ are all identically zero for $n>0$, A^0, B^0 are constant. The bound state eigenvalues are then given by the Bohr-Sommerfeld quantisation condition

$$S(d/2) - S(-d/2) = \int_{-d/2}^{d/2} \sqrt{\zeta^2 + q^2_0} \, dx \rightarrow \sqrt{q^2_0 - \eta^2} \, d = (\eta + \tfrac{1}{2})\pi. \tag{17}$$

The method has been applied in detail to the chirped square well. Consider first the potential

$$q = q_0 \exp\{i\beta(x)/\varepsilon\}, \qquad |x| < d/2$$
$$= 0, \qquad\qquad\qquad |x| > d/2. \tag{18}$$

Symmetry suggests that we look for solutions of the form

$$v_1 = C_1(x) \exp\{i(S + \beta/2)/\varepsilon\} + D_1(x) \exp\{-i(S - \beta/2)/\varepsilon\}$$
$$v_2 = C_2(x) \exp\{i(S - \beta/2)/\varepsilon\} + D_2(x) \exp\{-i(S + \beta/2)/\varepsilon\} \tag{19}$$

where

$$S = \int_{-d/2}^{x} \sqrt{(\zeta + \tfrac{\beta_x}{2})^2 + q^2_0} \, dx,$$

$$C_1 = \frac{C_0}{\sqrt{S_x}} \exp\{-\int_{-d/2}^{x} \frac{\beta_{xx}}{4S_x} dx\}, \qquad C_2 = \frac{i}{q_0}\{(\zeta + \frac{\beta_x}{2} + S_x\}C_1$$
$$D_1 = \frac{D_0}{\sqrt{S_x}} \exp\{+\int_{-d/2}^{x} \frac{\beta_{xx}}{4S_x} dx\}, \qquad D_2 = \frac{i}{q_0}(\zeta + \frac{\beta_x}{2}) - S_x\}D_1. \tag{20}$$

The expression for S is to be contrasted with that of the unchirped case. As is to be expected, the wavefunctions diverge at the classical turning points where $S_x=0$. Also it is easy to show that the coefficients satisfy conservation laws

$$C_1^2 + C_2^2 = \text{constant}$$
$$D_1^2 + D_2^2 = \text{constant}. \tag{21}$$

When β is small the potential is more easily treated by taking

$$q = q_0 \exp\{i\beta(x)\}, \qquad |x| < d/2$$
$$= 0, \qquad\qquad\qquad |x| > d/2 \tag{22}$$

but in this case it is necessary to go to higher order in the expansion 15 because S will always be given by the same expression as for the unchirped case 17, corresponding to the first term in a Taylor expansion with respect to β of the expression in 20. Closed form solutions for the transmission coefficient, a, are easily obtained for smooth β and compare favourably with exact calculations. Figures 1a-c show the variation of a with ζ constrained to be real $-\infty<\zeta<\infty$. The number of bound states (soliton solutions) is given by the total phase change divided by 2π [2]. Figure 1a shows the unchirped case for $q_0 d=3\pi/2$ on the boundary between one and two states. If we now modulate the potential with a phase $\beta=(x+d/2)^3$ results in the definite creation of two solitons in both the exact calculation, Fig. 1b, and the WKB solution, Fig. 1c.

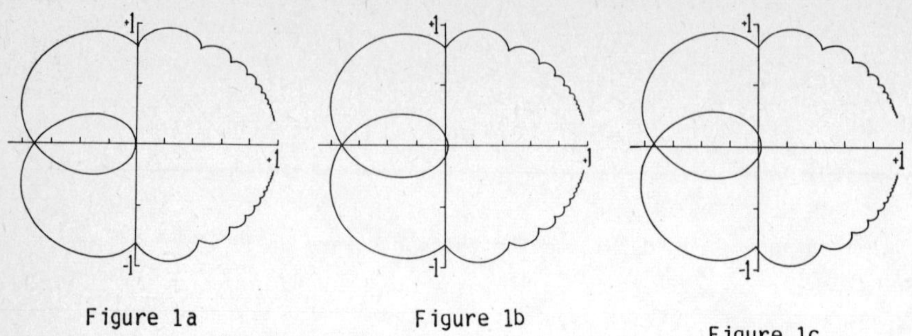

Figure 1a　　　　Figure 1b　　　　Figure 1c

We now consider a slightly different problem in which the NLS is perturbed rather than the initial data for the associated Z-S scattering equations. In particular we include the effects of linear absorption which yield the following equation

$$iq_t + q_{xx} + q|q|^2 = -i\Gamma q \tag{23}$$

Numerical integration of this equation [7] has given us some insight into the effects of loss and the stability of the soliton solutions. These investigations involve the study of some parameter such as the pulse width which can be used as an indication of the propagation of the solitons. For example the two soliton solution given by $q(t=0,x)=2\mathrm{sech}(x)$ has a pulse width which oscillates with period $\pi/4$ when $\Gamma=0$. In the presence of loss this period increases with t and eventually freezes out to behave as a linear pulse as the effect of the nonlinear term becomes negligible. However it is difficult to be precise about when the soliton nature of the propagation has become unimportant.

Let us consider a hybrid approach where we use the scattering equations to define the presence or absence of solitons but use the NLS to solve for the propagation of u. The Z-S problem is now written as

$$L(t)\phi_n(\zeta_n(t),x) = \zeta_n(t)\phi_n(\zeta_n(t),x) \tag{24}$$

and is considered as a separate problem for each t. Physically the $\zeta_n(t)$ define the solitons that would remain if the perturbation $i\Gamma q$ were switched off after a time t. We solve for $q(t,x)$ by numerical integration but now calculate $\zeta_n(t)$ from the scattering problem as q evolves. In Fig. 2 we show some preliminary results for the double soliton. The loss corresponds to $\Gamma=.05$. The scattering data decays monotonically toward zero as one would expect from the increase in the pulse width observed in the numerical simulation. The lower eigenvalue

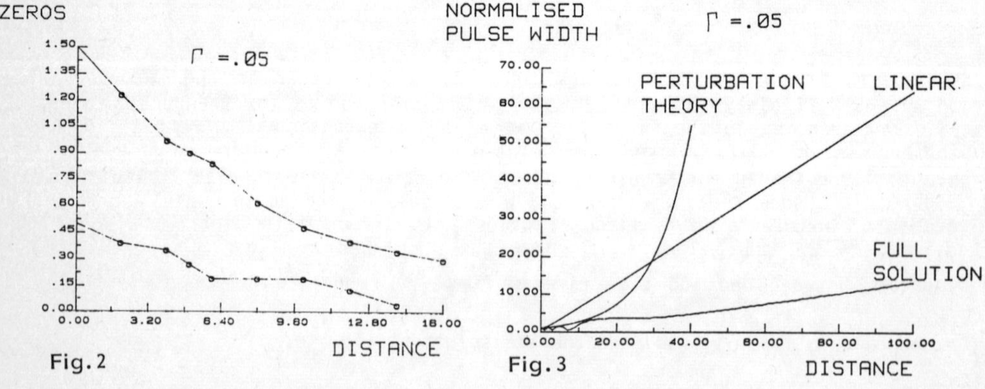

disappears when $\Gamma t \sim .7$ which corresponds well with the point at which the pulse oscillations freeze out (see Fig. 3 at $t \sim 14$). The upper eigenvalue appears to decrease exponentially which would imply that it never crosses the axis which is not what one would deduce from the pulse width evolution. This matter is still under investigation and will be reported elsewhere.

References
1. V.E. Zakharov and A.B. Shabat, Sov. Phys. JETP 34, 62 (1972)
2. M.J. Ablowitz, D.J. Kaup, A.C. Newell and H.Segur, SIAM 53, 249 (1974)
3. J. Satsuma and N. Yajima, Suppl. Prog. Theor. Phys. 55, 284 (1974)
4. J.N. Elgin and D.J. Kaup, Opt. Commun. 43, 233 (1982)
5. V.E. Zakharov, Funktional'nyi Analiz i Ego Prilozheniya 14, 15 (1980)
6. V.E. Zakharov, Physica 3D (1981)
7. N.J. Doran and K.J. Blow, IEEE J. Quant Elec. QE19, 1883 (1983)

The Zabolotskaya-Khokhlov Equation and the Inverse Scattering Problem of Classical Mechanics

John Gibbons

Department of Mathematics, Imperial College, London SW7 2BZ, Great Britain

Abstract   The Zabolotskaya-Khokhlov equation,

$$U_{TX} + (U^2/2)_{XX} = U_{YY}$$

which models beams of nonlinear, diffracting sound waves in two dimensions, may be written as the compatibility condition between two classical Hamiltonian flows, one evolving in Y with Hamiltonian $H_2 = P^2/2 + U(X,Y,T)$, and the other evolving in T, with Hamiltonian $H_3 = P^3/3 + P\,U(X,Y,T) + V(X,Y,T)$. Some solutions of the system may consequently be derived from solutions of the Benney hierarchy, a class of moment equations which arose originally as a description of long waves on a shallow perfect fluid, while the initial value problem is equivalent to the open problem of finding the potential $U(X,Y,T)$ in the Hamiltonian H_2, given the asymptotics of its orbits.

Section 1 Introduction

Let us consider a beam of sound waves, propagating almost parallel to the X-axis in the X-Y plane. The dispersion relation,

$$\omega^2 = k_X^{\ 2} + k_Y^{\ 2} \tag{1.1}$$

may then be approximated by:

$$\omega = k_X + k_Y^{\ 2}/2k_X \tag{1.2}$$

in which the first term describes propagation along the X-axis, and the second term describes diffraction effects. This gives us, after transforming to a frame moving in the X-direction at the speed of sound, and rescaling coordinates, the equation:

$$U_T = (\partial/\partial X)^{-1}\, U_{YY} \tag{1.3}$$

If we now include a UU_X term to describe a convective nonlinearity, we may obtain:

$$U_T + UU_X = (\partial/\partial X)^{-1}\, U_{YY} \tag{1.4}$$

which is the Zabolotskaya-Khokhlov equation [1]. Essentially it arises in similar circumstances to the Kadomtsev-Petviashvili equation,

$$u_t + uu_x - u_{xxx}/12 = (\partial/\partial x)^{-1}\, u_{yy} \tag{1.5}$$

when the u_{xxx} dispersion term may safely be neglected. This is the case, for

example, in some problems of sound propagation underwater. Since the dispersive
term is absent, it is not at all clear that there exist smooth global solutions,
since the UU_X term would tend to form shocks and the diffraction term might not be
able to prevent it.

In [2], Vinogradov and Vorob'ev constructed a class of solutions of (1.4),
using symmetry arguments. However few other solutions are known, if any. Since
(1.4) may be derived from (1.5) by a scaling transformation which removes its
dispersive term, one might hope that the appropriate limit of the linear problem
which solves the latter would lead to solutions of (1.4) itself. The present
paper is concerned with this approach.

In section 2, the semi-classical limit of the linear problem for (1.5) is
discussed. In section 3, the connection with the Benney hierarchy is used to
derive the 'hidden' symmetries of (1.4), while in section 4 we are concerned
with the initial value problem, which is connected with the problem of finding the
potential in a Hamiltonian, given the asymptotics of its orbits. The scattering
data in this problem is seen to be a deformation of the Radon transform of the
the potential.

<u>Section 2</u> The Semiclassical Limit

As Dryuma [3] has shown, (1.5) may be written as the compatibility condition
between two linear equations:

$$\left\{i\partial/\partial y + (i\partial/\partial x)^2/2 + u(x;y,t)\right\}\ \psi(x,y,t) = 0$$
$$\left\{i\partial/\partial t + (i\partial/\partial x)^3/3 + u(x,y,t)(i\partial/\partial x) + v(x,y,t)\right\}\ \psi(x,y,t) = 0$$

(2.1).

To pass from this system to (1.4), we must consider u and v to depend slowly on
x, y and t. Thus we introduce a small parameter ε, and write:

$$\begin{aligned} X &= \varepsilon x \\ Y &= \varepsilon y \\ T &= \varepsilon t \end{aligned} \qquad \begin{aligned} u(x,y,t) &= U(X,Y,T) \\ v(x,y,t) &= V(X,Y,T) \end{aligned}$$

(2.2).

We also take the wave function ψ to be of the form:

$$\psi(x,y,t) = \exp(-iS(X,Y,T)/\varepsilon)$$

(2.3).

In the scaled variables, (1.5) becomes:

$$\partial U/\partial T + U\ \partial U/\partial X - (\varepsilon^2/12)\ \partial^3 U/\partial X^3 = (\partial/\partial X)^{-1}\ \partial^2 U/\partial Y^2$$

(2.4),

which, on letting ε tend to zero, becomes (1.4). The analogous limits of the
linear equations (2.2) are:

$$\partial S/\partial Y + 1/2\ \left(\partial S/\partial X\right)^2 + U(X,Y,T) = 0$$
$$\partial S/\partial T + 1/3\ \left(\partial S/\partial X\right)^3 + U(X,Y,T)\ \partial S/\partial X + V(X,Y,T) - 0$$

(2.5),

which are a pair of Hamilton-Jacobi equations, the semiclassical limits of (2.1).
Such limits of completely integrable p.d.e.'s have been discussed, in different
formalisms, by Lax and Levermore [4], Lebedev and Manin [5] and Zakharov [6]. An

important example of such a system is Benney's equation, which will be discussed in the next section.

Equations (2.5) are equivalent to the following two pairs of canonical equations of motion:

$$\partial X/\partial Y = P \qquad\qquad \partial X/\partial T = P^2 + U$$
$$\partial P/\partial Y = -\partial U/\partial X \qquad \partial P/\partial T = -\partial(PU + V)/\partial X \qquad (2.6).$$

These pairs are generated respectively by the Hamiltonians:

$$H_2 = P^2/2 + U(X,Y,T) \qquad H_3 = P^3/3 + PU(X,Y,T) + V(X,Y,T) \qquad (2.7).$$

Hamiltonians of this form, satisfying the necessary compatibility condition $\partial U/\partial Y = \partial V/\partial X$, arise naturally in the theory of Benney's equations.

Section 3 Benney's Equations.

Benney's equations were introduced in [7] as a model of long waves on a perfect heavy fluid in two dimensions, with a free surface. In that paper, Benney showed that the moments of the horizontal component of the fluid velocity $P(X,Z,T)$ with respect to Z:

$$A_n(X,T) = \int_0^{h(X,T)} P^n(X,Z,T) \, dZ \qquad (3.1),$$

where h was the depth of the fluid, satisfied the following equations:

$$\partial A_n/\partial T + \partial A_{n+1}/\partial X + n \, A_{n-1} \, \partial A_0/\partial X = 0 \qquad (3.2).$$

Further, he showed that the system possessed an infinite set of conserved densities, polynomial in the moments A_n. Kupershmidt and Manin [8] found the Hamiltonian structure of the system, and were thus able to construct the commuting flows generated by these conserved densities. Zakharov [6], and later the present author [9], showed that such moment equations could also be derived from Vlasov equations. For example, (3.2) follows from:

$$\partial f(X,P,T)/\partial T + \{ P^2/2 + A_0 , f \} = 0 \qquad (3.3).$$

$$A_n(X,T) = \int_{-\infty}^{\infty} P^n \, f(X,P,T) \, dP$$

Here $\{ \, , \, \}$ denotes the canonical Poisson bracket on the X,P plane. As this system is one of an infinite collection of commuting flows, it is best to consider the solutions as depending on an infinite number of parameters, T_1, T_2, T_3 etc.; translations in T_1 are just translations in X, T_2 is just the T of (3.3), while T3 is the 'time' in the next equation in the hierarchy. This, when written as a Vlasov equation, takes the form:

$$\partial f(X,P;T_2,T_3)/\partial T_3 + \{ P^3/3 + P A_0 + A_1, f \} = 0 \qquad (3.4).$$

We may now see that the Hamiltonians appearing in the Poisson brackets in (3.3) and (3.4) are identical in form to those in (2.7). The compatibility condition between these, (3.3) and (3.4), is just (1.4); indeed a straightforward calculation shows that the 'density' A_0 satisfies:

$$\partial^2 A_0/\partial T_3 \partial X + \frac{1}{2} \partial^2 (A_0^2)/\partial X^2 = \partial^2 A_0/\partial T_2^2 \qquad (3.5),$$

which is the same as (1.4), provided the variables of this and the previous
sections are identified as follows:

$$A_0 \equiv U \qquad\qquad T_2 \equiv Y \qquad\qquad\qquad (3.6).$$
$$A_1 \equiv V \qquad\qquad T_3 \equiv T$$

The other flows in the hierarchy may all be generated by the following
construction, which is due to Lebedev and Manin [5]. If we denote by $\lambda(X,P)$ the
formal series:

$$\lambda = P + A_0/P + A_1/P^2 + A_2/P^3 + \ldots \qquad (3.7),$$

and denote by H_n that part of the series (λ^n/n) which is polynomial in P, then the
equations:

$$\partial f/\partial T_n + \left\{ H_n , f \right\} = 0 \qquad (3.8)$$

all commute. These, or rather the equations for A_0 which they, together with
(3.3) imply as consistency conditions, give the hidden symmetries of (1.4).
These equations may be written in the zero curvature form:

$$\partial H_2/\partial T_n - \partial H_n/\partial T_2 + \left\{ H_n, H_2 \right\} = 0 \qquad (3.9).$$

In particular, taking n=4, the simplest case, we obtain, after eliminating A_1, A_2
and A_3:

$$\partial A_0/\partial T_4 = 2 A_0 \, \partial A_0/\partial T_2 + \partial A_0/\partial X \, (\partial/\partial X)^{-1} \partial A_0/\partial T_2 + (\partial/\partial X)^{-2} \, \partial^3 A_0/\partial T_2^3 \qquad (3.10).$$

Although any solution of Benney's equations would yield a solution of (1.4),
very few of these have been constructed. Taking an ansatz for $f(X,P)$ of a
particularly simple form, for example, one may sometimes reduce the Vlasov
equations to systems involving a finite number of Riemann invariants. The 'cold
plasma' ansatz:

$$f(X,P;T_n) = h(X;T_n) \, \delta(P - P_0(X;T_n)) \qquad (3.11)$$

does this for example, the evolution in T_2 reducing to Riemann's classical shallow
water equations, which may be solved exactly. In general, although the hierarchy
(3.8) of Vlasov equations has been linearised exactly [10], the inverse
transformation seems not to possess any obvious closed-form solutions. This
problem is discussed in greater detail in [11], by the present author. The other
promising line of attack is to consider the asymptotics of the solutions as X and
the T_n all tend to infinity. This aspect of the problem is particularly relevant
to the scattering problem described below; however, it remains unsolved as yet.

Section 4   The Initial Value Problem.

The obvious formal approach to the initial value problem of (1.4) is to take the
inverse method of Manakov [12] for (1.5), and try to take its semi-classical limit.

The scattering data may, by this analogy, be defined fairly easily. Consider the
X,Y plane at T=0 (we are using the variables of sections 1 and 2). The initial
data consists of the function U(X,Y), which, we suppose, is smooth and tends to
zero faster than Y^{-1} along any straight line $X=X_0 + P_0Y$, as $Y \to \pm \infty$. We now need to
consider solutions of the first pair of equations (2.6), namely:

$$\partial X/\partial Y = P , \qquad \partial P/\partial Y = -\partial U/\partial X \qquad (4.1).$$

Solutions of this pair will be curves X(Y), P(Y), which have as their asymptotes
as $Y \to \pm\infty$ the lines:

$$X = X_{\pm} + P_{\pm}Y \qquad P = P_{\pm} \qquad (4.2).$$

The pairs of variables (X_+,P_+) and (X_-,P_-) are both canonical; the solutions of
(4.1) define a canonical map between them. It is this map (or rather, its
generating function) which is the semi-classical limit of Manakov's scattering
data. The time evolution of the scattering data is simple; if, at T=0, (X_-,P_-) is
mapped into (X_+,P_+), then for other times T, $(X_- + P_-^2T, P_-)$ is mapped into
$(X_+ +P_+^2T,P_+)$; this follows from the second pair of (2.6). Thus, if the map
from this scattering data back to the function U(X,Y) were known, the initial
value problem for (1.4) could be solved.

To see that, at least for some potentials, this inversion might be possible,
let us consider the case when U is small. Then the solutions of (4.1) will be
approximately straight lines, and we can calculate the deviations from the
straight trajectory perturbatively. We take U to be of order δ, with $\delta \ll 1$;
putting $U=\delta U_1$, $P = P_0 + \delta P_1 + O(\delta^2)$ and $X = X_0 + \delta X_1 + O(\delta^2)$ we get,
successively,

$$P_0 = P_-$$
$$X_0 = X_- + P_-Y \qquad (4.3),$$
$$P_1 = \int_{-\infty}^{Y} -\left(\frac{\partial U_1}{\partial X_-} (X_- + P_-Y',Y') \right) dY'$$

and so on. We thus obtain, as the first order approximation to P_+, the expression:

$$P_+ = P_- - \delta \frac{\partial}{\partial X_-} \int_{-\infty}^{\infty} U_1(X_- + P_-Y, Y) dY \qquad (4.4).$$

The integral here is the Radon transform of U1; it is well known that this
determines the function $U_1(X,Y)$ uniquely.

The problem remains open, however, as to how U(X,Y) may be recovered from the
scattering data in less trivial cases; the conceptual difficulty lies in working
out what the semiclassical analogues must be for the wavefunctions analytic in the
complex upper and lower half k-planes, which are fundamental to Manakov's method.
Nonetheless, since this inverse scattering problem is of considerable intrinsic
interest besides its relevance to the applications, the question deserves further
study.

<u>Section 5</u> Conclusions.

In this paper it has been shown how the Zabolotskaya-Khokhlov equation (1.4)
arises as the semiclassical limit of the Kadomtsev-Petviashvili equation (1.5),
for which the theory is relatively well understood. It inherits various algebraic
structures, such as an infinite number of symmetries, from the latter system, and
one may construct solutions of it from solutions of Benney's hierarchy. Further,
its initial value problem may be reduced to an inverse scattering problem, which
is the semiclassical limit of one already solved by Manakov. However, it does not
seem to be possible to carry his results over directly, and further work is
needed, even to establish such fundamental questions as whether there exist any
smooth global solutions.

<u>References</u>.

[1]   E.A. Zabolotskaya, R.V. Khokhlov, Sov. Phys. Acoustics <u>15</u> (1969) 35.

[2]   A.M. Vinogradov, E.M. Vorob'ev, Sov. Phys. Acoustics <u>22</u> (1976) 12.

[3]   V.S. Dryuma, JETP Lett. <u>19</u> (1974) 387.

[4]   P.D. Lax, C.D. Levermore, Proc. Nat. Acad. Sci. U.S.A. <u>76</u> (1979) 3602.

[5]   D.R. Lebedev, Yu.I. Manin, Phys. Lett. <u>74A</u> (1980) 154.

[6]   V.E. Zakharov, Funct. Anal. App. <u>14</u> (1980) 89.

[7]   D.J. Benney, Stud. Appl. Math. <u>52</u> (1973) 45.

[8]   B.A. Kupershmidt, Yu.I. Manin, Funct. Anal. App. <u>11</u> (1978) 188.

        B.A. Kupershmidt, Yu.I. Manin, Funct. Anal. App. <u>12</u> (1978) 20.

[9]   J. Gibbons, Physica <u>3D</u> (1981) 503.

[10]   J. Gibbons, Phys. Lett. <u>90A</u> (1982) 7.

[11]   J. Gibbons, preprint.

[12]   V.E. Zakharov, S.V. Manakov, Sov. Sci. Rev. – Phys. Rev. <u>1</u> (1979) 133.

        S.V. Manakov, Physica <u>3D</u> (1980) 420.

Fundamental Properties of the Binary Operators in Soliton Theory and Their Generalization

R. Hirota

Department of Applied Mathematics, Faculty of Engineering, Hiroshima University
Higashi-Hiroshima, Japan

We have introduced a binary operator to transform a class of nonlinear evolution
equations into the bilinear form

$$F(D_x, D_t, \ \ldots \ )f \cdot g = 0$$

The binary operator D_x^n operating on a pair of arbitrary functions f and g is defined
by

$$D_x^n f(x) \cdot g(x) = (\frac{\partial}{\partial x} - \frac{\partial}{\partial x'})^n f(x)g(x') \ \Big|_{\ x' \ = \ x}.$$

The usual differential operator $\frac{\partial}{\partial x}$ operating on a product of f and g is expressed
with

$$(\frac{\partial}{\partial x})^n \ f(x)g(x) = (\frac{\partial}{\partial x} + \frac{\partial}{\partial x'})^n \ f(x)g(x') \ \Big|_{\ x' \ = \ x}.$$

We introduce a generalized binary operator

$$D_{\alpha \cdot \beta, x}^n \ f(x) \cdot g(x) = (\alpha \frac{\partial}{\partial x} - \beta \frac{\partial}{\partial x'})^n \ f(x)g(x') \ \Big|_{\ x' \ = \ x}.$$

which includes both operators D_x^n and $(\partial / \partial x)^n$ as special cases.

The fundamental properties of the binary operators will be discussed and it will
be shown that a large number of nonlinear evolution equations which exhibit exact
solutions, are transformed into the bilinear form by virtue of the new operator.

I. Introduction

We have been developing a method[1] of finding exact solutions to a class of nonlinear evolution equations.

The method consists of the following steps.

$$L(u, u_t, u_x, u_{xx}, \ \ldots \ ) = 0 \tag{1.1}$$

where the subscripts indicate the partial differentiation with respect to indicated
variables.

First we assume that the dependent variable u is expressed with the ratio of two
functions g and f

$$u = g/f. \tag{1.2}$$

Let us introduce binary operators D_t, D_x, $\ldots$, operating on a pair of functions
$a(x,t)$ and $b(x,t)$, which are defined by

$$D_x^{\ n} D_t^{\ m} a(x,t) \cdot b(x,t)$$

$$= (\frac{\partial}{\partial x} - \frac{\partial}{\partial x'})^n (\frac{\partial}{\partial t} - \frac{\partial}{\partial t'})^m a(x,t) \, b(x',t') \Big|_{x' = x, \ t' = t} \qquad (1.3)$$

for non-negative integers n and m.

We have proved that higher order derivatives of the ratio $a(x,t)/b(x,t)$ are expressed with the binary operator. For example, we have

$$\frac{\partial}{\partial x} \frac{a}{b} = \frac{D_x a \cdot b}{b^2} , \qquad (1.4)$$

$$\frac{\partial^2}{\partial x^2} \frac{a}{b} = \frac{D_x^{\ 2} a \cdot b}{b^2} - \frac{a}{b} \frac{D_x^{\ 2} b \cdot b}{b^2} , \qquad (1.5)$$

$$\frac{\partial^3}{\partial x^3} \frac{a}{b} = \frac{D_x^{\ 3} a \cdot b}{b^2} - 3(\frac{D_x a \cdot b}{b^2}) \frac{D_x^{\ 2} b \cdot b}{b^2} . \qquad (1.6)$$

Substituting eq. (1.2) into eq. (1.1) and using the above relations, we transform
eq. (1.1) into a homogeneous equation of f and g, which can be split into a set
of bilinear equations. Suppose that we have transformed eq. (1.1) into the following bilinear equation

$$F_1(D_t, D_x, \ \cdots \ ) g \cdot f = 0, \qquad (1.7)$$

$$F_2(D_t, D_x, \ \cdots \ ) f \cdot f + F_3(D_t, D_x, \ \cdots) g \cdot g = 0. \qquad (1.8)$$

We expand f and g in a formal power series in ε

$$f = 1 + \varepsilon f_1 + \varepsilon^2 f_2 + \ \cdots \ , \qquad (1.9)$$

$$g = g_0 + \varepsilon g_1 + \varepsilon^2 g_2 + \ \cdots \ . \qquad (1.10)$$

Substituting eqs. (1.9) and (1.10) into eqs. (1.7) and (1.8) and collecting terms
with the same power of ε, we obtain a set of linear equations for f_n and g_n (n = 1,
2, $\cdots$), which is to be solved successively. If the solution series terminates
we have an exact solution to the nonlinear evolution equation (1.1).

The method has been successfully applied to find explicit solutions to many nonlinear evolution equations, especially to soliton equations[2], [3].

However, the method is entirely dependent on the assumption that the solution is
expressed with the ratio of two functions and each of which consists of finite terms.
This implies that we cannot obtain the explicit solution by the present method to
the differential equation whose solution u has, for example, the following form

$$u = \frac{g^\alpha}{f^\beta} \qquad (1.11)$$

where α and β are not integers, even if f and g consist of finite terms.

In order to overcome the defect, we shall introduce a generalized binary operator
$D_{\alpha \cdot \beta, x}^n$ operating on a pair of smooth functions a and b, which are now defined by

$$D^n_{\alpha \cdot \beta, x}\, a(x) \cdot b(x) = (\alpha \frac{\partial}{\partial x} - \beta \frac{\partial}{\partial x'},)^n\, a(x)\, b(x') \Big|_{x' = x} \qquad (1.12)$$

where α and β are constants.

With the use of the new operator we show in section II that higher order derivatives of the ratio a^{α}/b^{β} are transformed into the form similar to that of eqs.(1.4), (1.5) and (1.6). Accordingly the nonlinear evolution equation whose solution is of type (1.11), is converted into the bilinear form.

We have a number of nonlinear evolution equations which are transformed into the bilinear form by virtue of the new operator and exhibit exact solutions. Among them we present in section III two examples of physical importance: One is the generalized Ginzburg-Landau equation[4] and another a couple of nonlinear diffusion equations.

The Bäcklund transformation which connects solutions of two evolution equations, has been used to construct N-soliton solutions, to obtain the higher order conservation laws and to derive the inverse scattering transformation[2] . In section IV, it is shown that the bilinear form of the Bäcklund transformation described in Ref.2 can be extended, with the help of the new operator, to connect solutions of two different soliton equations[5], and they can be used to generate third soliton equations.

II. Generalized Binary Operator

We introduce a generalized binary operator $D^n_{\alpha \cdot \beta, x}$ operating on a pair of smooth function a(x) and b(x);

$$D^n_{\alpha \cdot \beta, x}\, a(x) \cdot b(x) = (\alpha \frac{\partial}{\partial x} - \beta \frac{\partial}{\partial x'},)^n\, a(x)b(x') \Big|_{x' = x}, \qquad (2.1)$$

where n is a natural number, α and β are constants.

Suppose a function u(x) is expressed with the ratio of two functions $a^{\alpha}(x)$ and $b^{\beta}(x)$

$$u(x) = a^{\alpha}(x)/b^{\beta}(x). \qquad (2.2)$$

Then we find the following relation, which will be shown in the following

$$\exp(\varepsilon \frac{\partial}{\partial x})[\frac{a^{\alpha}(x)}{b^{\beta}(x)}] = [\frac{a^{\alpha}(x)}{b^{\beta}(x)}] [\frac{\exp(\varepsilon D_{\alpha \cdot \beta, x})\, a(x) \cdot b(x)}{a(x)b(x)}]$$

$$\times \exp \{\sum_{n=2}^{\infty} \frac{\varepsilon^n}{n!} [(\alpha - \alpha^n)(\log a)_{n,x} - (\beta + (-\beta)^n)(\log b)_{n,x}]\} , \qquad (2.3)$$

where the subscript n,x indicates the n-th derivative with respect to x.

The left hand side of eq. (2.3) is nothing but

$$a^{\alpha}(x + \varepsilon)/b^{\beta}(x + \varepsilon), \qquad (2.4)$$

which is rewritten, introducing functions $a(x + \alpha \varepsilon)$ and $b(x - \beta \varepsilon)$ as

$$a(x + \alpha \varepsilon)b(x - \beta \varepsilon)[a^{\alpha}(x + \varepsilon)/a(x + \alpha \varepsilon)] [b^{\beta}(x + \varepsilon)b(x - \beta \varepsilon)]^{-1}. \qquad (2.5)$$

The product $a(x + \alpha \varepsilon)b(x - \beta \varepsilon)$ is expressed with the binary operator $D_{\alpha \cdot \beta, x}$

$$a(x + \alpha\varepsilon)b(x - \beta\varepsilon) = \exp(\varepsilon D_{\alpha\cdot\beta,x})a(x)\cdot b(x). \tag{2.6}$$

On the other hand, the Taylor expansions

$$\log\,[a^{\alpha}(x + \varepsilon)/a(x + \alpha\varepsilon)] = \alpha\sum_{n=0}^{\infty}\frac{\varepsilon^n}{n!}\,[\log a]_{n,x} - \sum_{n=0}^{\infty}\frac{(\alpha\varepsilon)^n}{n!}\,[\log a]_{n,x} \tag{2.7}$$

$$= (\alpha - 1)\log a + \sum_{n=2}^{\infty}\frac{\varepsilon^n}{n!}\,(\alpha - \alpha^n)[\log a(x)]_{n,x} \tag{2.8}$$

and

$$\log\,[b^{\beta}(x + \varepsilon)b(x - \beta\varepsilon)]$$

$$= \beta\sum_{n=0}^{\infty}\frac{\varepsilon^n}{n!}\,[\log b]_{n,x} + \sum_{n=0}^{\infty}\frac{(-\beta\varepsilon)^n}{n!}\,[\log b]_{n,x} \tag{2.9}$$

$$= (\beta + 1)\log b + \sum_{n=2}^{\infty}\frac{\varepsilon^n}{n!}\,(\beta + (-\beta)^n)[\log b]_{n,x} \tag{2.10}$$

give that the term

$$[a^{\alpha}(x + \varepsilon)/a(x + \alpha\varepsilon)][b^{\beta}(x + \varepsilon)b(x - \beta\varepsilon)]^{-1} \tag{2.11}$$

is converted into

$$[a(x)/b(x)][a(x)b(x)]^{-1}$$

$$\times \exp\left\{\sum_{n=2}^{\infty}\frac{\varepsilon}{n!}\,[(\alpha - \alpha^n)(\log a)_{n,x} - (\beta + (-\beta)^n)(\log b)_{n,x}]\right\}\,. \tag{2.12}$$

Since all factors in eq. (2.5) are transformed into new expressions, we get the relation (2.3).

Expanding eq. (2.3) we find that higher order derivatives of $u(x)$ $(=a^{\alpha}/b^{\beta})$ with respect to x are transformed into new expressions with $D_{\alpha\cdot\beta,x}^n$ as follows.

$$\frac{\partial}{\partial x}\,\frac{a^{\alpha}}{b^{\beta}} = \frac{a^{\alpha}}{b^{\beta}}\,\frac{D_{\alpha\cdot\beta,x}\,a\cdot b}{ab}\,, \tag{2.13}$$

$$\frac{\partial^2}{\partial x^2}\,\frac{a^{\alpha}}{b^{\beta}} = \frac{a^{\alpha}}{b^{\beta}}\left\{\frac{D_{\alpha\cdot\beta,x}^2\,a\cdot b}{ab} + (\alpha - \alpha^2)(\log a)_{xx} - (\beta + \beta^2)(\log b)_{xx}\right\}\,, \tag{2.14}$$

$$\frac{\partial^3}{\partial x^3}\,\frac{a^{\alpha}}{b^{\beta}} = \frac{a^{\alpha}}{b^{\beta}}\left\{\frac{D_{\alpha\cdot\beta,x}^3\,a\cdot b}{ab} + 3\left(\frac{D_{\alpha\cdot\beta,x}\,a\cdot b}{ab}\right)\right.$$

$$\times\,[(\alpha - \alpha^2)(\log a)_{xx} - (\beta + \beta^2)(\log b)_{xx}]$$

$$\left. + (\alpha - \alpha^3)(\log a)_{xxx} - (\beta - \beta^3)(\log b)_{xxx}\right\}\,, \tag{2.15}$$

$$\cdots\,.$$

If one notices the relation

$$(\log b)_{xx} = \frac{D_x^2\,b\cdot b}{2b^2}\,, \tag{2.16}$$

one find that eqs. (2.13), (2.14) and (2.15) reduce to eqs. (1.8), (1.9) and (1.10), respectively when $\alpha = \beta = 1$.

III. Nonlinear Evolution Equations

There are a number of nonlinear evolution equations which are transformed into the bilinear forms in terms of the new binary operators and exhibit exact solutions. Among them we present two examples of physical importance.

(i) Generalized Ginzburg-Landau equation

We have the generalized Ginzburg-Landau equation

$$i\psi_t + p\psi_{xx} + q|\psi|^2\psi = i\gamma\psi \tag{3.1}$$

which describes phase-transitions and wave propagations in many non-equilibrium systems. In eq. (3.1) $p = p_r + ip_i$, $q = q_r + iq_i$, $p_i < 0$ and $\gamma q_i > 0$.

Nozaki and Bekki[4] have obtained exact solutions to eq. (3.1). They used a dependent variable transformation

$$\psi(x,t) = \exp[i(kx - \Omega t)]G(x,t)/F^\beta(x,t) \tag{3.2}$$

where k, Ω and $F(x,t)$ are assumed to be real, and β is a complex parameter satisfying the relation $\beta + \beta^* = 2$. Substituting eq. (3.2) into eq. (3.1) and using eqs. (2.13) and (2.14), we obtain the following coupled bilinear equations

$$(\Omega - pk^2 - \lambda + iD_{\beta,t} + 2ikpD_{\beta,x} + pD_{\beta,x}^2) \; G \cdot F = HG , \tag{3.3}$$

$$[p(\beta + \beta^2)/2D_x^2 + i\gamma - \lambda] \; F \cdot F - q \; |G|^2 = HF \tag{3.4}$$

where λ and H are arbitrary constant and function respectively, introduced to decouple the equation. In eqs. (3.3) and (3.4) operators $D_{\beta,t}$, $D_{\beta,x}$ and D_x are $D_{1\cdot\beta,t}$, $D_{1\cdot\beta,x}$ and $D_{1\cdot1,x}$ in the present notation, respectively.

Solving eqs. (3.3) and (3.4) Nozaki and Bekki have obtained a solitary wave solution, a hole solution, shock-type solution, and two shock-wave solution. For details see reference 4.

(ii) Coupled nonlinear diffusion equations

We consider coupled nonlinear diffusion equations of the form

$$u_t - d_1 u_{xx} = h_1(u,v) , \tag{3.5}$$

$$v_t - d_2 v_{xx} = h_2(u,v) \tag{3.6}$$

which describe a model of two competing and diffusion species in the field of population dynamics[6], where u and v are the population densities of the two species. It is rather impossible to find explicit solutions to eqs. (3.5) and (3.6) for a general form of $h_1(u,v)$ and $h_2(u,v)$. However if we restrict the form of $h_1(u,v)$ and $h_2(u,v)$ to

$$h_1(u,v) = [2(d_2 - d_1) + \alpha(5d_1 - d_2) v - d_1(\alpha + \alpha^2) u(1 - v)^{2-\alpha}]u$$

$$- 2d_1(1 - v)^{\alpha-2}v^2 , \tag{3.7}$$

$$h_2(u,v) = 2d_2u(1 - v)^{3-\alpha} , \tag{3.8}$$

where d_1, d_2 and α are constants, we can transform eqs. (3.5) and (3.6) into the bilinear form and obtain an explicit solution to them. The solution represents two travelling waves colliding with each other and is written as

$$u = (\exp \eta_1 - \exp \eta_2)^2/(1 + \exp \eta_1 + \exp \eta_2)^\alpha \ , \tag{3.9}$$

$$v = (\exp \eta_1 + \exp \eta_2)/(1 + \exp \eta_1 + \exp \eta_2) \qquad \text{where} \tag{3.10}$$

$$\eta_1 = d_2 t + x + \text{const.} \tag{3.11}$$

$$\eta_2 = d_2 t - x + \text{const.} \tag{3.12}$$

IV. Bilinear Bäcklund Transformation

We shall describe two examples of non-auto Bäcklund transformation of bilinear form which connect solutions of two different equations of bilinear form.

First we describe the Bäcklund transformation derived by Fordy and Gibbons[5]. They have shown that the Sawada-Kotera equation

$$u_t + 45u^2 u_x + 15(u_x u_{xx} + u_{3x} u) + u_{5x} = 0 \tag{4.1}$$

and the Kaup-Kupershmidt equation

$$w_t + 180w^2 w_x + 30(\ (5/2) \ w_x w_{xx} + w_{3x} w) + w_{5x} = 0 \tag{4.2}$$

are related to each other by the Miura transformation

$$u = v_x - v^2, \tag{4.3}$$

$$w = -v_x - \frac{1}{2} v^2. \tag{4.4}$$

They also found that v satisfies the following equation

$$v_t - 5(v_x v_{3x} + v_{xx}^2 + v_x^3 + 4vv_x v_{xx} + v^2 v_{xxx} - v^4 v_x) + v_{5x} = 0 \tag{4.5}$$

which we call modified Kaup-Kupershmidt equation.

The dependent variable transformations

$$u = 2(\log h)_{xx} \tag{4.6}$$

and

$$w = (1/2)(\log f)_{xx} \tag{4.7}$$

transform eqs. (4.1) and (4.2) into the bilinear forms

$$D_x(D_t + D_x^5)h \cdot h = 0 \tag{4.8}$$

and

$$D_x(D_t + (1/16)D_x^5) \ f \cdot f + (15/16)D_x^2 \ f \cdot g = 0, \tag{4.9}$$

$$D_x^4 \ f \cdot f = fg \tag{4.10}$$

respectively.

We note that modified Kaup-Kupershmidt equation is transformed, by the dependent variable transformation

$$v = -3[\log (f/h^2)]_x \, , \tag{4.11}$$

into the bilinear form[7)]

$$(D_{2,t} + (1/6)D_{2,x}^5)f \cdot h = 0, \tag{4.12}$$

$$D_{2,x}^2 f \cdot h = 0 \tag{4.13}$$

where we have used $D_{2,x}$ in place of $D_{1 \cdot 2,x}$ for simplicity.

One of the bilinear forms, eq. (4.13) has been pointed out by Fordy and Gibbons[5)]

Since eqs. (4.12) and (4.13) are linear equations for each of f and h, we can easily find the compatibility condition for each of f and h. We have found[7)] that the Sawada-Kotera equation is the compatibility condition for f and the Kaup-Kupershmidt equation is that for h.

Second, we consider the following bilinear equations

$$D_{2,x}(D_{2,x}D_{2,\tau} - 6)f \cdot h = 0, \tag{4.14}$$

$$D_{2,x}^2 f \cdot h = 0. \tag{4.15}$$

They are transformed into a set of equations

$$E_\tau = P, \tag{4.16}$$

$$P_x = -E(P-N), \tag{4.17}$$

$$N_x = 2EP, \tag{4.18}$$

through the dependent variable transformation

$$\phi = \log (f/h^2), \tag{4.19}$$

$$\rho = \log (fh^4), \tag{4.20}$$

$$E = \phi_x \, , \tag{4.21}$$

$$P = \phi_{\tau x} \, , \tag{4.22}$$

$$N = -\rho_{x\tau} + 3. \tag{4.23}$$

These equations are similar to but are different from the equations describing the self-induced transparency in nonlinear optics (see for example Ref. 8).

$$\frac{\partial E}{\partial \xi} = P, \tag{4.24}$$

$$\frac{\partial P}{\partial t} = EN, \tag{4.25}$$

$$\frac{\partial N}{\partial t} = -EP. \tag{4.26}$$

Equations (4.14) and (4.15) are linear equations for each of f and h. Hence we can derive the new nonlinear equations from the compatibility conditions for f and h as was done for the previous bilinear equations. In fact the compatibility condition for f gives

$$w_{\tau xx} = 3(1-w_\tau)w_x \ ,\tag{4.27}$$

where

$$w = 2(\log h)_x\tag{4.28}$$

and that for h gives

$$Z_{xx} - 3(1-Z_\tau)Z_x + \frac{3}{4}\ \frac{Z_{\tau x}^2}{1-Z_\tau} = 0,\tag{4.29}$$

where

$$Z = (\log f)_x \ .\tag{4.30}$$

Finally we note that the nonlinear evolution equations described in this section exhibit N-soliton solutions[7].

References:

[1] Ryogo Hirota: "Direct Methods of Finding Exact Solutions of Nonlinear Evolution Equations", in Bäcklund Transformations, ed. by R.M. Miura, Lecture Notes in Mathematics (Springer, Berlin, Heidelberg, New York 1976) Vol. 515.
[2] Ryogo Hirota and Junkichi Satsuma: Prog. Theor. Phys. Suppl. No. 59 (1976)64.
[3] Ryogo Hirota: "Direct Methods in Soliton Theory", in Solitons, ed. R.K. Bullough and P.J. Caudrey(Topics in Current Physics 17, Springer-Verlag, 1980).
[4] Kazuhiro Nozaki and Nozaki Bekki:J.Phys. Soc. Jpn. 53 (1984) 1581.
[5] Allan P. Fordy and John Gibbons: Physics Letters 75A (1980) 325.
[6] Yuzo Hosono and Masayasu Mimura: J. Math. Kyoto Univ. 22-3 (1982) 435.
[7] Ryogo Hirota in preparation.
[8] G.L. Lamb, Jr. and D.W. McLaughlin: "Aspects of Soliton Physics", in Solitons, ed. R.K.Bullough and P.J. Caudrey (Topics in Current Physics 17, Springer-Verlag, 1980).

On an Exactly Solvable Nonlinear Diffusion Equation

Junkichi Satsuma

General Education, Miyazaki Medical College, Kiyotake, Miyazaki 889-16, Japan

Synopsis

A nonlinear integro-differential equation describing a diffusion process with non-local mutual interaction is considered. It can be exactly linearized by a dependent variable transformation. A perturbational approach is employed to understand the structure of solutions of the equation. First, a series of differential equations approximating the integro-differential equation is derived. Then the effect of higher order corrections on the equilibrium solution is investigated.

I Introduction

We owe the success of soliton theory to the finding of the analytical methods to obtain the exact solutions of nonlinear equations. Those solutions have revealed the importance of solitons in various nonlinear dispersive systems. However, it is not the case for the nonlinear dissipative systems. There are only few examples for which the exact treatment is possible [1]-[4]. The most important and famous example is Burgers' equation. It is transformed into the heat equation by means of the dependent variable transformation. The properties of solutions of Burgers' equation have been studied in detail through the linearization technique.

In [2], the author has proposed a nonlinear integro-differential equation which describes a diffusion process with nonlinear and nonlocal mutual interaction. It is an extension of Burgers' equation and can be solved exactly by transforming into a linear diffusion equation. We can obtain some explicit solutions including equilibrium state and time-dependent solutions. An interesting feature of the time-dependent solution is to exhibit a blowing up depending on the initial values, which does not happen in the solutions of Burgers' equation.

In this paper we employ a perturbational approach to understand the structure of solutions of the integro-differential equation. First in Sec. II, we present the equation and survey the method of exact solutions. Then in Sec. III, by expanding the integral term with respect to the parameter contained in the term, we derive a series of partial differential equations which approximate the integro-differential equation. Especially in the lowest order of the expansion, we obtain Burgers' equation. Finally in Sec. IV, we study how the higher order corrections affect the equilibrium solution and discuss the relation between the exact solutions of the integro-differential equation and the solutions of the approximated equations.

50

II Model Equation and Exact Solutions

The equation we consider is

$$u_t - u_{xx} + (Tu \cdot u)_x = 0, \qquad \text{where} \tag{1}$$

$$Tu(x) = P \int_{-\infty}^{\infty} \coth \frac{\pi(y-x)}{2\delta} \, u(y) dy, \tag{2}$$

and where P means the principal value and δ is a positive parameter. We impose the following conditions on u:

$$u(t,x) \geq 0 \qquad \text{for } -\infty < x < \infty, \qquad \text{and} \tag{3}$$

$$I = \int_{-\infty}^{\infty} u(t,x) dx < \infty. \tag{4}$$

In order to obtain the exact solution of (1), we introduce a dependent variable transformation,

$$u(t,x) = - \frac{i}{2\delta} \frac{\partial}{\partial x} \log[f^-(t,x)/f^+(t,x)]. \tag{5}$$

Substituting (5) into (1) and noticing the analytical properties of $f^{\pm}(t,x)$, we obtain

$$f^+(f_t^- - f_{xx}^- + If_x^-) - f^-(f_t^+ - f_{xx}^+ + If_x^+) = 0. \tag{6}$$

Hence if one can find the functions $f^{\pm}(t,x)$ such that

$$f_t^{\pm} = f_{xx}^{\pm} - If_x^{\pm} \tag{7}$$

and $\partial \log f^{\pm}(t,x)/\partial x$ are analytic in their corresponding complex planes, then we get a solution u(t,x) of (1) through the relation (5). This procedure solves in principle the initial value problem of (1) for the initial data u(0,x). (See [2] and [5] for the detail.)

Let us show two special solutions:

A) Equilibrium solution (One-pulse solution)

$$u(t,x) = u(0,x)$$

$$= \frac{p}{2\delta} \frac{\sin p\delta}{\cosh px + \cos p\delta}, \tag{8}$$

where p is a real constant satisfying

$$0 < p < \pi/\delta . \tag{9}$$

It is noted that $I = p$ for this equilibrium solution.

B) Two-pulse solution

Take an initial function as

$$u(0,x) = \frac{p}{2\delta} \sin p\delta \left[\frac{1}{\cosh p(x-\alpha) + \cos p\delta} + \frac{1}{\cosh p(x+\alpha) + \cos p\delta} \right], \tag{10}$$

where α is a nonnegative constant. This function represents two stationary pulses with the same amplitude locating at a distance 2α symmetrically with respect to x = 0. The exact solution tells us that

i) if $0 < p\delta < \pi/2$, then

$$\lim_{t \to \infty} u(t,x) = \frac{p}{\delta} \frac{\sin 2p\delta}{\cosh 2px + \cos 2p\delta} \, , \tag{11}$$

ii) if $p\delta = \pi/2$, then u grows up,

and

iii) if $\pi/2 < p\delta < \pi$, then u blows up at

$$t = (1/p^2) \log [\cosh p\alpha/\cos(\pi-p\delta)]. \tag{12}$$

Namely if

$$\delta I = 2p\delta < \pi, \tag{13}$$

the initial two pulses gather to make one pulse and, otherwise, the solution grows or blows up.

III Perturbational Analysis

In order to approximate (1), we first expand Tu(x) with respect to δ. It is easy to see that Tu is written as

$$Tu(x) = \int_x^\infty [1 + \frac{2\ e^{-\pi(y-x)/\delta}}{1 - e^{-\pi(y-x)/\delta}}]u(y)dy - \int_{-\infty}^x [1 + \frac{2\ e^{\pi(y-x)/\delta}}{1 - e^{\pi(y-x)/\delta}}]u(y)dy \ . \tag{14}$$

Then noting (4) and changing the integral variables, we have

$$Tu(x) = I - 2 \int_{-\infty}^x u(y)dy + \frac{2\delta}{\pi} \int_0^\infty \frac{e^{-z}}{1-e^{-z}} [u(x + \frac{\delta}{\pi} z) - u(x - \frac{\delta}{\pi} z)]dz. \tag{15}$$

Expanding $u(x \pm \delta z/\pi)$ as Taylor series, we obtain

$$Tu(x) = I - 2 \int_{-\infty}^x u(y)dy + 4 (\frac{\delta}{\pi})^2 \sum_{n=0}^\infty (\frac{\delta}{\pi})^{2n} \zeta(2n+2)\ u^{(2n+1)}(x), \tag{16}$$

where $u^{(n)}$ denotes the n-th derivative of $u(x)$ and $\zeta(x)$ is the Zeta function of Riemann defined by the series

$$\zeta(z) = \sum_{k=1}^\infty 1/k^z. \tag{17}$$

Especially, we have $\zeta(2) = \pi^2/6$, $\zeta(4) = \pi^4/90$, $\zeta(6) = \pi^6/945$ and so on. Hence (1) is reduced to

$$u_t - u_{xx} + [\{I - 2 \int_{-\infty}^x u(y)dy + 4(\frac{\delta}{\pi})^2 \sum_{n=0}^\infty (\frac{\delta}{\pi})^{2n} \zeta(2n+2)\ u^{(2n+1)}\}u]_x = 0. \tag{18}$$

Let w(x) be

$$w(x) = \int_{-\infty}^x u(y)dy. \tag{19}$$

The conditions (3) and (4) give

$$w(x) \to 0 \quad \text{as} \quad x \to -\infty \quad \text{and}$$

$$w(x) \to I \quad \text{as} \quad x \to \infty.$$

Equation (18) is rewritten in terms of w as

$$w_t - w_{xx} + Iw_x - 2ww_x + \frac{2}{3} \delta^2 w_x w_{xx} + \frac{2}{45} \delta^4 w_x w_{xxxx} + \frac{4}{945} \delta^6 w_x w_{xxxxxx} + \cdots = 0. \tag{20}$$

In the lowest order of expansion, we have Burgers' equation,

$$w_t - w_{xx} + Iw_x - 2ww_x = 0. \tag{21}$$

If we include the term of order δ^2, we get

$$w_t - (1 - \tfrac{2}{3}\delta^2 w_x)w_{xx} + Iw_x - 2ww_x = 0. \tag{22}$$

Since w_x is always positive, we may say that the correction term plays a role in decreasing the diffusion coefficient in Burgers' equation.

IV Equilibrium Solution

In this section, we discuss the equilibrium solution of the perturbed equation (20). For this purpose, it is convenient to introduce

$$\phi = w - \frac{I}{2} . \tag{23}$$

The stationary solution of (20) is governed by the ordinary differential equation,

$$\phi_x = -(\phi^2 - I^2/4) + \frac{1}{3}\delta^2\phi_x^2 + \frac{2}{45}\delta^4(\phi_{xxx}\phi_x - \frac{1}{2}\phi_{xx}^2)$$

$$+ \frac{4}{945}\delta^6(\phi_{xxxxx}\phi_x - \phi_{xxxx}\phi_{xx} + \frac{1}{2}\phi_{xxx}^2) + \cdots . \tag{24}$$

In the lowest order of expansion, (24) gives

$$\phi_x = -(\phi^2 - I^2/4), \tag{25}$$

which yields the equilibrium solution of Burgers' equation,

$$\phi = \tanh Ix/2. \tag{26}$$

Let us examine the effect of the term of order δ^2 on (26). The perturbed equation is written as

$$\phi_x = -(\phi^2 - \frac{I^2}{4}) + \frac{1}{3}\delta^2\phi_x^2 . \tag{27}$$

From (27) we have

$$\frac{\phi_x}{\phi^2 - I^2/4} + \frac{1}{3}\delta^2\phi_x = -1. \tag{28}$$

Integration of (28) gives

$$\frac{1}{I}\log\left|\frac{I-2\phi}{I+2\phi}\right| + \frac{1}{3}\delta^2\phi = -x, \tag{29}$$

which is the implicit form of the equilibrium solution of (27). The term $\frac{1}{3}\delta^2\phi$ in (29) sharpens the gradient of the solution near $x = 0$. By calculating the magnitude of ϕ_x at $\phi = 0$, we can get the value of δI for which the analyticity of the solution is broken. In this approximation, it becomes $2\sqrt{3} \approx 3.46$.

If the term of order δ^4 in (24) is included, the value is found to be $\delta I \approx 3.20$. Furthermore, if the term of order δ^6 is also included, it is calculated as $\delta I \approx 3.16$. In Sec. II, we have shown that the two-pulse solution blows up if δI exceeds π. The perturbational analysis in this section shows a good agreement with the exact result.

Acknowledgement

The author expresses his sincere thanks to Professor Nobuo Yajima for useful advice
and discussions. He is also grateful to Professor Shoji Yotsutani for valuable
comments.

References
1.   J.M.Burgers: Adv. Appl. Mech. $\underline{1}$, 171 (1948)
2.   J.Satsuma: J. Phys. Soc. Jpn. $\underline{50}$, 1423 (1981)
3.   T.Kawahara and M.Tanaka: Phys. Lett. $\underline{97A}$, 311 (1983)
4.   K.Nozaki and N.Bekki: J. Phys. Soc. Jpn. $\underline{53}$, 1581 (1984)
5.   J.Satsuma and M.Mimura: preprint, "Exact Treatment of Nonlinear Diffusion
     Equations with Singular Integral Terms"

Hirota's Method and the Painlevé Property

J.D. Gibbon

Department of Mathematics, Imperial College, London SW7 2BZ, Great Britain

M. Tabor

Department of Applied Physics & Nuclear Engineering, Columbia University
New York, NY 10027, USA

1. Introduction

Given a system of nonlinear ordinary or partial differential equations a most
challenging problem is to find an analytical test to determine whether the given
system is *integrable*.   In the case of systems of o.d.e's integrability (in the
classical sense of "integration by quadratures" [1]) requires one to find as many
integrals of the motion as the order of the system.   However, in the case of
Hamiltonian systems, owing to the special symplectic structure of phase space, a
complete integration can be effected by the identification of as many involutive
integrals as there are degrees of freedom (N).   For integrable Hamiltonian systems
the flow is confined to N-dimensional tori embedded in the 2N-dimensional phase
space.

For partial differential equations "complete integrability" is normally associated
with the identification of an infinite number of non-trivial conservation laws
(integrals).   It would appear that such systems can be solved by the Inverse
Scattering Transform (IST) method, the scattering data of which may be shown to be
related to the integrals [2].   As is well known, many systems soluble by IST exhibit
n-soliton solutions.   For an arbitrary given nonlinear p.d.e, finding a spectrum
preserved operator, if it exists, and/or identifying all (if any) the conserved
quantities can be a daunting task.   An important contribution, which we shall discuss at length later, has been made by HIROTA [3] who invented a novel means of
calculating n-soliton solutions with its attendant implication of integrability
without having to resort to the IST.   Direct tests of integrability have proved
elusive but in the case of o.d.e.'s the answer may lie in the classic work of
SOFYA KOVALEVSKAYA [4].   She was able to identify the integrable cases of the Euler
-Poisson equations by determining those parameter values for which the only movable
singularities, exhibited by the solutions in the complex time plane were ordinary
poles.   Subsequent work by PAINLEVÉ and co-workers [5] on the classification of
second order o.d.e's with this property has led to the terminology "Painlevé
property".   The six Painlevé Transcendents are, of course, the most famous examples
of such equations.   In the last few years the Painlevé property has been used
successfully to identify integrable cases of the Lorenz system [6], the Henon-Heiles
system [7], the three particle Toda Lattice etc. [8].   In the case of completely
integrable p.d.e's, a number of workers [9-11] have observed that when these
equations are reduced by means of a similarity transformation or travelling wave
reduction, the resulting o.d.e's have the Painlevé property.   In [9] it was conjectured that a p.d.e. will be completely integrable if all the o.d.e's derived
from it by all possible reductions have the Painlevé property.   In [10] it was
shown, under certain restrictions, that if G is an analytic, regular symmetry group
of a completely integrable p.d.e, then the reduced o.d.e for the G-invariant
solutions is of Painlevé type.   Naturally it would be desirable to extend the idea
of the Painlevé property to p.d.e's in a way that does not require the reduction to
o.d.e's.   The problem is that, unlike the latter, the singularities of the former
are no longer isolated points but lie on analytic manifolds (the "singular manifolds") in the complex hyperspace.   WEISS, TABOR and CARNEVALE [12] have suggested
that the natural extension of the Painlevé property to p.d.e's is that the solutions
be single valued about movable singular manifolds.   It would appear that many
integrable p.d.e's have this generalized Painlevé property.   Furthermore, for those

systems that do, the Painlevé analysis contains much valuable information. WEISS
[13] has developed a general procedure, involving the Schwarzian derivative of the
singular manifold for finding the Lax pair of the system. GIBBON ET AL [14] have
recently shown a direct connection between the Bäcklund transformation that can be
derived from the Painlevé analysis, and Hirota's method. Connections with pole
dynamics [15] and rational solutions [14,16] have also been found. An extension
of this approach to discrete systems, e.g. the Toda chain, has also been found [17].

2. The Painlevé Property

If $u = u(z_1,\ldots,z_n)$ is a meromorphic function of n complex variables, the singularities
of u lie on 2n-2 dimensional manifolds determined by equations of the form
$\varphi(z_1,\ldots,z_n) = 0$ where φ is analytic in a neighbourhood of the manifold [12].
WEISS, TABOR AND CARNEVALE define the Painlevé property for a p.d.e to be that the
solutions are "single valued" about the movable singular manifolds. This is determined by making the ansatz that the dependent variable(s) can be expanded as a
generalized Laurent series of the form

$$u = \varphi^{\alpha} \sum_{j=0}^{\infty} u_j \varphi^j \qquad (2.1)$$

where $u_j = u_j(z_1,\ldots,z_n)$ and $\varphi = \varphi(z_1,\ldots,z_n)$ are analytic functions of the independent variables. (In the case of one independent variable φ can be reduced to $(z-z_0)$,
where z_0 is the arbitrary pole position, and (2.1) becomes the standard Laurent
series). Direct substitution of (2.1) into the given p.d.e determines the possible
values of α and defines a set of recursion relations between the u_j. Note that
these recursion relations are now systems of p.d.e's involving φ rather than just
a set of algebraic equations that one would find for the standard Laurent series.
For certain values of j, the recursion relations are found to be undefined. These
points, termed "resonances", correspond to the arbitrariness of the associated u_j.
At each resonance one finds certain conditions on the preceeding u_j and φ
("compatibility conditions") which must be satisfied in order to ensure that the
corresponding u_j is indeed arbitrary. In keeping with the Cauchy-Kovalevskaya
theorem the expansion (2.1) must have the requisite number of arbitrary functions
in order to be a valid general solution. The method is well illustrated by the
KdV equation

$$u_t + 12uu_x + u_{xxx} = 0 \qquad (2.2)$$

Making the substitution (2.1), the leading order is easily determined to be $\alpha = -2$
and the resonances found to occur at $j = -1, 4, 6$. The resonance at $j = -1$ corresponds to the arbitrariness of φ (c.f. the arbitrary pole position z_0 in the 1-
dimensional case). For the first few j one finds the following:-

$$j = 0: \qquad u_0 = -\varphi_x^2 \qquad (2.3a)$$

$$j = 1: \qquad u_1 = \varphi_{xx} \qquad (2.3b)$$

$$j = 2: \qquad \varphi_x\varphi_t + 12u_2\varphi_x^2 + 4\varphi_x\varphi_{xxx} - 3\varphi_{xx}^2 = 0 \qquad (2.3c)$$

$$j = 3: \qquad \varphi_{xt} + 12u_2\varphi_{xx} - u_3\varphi_{xx}^2 + \varphi_{xxxx} = 0 \qquad (2.3d)$$

$$j = 4: \qquad u_4 \text{ arbitrary subject to the compatibility condition}$$

$$\frac{\partial}{\partial x} (\varphi_{xt} + 12u_2\varphi_{xx} - u_3\varphi_x^2 + \varphi_{xxxx}) = 0 \, .$$

Clearly by the results at $j = 3$ this condition is always satisfied. The compatibility condition at $j = 6$ is also satisfied thereby ensuring the arbitrariness of
u_6. Thus the expansion (2.1) is valid and has the requisite number of arbitrary
functions (φ, u_4 and u_6) to correspond to a general solution to (2.2). The KdV
equation can therefore be said to possess the "Painlevé property" in the sense of
WEISS ET AL [12].

A Bäcklund transformation can now be derived in the following way.   The arbitrary
functions u_4 and u_6 are set to zero and by requiring also that $u_3 = 0$, it is easy
to demonstrate that all $u_j = 0$, for $j \geqslant 3$, provided that u_2 is itself a solution
to the KdV equation (2.2).   Combining (2.3a) and (2.3b) as a second logarithmic
derivative we have

$$u = \frac{\partial^2}{\partial x^2} \log \varphi + u_2 \tag{2.4a}$$

$$\varphi_x \varphi_t + 12 u_2 \varphi_x^2 + 4 \varphi_x \varphi_{xxx} - 3 \varphi_{xx}^2 = 0 \tag{2.4b}$$

$$\varphi_{xt} + 12 u_2 \varphi_{xx} + \varphi_{xxxx} = 0 \tag{2.4c}$$

where both u and u_2 are, by definition, solutions of (2.2).   The set of equations
(2.4) now define a Bäcklund transformation for the KdV equation.   Furthermore one
may derive the Lax pair for the KdV equation from equations (2.4) by two distinctly
different routes.

The first involves solving (2.4b) for φ_t, differentiating w.r.t x, using (2.4c)
and making the substitution $\varphi_x = \psi^2$.   This results in (2.4b) and (2.4c) being
transformed into the Lax pair

$$\psi_{xx} + (2u_2 - \lambda)\psi = 0 \tag{2.5a}$$

$$\psi_t + (6u_2 + \lambda)\psi_x + \psi_{xxx} = 0 \tag{2.5b}$$

where we have identified the singular manifold with the squared eigenfunction of
the scattering problem.   The second route, developed by WEISS [13], involves eliminating u_2 from (2.4b) and (2.4c) to obtain a single equation for φ of the form

$$\frac{\varphi_t}{\varphi_x} + \left\{ \varphi ; x \right\} = \lambda \tag{2.6a}$$

where

$$\left\{ \varphi ; x \right\} = \frac{\varphi_{xxx}}{\varphi_x} - \frac{3}{2} \left(\frac{\varphi_{xx}}{\varphi_x} \right)^2 \tag{2.6b}$$

is the Schwarzian derivative of φ.   The invariance of (2.6b) under the Mobius group
suggests that (2.6a) can be linearized by the substitution $\varphi = \psi_1/\psi_2$ where (ψ_1, ψ_2)
satisfy the same system of linear equations.   By this procedure one is indeed able
to obtain the Lax pair for the KdV equations.

A large number of integrable p.d.e's have been shown to pass the Painlevé test
[12-16] and their Bäcklund transformations found from the truncated expansions.
The derivation of Lax pairs from the Painlevé analysis using Weiss's Schwarzian
derivative procedure seems to work best for those systems with scalar scattering
problems.   For those systems with multicomponent scattering schemes (i.e. the AKNS
method) the connection between the singular manifold and the scattering eigenfunctions is more subtle.   In fact this connection is conveniently displayed when
we demonstrate the relationship between the Painlevé property and Hirota's method.
Before we do this it is useful to carry out the Painlevé analysis for the nonlinear
Schrödinger (NLS) equation

$$iu_t + u_{xx} + 2u|u|^2 = 0 \quad . \tag{2.7}$$

This must be treated as a two component system

$$iu_t + u_{xxx} + 2u^2 v = 0 \tag{2.8a}$$

$$-iv_t + v_{xx} + 2v^2 u = 0 \tag{2.8b}$$

where $v = u^*$.   Now both u and v have expansions about the same singular manifold
of the form

$$u = \varphi^{-1} \sum_{j=0}^{\infty} u_j \varphi^j \quad ; \qquad v = \varphi^{-1} \sum_{j=0}^{\infty} v_j \varphi^j \qquad\qquad (2.9)$$

with resonances at $j = -1, 0, 3, 4$ and one may indeed show that (2.8) passes the Painlevé test [15].   The truncated expansions

$$u = \frac{u_0}{\varphi} + u_1 \quad \text{and} \quad v = \frac{v_0}{\varphi} + v_1 \qquad\qquad (2.10)$$

are found to take the form

$$uv = \frac{\partial^2}{\partial x^2} \log \varphi + u_1 v_1 \qquad\qquad (2.11)$$

where (u,v) and (u_1,v_1) both satisfy the NLS equation (2.8).

3. Hirota's Method and the Painlevé Property

An important technique in the study of nonlinear p.d.e's is the method developed by HIROTA [3] to find n-soliton solutions.   This is a direct method in the sense that it does not require a solution of the IST problem and for some systems n-soliton solutions were found before any Lax pair was known.   A general rule of thumb seems to be that if n-soliton solutions with $n \geqslant 3$ can be constructed, then they can be found for all n.   This property coincides with but does not logically imply the existence of an infinite set of conservation laws and hence complete integrability.   Certain aspects of Hirota's procedure have, to date, seemed somewhat mysterious but now we believe much of this can be resolved when the connection with the Painlevé property is demonstrated.

Hirota's method and its many applications have been well documented so here we will only sketch out the barest details that are sufficient for our purpose.   For the KdV equation (2.2), the method proceeds by making the substitution

$$u(x,t) = \frac{\partial^2}{\partial x^2} \log f(x,t) \qquad\qquad (3.1)$$

which yields an equation for f which can be expressed in terms of Hirota's bilinear operators as

$$(D_x^{\,4} + D_x D_t) f.f \;=\; 0 \qquad\qquad (3.2)$$

where $D_x^{\,n} = (\frac{\partial}{\partial x} - \frac{\partial}{\partial x'})^n f(x,t) f(x',t)/x=x'$ etc.   The next stage is to attempt to construct an n-soliton solution, $f^{(n)}$, by making a formal expansion of the $f^{(n)}$.

As is well known, this expansion has the apparently magical property of "self-truncating" for integrable systems [3].

Hirota's method and the Painlevé property are connected in the following way. We can think of u and u_2, in the Bäcklund transformation (2.4a), as an adjacent pair of solutions in a set $\{u^{(n)}\}$ which are related to Hirota's f-function $f^{(n)}$ by

$$u^{(n)} = [\log f^{(n)}]_{xx} \qquad\qquad (3.3)$$

Relabelling the u, u_2 and φ in (2.4a) as $u^{(n)}$, $u^{(n-1)}$ and φ_{n-1} respectively, we can prove the following theorem [14]:

Theorem I

If the functions $f^{(n)}$ $(n=1,2,3,\ldots)$ satisfy the Hirota equation (3.2), for every n, and if

$$f^{(n)} = \varphi_{n-1} f^{(n-1)} \qquad\qquad (3.4)$$

58

then the resulting equation in φ_{n-1} and $u^{(n-1)}$ is satisfied by the Painlevé relations (2.4b,c). Furthermore

$$f^{(n)} = \prod_{i=0}^{n-1} \varphi_i \quad . \tag{3.5}$$

The proof of the theorem is straightforward [14]. What it tells us is the following. At each iteration of the B.T. (2.4), a new singular manifold φ_i is produced - the product of these giving the $f^{(n)}$ (3.6). Since the B.T. (2.4) is only self-consistent if the system (in this case the KdV equation) possesses the Painlevé property then the self-truncation of Hirota's equation (3.2) will, by theorem 1, only occur if the system has this property. This connection between Hirota's method and the Painlevé property has been demonstrated for a large number of systems [14]. Equally, systems with non-truncating Hirota expansions are found *not* to pass the Painlevé test. There are a number of completely integrable systems, such as the modified KdV and the NLS equations, for which Hirota's method requires a different substitution from (3.1). Here one starts with $u = g/f$ and, in the case of NLS, obtains the pair of bilinear equations

$$(D_x^2 + i\, D_t)g.f = 0 \tag{3.6a}$$
$$D_x^2(f.f) = |g|^2 \tag{3.6b}$$

Now one works with formal self-truncating ε-expansions which are odd in g and even in f [3]. The connection with the Painlevé property may again be demonstrated through the following theorem (labelling the $u(v)$, $u_1(v_1)$ and φ of (2.10) as $u_n(v_n)$, $u_{n-1}(v_{n-1})$ and φ_{n-1} respectively) [14]:

Theorem II

If $f^{(n)}$ and $g^{(n)}$ satisfy

$$(D_x^2 + iD_t)g^{(n)}.f^{(n)} = 0 \tag{3.7a}$$
$$D_x^2(f^{(n)}.g^{(n)}) = |g|^2 \tag{3.7b}$$

and if

$$f^{(n)} = \varphi_{n-1} f^{(n-1)} \tag{3.7c}$$
$$g^{(n)} = u_0 f^{(n-1)} + \varphi_{n-1} g^{(n-1)} \tag{3.7d}$$

then the resulting equations in φ_{n-1}, u_0 and u_{n-1} are satisfied by the conditions found from the Painlevé analysis for the NLS equation. Furthermore

$$f^{(n)} = \prod_{i=0}^{n-1} \varphi_i \tag{3.8}$$

An almost identical theorem can be proved for the modified KdV equation.

4. Connections with other properties of completely integrable systems

The Bäcklund transformations derived in §2 from the Painlevé analysis are of the pseudopotential type [18,19]. In the case of the KdV equation we can also show the connection with the "standard" B.T. with the help of Hirota's formalism. Hirota's form of the B.T. [3], on using the result of Theorem I, transforms to

$$[\frac{\partial^2}{\partial x^2} + 2u^{(i-1)}]\varphi_{i-1} = k^2 \varphi_{i-1} \tag{4.1}$$

Equation (4.1) is, of course, just a Schrödinger equation with the eigenfunction

φ_{i-1} and potential $u^{(i-1)}$.   This result is an example of Crum's theorem [20].   In the case of multicomponent scattering problems such as the modified KdV, sine-Gordon, NLS equations, etc., we find the following connection between the singular manifold φ_n and the AKNS eigenfunctions v_1, v_2:

$$\varphi_n = (v_1^{(n)} - iv_2^{(n)})/(v_1^{(n)} + iv_2^{(n)}) \qquad , \tag{4.2}$$

where the $v_i^{(n)}$ are the eigenfunctions associated with the nth solution to the scattering problem.   The details of this analysis are given in [14].   The Painlevé analysis can also be used to obtain rational solutions [14] and pole dynamics [15]. In the former case, for the KdV equation in particular, we find a direct connection with the work of ADLER AND MOSER [21].   These authors give the following theorem:

Theorem III: If $\Theta^{(n)}(\tau_1, \tau_2, \dots, \tau_n)$ with $\tau_1 = x$ is a sequence of polynomials in n-variables and satisfies the recursion relations $D_x(\Theta^{(n+1)}, \Theta^{(n-1)}) = (2n+1)(\Theta^{(n)})2$; $\Theta^{(0)} = 1$, $\Theta^{(1)} = \tau_1 = x$ then the most general potential $u^{(n)}$, obtained by n applications of the Crum transformation to $u^{(0)} = 0$ with $\lambda = 0$ is of the form $u^{(n)} = (\log \Theta^{(n)})_{xx}$, $n = 0,1,2,\dots$ Moreover the eigenfunctions Φ_n at the nth step are given by

$$\Theta^{(n+1)} = \Theta^{(n)} \Phi_n \text{ with } (-\frac{\partial}{\partial x^2} + 2u^{(n)})\Phi_n = 0$$

(we note that the higher τ_i are associated with the time variables of the set of higher KdV equations.   In view of our own results, this theorem is easy to understand.   The $\Theta^{(n)}$ are just Hirota's solutions $f^{(n)}$ and the Φ_n are our singular manifolds φ_n.

Our discussion of rational and polar solutions leads us to a few remarks about the so-called τ-function.   The solutions to the Painlevé transcendents may all be expressed as second logarithmic derivatives of a certain entire function called the τ-function [22].   It would indeed seem, since the solutions to completely integrable systems are meromorphic, that Hirota's functions $f^{(n)}$ (which are just made up of (entire) exponential, polynomial or theta functions) are themselves multi-dimensional τ-functions.   Furthermore, from Theorem I we have the relation $\varphi_{n-1} = f^{(n)}/f^{(n-1)}$ and since the φ_n are meromorphic in the generalised Painlevé sense (as demonstrated for the KdV equation by WEISS [13]) it follows that the $f^{(n)}$ must be entire (the ratio of two entire functions is meromorphic).   We also note that the zero divisor of the functions $f^{(n)}$ defines the singular manifolds $\varphi_n = 0$. Finally we briefly mention some recent results for discrete systems.   A number of nonlinear chains such as the Toda chain

$$\ddot{Q}_n = \exp(Q_{n-1} - Q_n) - \exp(Q_n - Q_{n+1}) \tag{4.3}$$

and the Mikhailov, Fordy, Gibbons chain [23,24]

$$Q_{n,xt} = \exp(Q_{n-1} - Q_n) - \exp(Q_n - Q_{n+1}) \tag{4.4}$$

have been demonstrated to be completely integrable systems.   In terms of the variable $Q_n = S_{n-1} - S_n$ these equations take a slightly more compact form, e.g. (4.4) becomes

$$1 + S_{n,xt} = \exp(S_{n+1} + S_{n-1} - 2S_n) \qquad . \tag{4.5}$$

Motivated by the results of the Painlevé analysis for the Boussinesq equation (a continuum limit of the Toda chain), GIBBON AND TABOR [17] have found that the ansatz $S_n = \log \varphi_n + \bar{S}_n$, where S_n and $\bar{S}_n$ are separated solutions to (4.5) and φ_n a singular manifold, leads to the following equation for φ_n

$$\varphi_n \varphi_{n,xt} - \varphi_{n,x} \varphi_{n,t} = (1 + \bar{S}_{n,xt})(\varphi_{n+1}\varphi_{n-1} - \varphi_n^2) \tag{4.6}$$

from which one may deduce the Lax pair [24],

$$\varphi_{n,x} = -\bar{Q}_{n+1,x} \varphi_n + \lambda \varphi_{n+1}$$

$$\varphi_{n,t} = -\frac{1}{\lambda}\,\varphi_{n-1}\,\exp(\bar{Q}_n - \bar{Q}_{n+1}) \tag{4.7}$$

as well as the standard Bäcklund transformation.   Similar results for the Toda
chain may also be derived using this approach.

Acknowledgements

Michael Tabor is supported in part by U.S. Department of Energy grant
DEFG02-84ER13190 and an Alfred P. Sloan Research Fellowship.

References

1   E.T. Whittaker; Treatise on Analytical Dynamics, C.U.P., 1904.
2   V.E Zakharov and L. Faddeev; Funct. Anal. Applns. 5, 280 (1972).
3   R. Hirota; in "Solitons", eds: R.K. Bullough and P.J. Caudrey, Springer, Berlin
      (1980).
4   S. Kovalevskaya; Acta Mathematica, 12, 177, (1889), (Sur le problème de la
      rotation d'un corps solide autour d'un point fixe).
5   E.L. Ince; "Ordinary Differential Equations", Dover, New York (1956).
6   M. Tabor and J. Weiss; Phys. Rev. A 24, 2157 (1981).
7   Y.F. Chang, M. Tabor and J. Weiss; J. Math. Phys., 23, 531 (1982).
8   T. Bountis, H. Segur and F. Vivaldi; Phys. Rev., 25A, 1257 (1982).
9   M.J. Ablowitz, A. Ramani and H. Segur; J. Math. Phys., 21, 715 (1980).
10 J.B. McLeod and P.J. Olver; SIAM J. Math Anal., 14, 488, (1983).
11 R. Nakach; Plasma Physics, Proc. 36th Nobel Symposium (ed. Wilhelmsson), Plenum,
      N.Y., (1977).
12 J. Weiss, M. Tabor and G. Carnevale; J. Math. Phys., 24, 522, (1983).
13 J. Weiss; J. Math. Phys., 24, 1405 (1983).
14 J.D. Gibbon, P. Radmore, D. Wood and M. Tabor; to appear in Stud. in Appl. Math.,
      (1984).
15 D.V. Chudnovsky, G.V. Chudnovsky and M. Tabor; Phys. Lett., 97A, 268 (1983).
16 J. Weiss, J. Math. Phys., 25, 13 (1984).
17 J.D. Gibbon and M. Tabor; preprint.
18 H. Wahlquist in "Bäcklund Transformations"; 515, R.M. Miura (Ed), Springer,
      Heidelberg (1976).
19 D.V. Chudnovsky and G.V. Chudnovsky; Proc. Nat. Acad. Sci., 80, 1774 (1983).
20 M.M. Crum; Quart. J. Math. Oxford (2), 6, 121 (1955).
21 M. Adler and J. Moser; Commun. Math. Phys. 41, 1 (1978).
22 K. Okomoto; Physica 2D, 525 (1981).
23 A. Mikhailov; JETP Letts., 30, 414 (1979).
24 A.P. Fordy and J. Gibbons; Commun. Math. Phys., 77, 21-30 (1980).

A New Approach to Completely Integrable Partial Differential Equations by Means of the Singularity Analysis

Hitoshi Harada and Shin'ichi Oishi

Department of Electronics and Communication Engineering, School of Science and Engineering, Waseda University, Tokyo 160, Japan

§1. INTRODUCTION

Several years ago, Ablowitz, Ramani and Segur made the Painlevé conjecture in [1]: every nonlinear ordinary differential equation (ODE) obtained by an exact reduction of a completely integrable nonlinear partial differential equation (PDE) is of Painlevé type. Here an ODE is said to be of Painlevé type if its general solution has no movable singularities other than poles. Moreover they presented an explicit algorithm to test whether a given ODE satisfies certain necessary conditions for it to be of Painlevé type [1,2].

Recently, Weiss et al. proposed the PDE version of this test based on their own definition of the Painlevé property (P-property) in [3]. Hereafter we refer to it as the Weiss test. This test seems to be useful in investigating the soliton equations because it provides us with the Bäcklund transformations, the linearized equations and so on for the PDE's confirmed to possess the P-property. Various soliton equations are investigated by the Weiss test [3-6].

Here we present a method in order to find candidates for completely integrable PDE's. Our method is based on the Weiss test. We analyse a given nonlinear PDE with parameters as coefficients, values of which we will determine so that the PDE should possess the generalized P-property defined in the next section. Our approach aims at the systematic study of completely integrable PDE's from the viewpoint of the analyticity of their general solutions. We expect that investigations of various forms of nonlinear PDE's will be of some help to the systematic study of the complete integrability.

§2. THE GENERALIZED P-PROPERTY

Let us explain our position on the singularity analysis before introducing our procedure. For the sake of simplicity and definiteness, we limit ourselves to considering a quasi-linear PDE of order N in $u = u(x, t_1, \ldots, t_n)$ in the following form:

$$F(u) = \partial_x^N + \tilde{F}(u) = 0, \tag{1}$$

where $\partial = \partial/\partial x$ and $\tilde{F}(u)$ is assumed to be a polynomial in partial derivatives of u of order at most N-1. We consider (1) in the complex domain.

We say that the equation possesses the generalized P-property if it has a formal solution satisfying the following two conditions:
(I) The solution is expressible in terms of a dependent variable $\phi = \phi(x, t_1, \ldots, t_n)$ in the following form of the "Laurent series about $\phi = 0$":

$$u = \sum_{j=0}^{\infty} u_j \, \phi^{j-\alpha}, \tag{2}$$

where α is a positive integer; ϕ and $u_j=u_j(x,t_1,\ldots,t_n)$ are all analytic functions in the neighborhood of the singularity manifold, i.e., the manifold of singularities of the u defined by the condition $\phi(x,t_1,\ldots,t_n)=0$.

(II) Then the solution (2) contains N arbitrary functions: one of them is the ϕ, and others are among the u_j's.

Our conditions claim that the u is the formal general solution and has a "movable pole at $\phi=0$", by the Cauchy-Kovalevskaya theorem (see [7]). Many of soliton equations are found to possess the generalized P-property [3-6].

§3. SINGULARITY ANALYSIS

We now introduce our procedure. We will explain it by analysing as an example the fifth order KdV type equations in the following form:

$$u_{xxxxx}+a_1 uu_{xxx}+a_2 u_x u_{xx}+a_3 u^2 u_x+a_4 u_t=0, \tag{3}$$

where subscripts denote partial differentiations; a_1,a_2,a_3 and a_4 are parameters.

Let us consider (1), in which we now suppose that a "relevant" number of parameters for our analysis are involved as coefficients. Before starting our analysis, we should consider freedom of the scale transformations.

[Example] For (3), we consider scale transformations

$$\partial/\partial x \to \xi(\partial/\partial x), \quad \partial/\partial t \to \tau(\partial/\partial t), \quad u \to cu, \tag{4}$$

where ξ, τ and c are nonzero scaling factors. By this, (3) is transformed into

$$u_{xxxxx}+(a_1 b_1)uu_{xxx}+(a_2 b_1)u_x u_{xx}+(a_3 b_1^2)u^2 u_x+(a_4 b_2)u_t=0, \tag{5}$$

where we have set $b_1=c\xi^{-2}$, $b_2=\tau\xi^{-5}$. When $a_4=0$, (5) reduces to the ODE case. Therefore we set $b_2=1/a_4$ for the time being and rewrite (5) as follows:

$$u_{xxxxx}+A_1 uu_{xxx}+A_2 u_x u_{xx}+A_3 u^2 u_x+u_t=0. \tag{6}$$

If $a_3\neq0$, or $A_3\neq0$, then A_3 may be used as a scaling factor. We note that, even after setting any value to A_3, there remains freedom of the scale transformations that changes the signs of A_1 and A_2 simultaneously.

§3.1. LEADING ORDER ANALYSIS

In this step, we begin with determining the first term of the Laurent series (2). Let us introduce degrees for monomials of partial derivatives of u by setting $\deg(u)=\alpha$ and $\deg(\partial/\partial x)=\deg(\partial/\partial t_i)=1$ $(i=1,\ldots,n)$. They are linear functions in the α. Then substitution of (2) into each of the monomials appearing in F(u) shows that, if $A\phi^{-d}$ is the dominant term of this monomial as $\phi\to0$, then d coincides with the degree of this monomial. (We suppose α to be positive and regard the partial derivatives of the ϕ as independent of the ϕ. Therefore we equate the equation obtained by substituting (2) into (1), with respect to each power of the ϕ.)

Because of (II), the value of the α must be so chosen that the degree of the highest derivatives ∂_x^N in (1) should be equal to one or more degrees of the monomials in $\widetilde{F}(u)$ and higher than the rest of the

degrees. For the α, the homogeneous terms with the highest degree
are called the leading terms. Then if the value of the α is not an
integer, we must eliminate those terms which lead to such a choice,
by setting corresponding parameters to be zero. In this manner we
can determine the possible value of the α of an integer.

For the α, an algebraic equation for the u_0 in (2) is usually obtained from the dominant power $\phi^{-(\alpha+N)}$. This equation, which may be
of some order, will determine the u_0, but it must not be solved at
this stage because only one value of the roots of it is suitable.
[Example] In (6), the degrees of the monomials u_{xxxxx}, uu_{xxx}, $u_x u_{xx}$,
$u^2 u_x$ and u_t are $\alpha+5$, $2\alpha+3$, $2\alpha+3$, $3\alpha+1$ and $\alpha+1$, respectively. One can
easily find that $\alpha+5=2\alpha+3=3\alpha+1=7$ and $\alpha+1=3(<7)$ when $\alpha=2$. The leading
terms corresponding to $\alpha=2$ are the first four terms in (6). Then the
u_0 should be determined by

$$A_3 B^2 - 6(2A_1 + A_2)B + 360 = 0, \tag{7}$$

where we have set $u_0 = -B\phi_x^2$.

§3.2 RESONANCE STRUCTURE ANALYSIS

We say that a "resonance" occurs at j when the coefficient u_j of the
term $\phi^{j-\alpha}$ in (2) is arbitrary. The resonances can be found as the
roots of the "indicial equation". In this step we will determine the
values of parameters involved in the leading terms.

First we set up the indicial equation for (1). For this purpose,
we substitute

$$u = u_0 \phi^{-\alpha} + u_j \phi^{j-\alpha}, \tag{8}$$

into the polynomial consisting of the leading terms only, and ex-
tract the coefficient $\tilde{Q}(j) = Q(j)u_j$ of the power $\phi^{j-\alpha-N}$. Then $Q(j)=0$
is the indicial equation, which is found to be a Nth degree algebraic
equation for j because the leading terms involve the highest derivative. We set the leading coefficient of this equation to be unity
for definiteness.

Let $r_1,\ldots,r_N$ be N roots of the indicial equation. Since one root
is always -1, which corresponds to the arbitrariness of the ϕ, we denotes it by r_N. At this point we have to impose a restriction on (1)
that the leading terms do not involve any N-1th derivative, i.e.,
$\alpha\neq1$, in order to carry out this step determinatively. Under this,
the coefficient of j^{N-1} in the indicial equation is $-N(N+2\alpha-1)/2$.
Thus we have

$$r_1 + r_2 + \ldots + r_{N-1} = 1 + N(N+2\alpha-1)/2. \tag{9}$$

We can immediately write down all the possible combinations of values
of the roots $r_1,\ldots,r_{N-1}$ satisfying (9), because they must be nonnegative distinct integers by (II). (Besides they must be positive unless u_0 is arbitrary.) Then, solving the equations obtained by substituting these values into the indicial equation with the one for u_0
obtained in the previous step, one can determine values of the u_0 and
the parameters in question for each of the possible sets.

[Example] The indicial equation for (6) is found, by using (7), to be as follows:

$$(j+1)(j-6)[j^3-15j^2+(86-A_1B)j-\{240-2(2A_1+A_2)B\}]$$
$$=(j-r_1)(j-r_2)(j-r_3)(j-r_4)(j-r_5)=0, \tag{10}$$

where $r_4=6$, $r_5=-1$. Then, from (7) and (10), we have

$$r_1+r_2+r_3=15, \tag{11}$$
$$B=\pm\{(360-3r_1r_2r_3)/A_3\}^{1/2}, \tag{12}$$
$$A_1=\{86-(r_1r_2+r_2r_3+r_3r_1)\}/B, \tag{13}$$
$$A_2=\{2(r_1r_2+r_2r_3+r_3r_1)-r_1r_2r_3/2-52\}/B. \tag{14}$$

Here the double sign in (12) corresponds to the remaining freedom of the scale transformations as noted earlier, and we choose the plus sign for the definiteness. Then, there are nine cases of the possible combinations of the values of r_1, r_2, r_3, as in the following table. (We used the A_3 as the scaling factor and assigned appropriate values to A_3, because if $A_3=0$, (6) is found not to possess the generalized P-property.) The cases (C4), (C7), (C9) are the Sawada-Kotera equation, the fifth order KdV equation of Lax and the Kaup equation, respectively.

<u>Table</u> Values of A_1, A_2, A_3, B for the possible values of r_1, r_2, r_3

Cases	(r_3,r_2,r_1)	A_1	A_2	A_3	B
(C1)	(1, 2,12)	4	1	2	12
(C2)	(1, 3,11)	26	17	116	3/2
(C3)	(1, 4,10)	8	9	15	4
(C4)	(2, 3,10)	15	15	45	2
(C5)	(1, 5, 9)	18	29	100	3/2
(C6)	(2, 4, 9)	2	3	1	12
(C7)	(2, 5, 8)	10	20	30	2
(C8)	(3, 4, 8)	3	6	2	6
(C9)	(3, 5, 7)	15	75/2	45	1

3.3 <u>RESONANCE CRITERIA ANALYSIS</u>

We explain in this step the way to determine values of the undetermined parameters for a fixed set of the resonances. Let r_1 be the greatest number in the resonances. Then we substitute

$$u=\sum_{j=0}^{r_1} u_j\phi^{j-\alpha}, \tag{15}$$

into the original equation (1) and, for $j=1,\ldots,r_1$, require

$$Q(j)u_j+R_j=0, \tag{16}$$

where the left hand side of (16) is the coefficient of $\phi^{j-\alpha-N}$. Since $Q(j)=0$ is the indicial equation, R_j should identically vanish for any resonance j. When this identity holds, u_j is arbitrary. On the contrary, we should consider a coefficient in the form $a_j+b_j\ln\phi$, instead of u_j, if not so. We refer to this as the resonance criterion. Of course, u_j's can be determined by (16) unless the resonances occur at j's.

If there exists some resonance j   such that the identity $R_j=0$ will
not be able to be satisfied for any choices of the undetermined
parameters, then the choice of the parameters already determined is
not suitable for the generalized P-property. In this manner we can
determine the suitable choice of the parameters, namely the equations
possessing the generalized P-property.
[Example] The result of the resonance criteria analysis is that only
three cases (C4), (C7), (C9) have satisfied all the criteria. Thus
we have determined all the fifth order KdV type equations which possess the generalized P-property. This result coincides with the conclusion from the Sato theory (see [8]).

REFERENCES

1. M.J. Ablowitz, A. Ramani and H. Segur: Lett. Nuovo Cimento $\underline{23}$, 333
   (1978)
2. M.J. Ablowitz, A. Ramani and H. Segur: J. Math. Phys. $\underline{21}$, 715
   (1978)
3. J. Weiss, M. Tabor and G.Carnevale: J. Math. Phys. $\underline{24}$, 522 (1983)
4. J. Weiss: J. Math. Phys. $\underline{24}$, 1405 (1983)
5. J. Weiss: J. Math. Phys. $\underline{25}$, 13 (1984)
6. M. Jimbo, M.D. Kruskal and T. Miwa: Phys. Lett. $\underline{92A}$, 59 (1982)
7. I.G. Petrovsky: Lectures on Partial Differential Equations
   (Interscience, London 1954)
8. M. Jimbo and T. Miwa: Publ. RIMS, Kyoto Univ. $\underline{19}$, 943 (1983)

Part II

Field Theory and Statistical Mechanics of Solitons

Quantum Inverse Scattering Method

Miki Wadati

Institute of Physics, College of Arts and Sciences, University of Tokyo, Komaba 3-8-1
Tokyo 153, Japan

1. Introduction

The inverse scattering method is one of the most important discoveries in modern
mathematical physics. It is a unique method whereby the initial value problem of a
nonlinear evolution equation is solved. We now have its extension to the quantum
theory which we call the quantum inverse scattering method. It provides not only
a powerful method for studying completely integrable quantum systems, but also a
unified viewpoint on the structure of solvable models in quantum field theory and
statistical mechanics.

In this lecture I will review the developments of the quantum inverse scattering method by considering the quantum nonlinear Schrödinger model:

$$i\phi_t + \phi_{xx} - 2\kappa\phi^\dagger\phi\phi = 0,$$

$$\kappa = \varepsilon\alpha^2 , \qquad \alpha = |\kappa|^{1/2} , \qquad \varepsilon = \pm 1 . \tag{1.1}$$

Here $\phi(x)$ is a nonrelativistic boson field with equal time commutation relations

$$[\phi(x), \phi(y)] = [\phi^\dagger(x), \phi^\dagger(y)] = 0, \qquad [\phi(x), \phi^\dagger(y)] = \delta(x-y) . \tag{1.2}$$

The Hamiltonian of the system is given by

$$H = \int dx\ \phi_x^\dagger\phi_x + \kappa\int dx\ \phi^\dagger\phi^\dagger\phi\phi . \tag{1.3}$$

With the Hamiltonian (1.3) and the commutation relations (1.2), the equation of
motion $i\phi_t = [\phi, H]$ is (1.1). Note that the units $n = 1$ and $2m = 1$ are chosen.

The nonlinear Schrödinger model has been studied extensively both in classical
and quantum theories. The classical nonlinear Schrödinger model can be solved by
the inverse scattering method [1]. Bright and dark solitons exist for the attractive case $\kappa < 0$ and the repulsive case $\kappa > 0$, respectively. The quantum nonlinear
Schrödinger model describes the second quantization of the δ-function bose gas [2,
3]. The ground state and elementary excitations are obtained by the Bethe ansatz
method. This knowledge provides a useful test for the newly developed method.

2. Quantum Inverse Scattering Method

We formulate the quantum version of the inverse scattering method. The starting
point is an operator version of the auxiliary linear problem (Zakharov-Shabat eigenvalue problem):

$$\psi_{1x} + \frac{i}{2}\lambda\psi_1 = i\alpha\phi^\dagger\psi_2 , \qquad \psi_{2x} - \frac{i}{2}\lambda\psi_2 = -i\varepsilon\alpha\psi_1\phi . \tag{2.1}$$

We assume that $\phi(x) \to 0$ weakly as $|x| \to \infty$. Jost functions and scattering data are
defined as in Table 1.

Table 1.   Jost functions and scattering data

	$x = -\infty$	$x = \infty$
$\psi(x,\lambda)$	$\begin{pmatrix} 1 \\ 0 \end{pmatrix} e^{-\frac{i}{2}\lambda x}$	$\begin{pmatrix} A(\lambda)e^{-\frac{i}{2}\lambda x} \\ B(\lambda)e^{\frac{i}{2}\lambda x} \end{pmatrix}$
$\chi(x,\lambda)$	$\begin{pmatrix} -\varepsilon B^{\dagger}(\lambda)e^{-\frac{i}{2}\lambda x} \\ A(\lambda)e^{\frac{i}{2}\lambda x} \end{pmatrix}$	$\begin{pmatrix} 0 \\ 1 \end{pmatrix} e^{\frac{i}{2}\lambda x}$

For real λ, when $\psi(x,\lambda)$ satisfies (2.1), ${}^{t}\bar{\psi}(x,\lambda) = (\varepsilon\psi_2^{\dagger}(x,\lambda),\ \psi_1^{\dagger}(x,\lambda))$ also satisfies (2.1). Similarly, ${}^{t}\bar{\chi}(x,\lambda) = (\chi_2^{\dagger}(x,\lambda),\ \varepsilon\chi_1^{\dagger}(x,\lambda))$ is introduced. For complex λ, we have

$$\bar{\psi}(x,\lambda) = \begin{pmatrix} \varepsilon\psi_2^{\dagger}(x,\lambda^*) \\ \psi_1^{\dagger}(x,\lambda^*) \end{pmatrix} \qquad \bar{\chi}(x,\lambda) = \begin{pmatrix} \chi_2^{\dagger}(x,\lambda^*) \\ \varepsilon\chi_1^{\dagger}(x,\lambda^*) \end{pmatrix} . \qquad (2.2)$$

While ψ and χ are analytic in the upper half λ-plane, $\bar{\psi}$ and $\bar{\chi}$ are analytic in the lower half λ-plane.

Scattering data can be expressed in terms of the field variables. An auxiliary linear problem with the boundary condition (Table 1) is cast into the integral equations. Iterating the equations and taking the limit $x \to \infty$, we obtain

$$A(\lambda) = 1 + \sum_{n=1}^{\infty} \kappa^n \int \prod_{j=1}^{n} dx_j \prod_{j=1}^{n} dy_j \theta(x_1 > y_1 > \cdots\cdots > x_n > y_n)$$

$$\times\ e^{i\lambda(x_1 + \cdots + x_n - y_1 \cdots - y_n)} \phi^{\dagger}(x_1)\cdots\phi^{\dagger}(x_n)\phi(y_1)\cdots\phi(y_n) , \qquad (2.3a)$$

$$B(\lambda) = -i\varepsilon\alpha \sum_{n=0}^{\infty} \kappa^n \int \prod_{j=1}^{n} dx_j \prod_{j=1}^{n+1} dy_j \theta(y_1 > x_1 > \cdots\ x_n > y_{n+1})$$

$$\times\ e^{i\lambda(x_1 + \cdots + x_n - y_1 - \cdots - y_{n+1})} \phi^{\dagger}(x_1)\cdots\phi^{\dagger}(x_n)\phi(y_1)\cdots\phi(y_{n+1}) , \qquad (2.3b)$$

where $\theta(x_1 > x_2 > \cdots > x_n) = \theta(x_1 - x_2)\theta(x_2 - x_3)\cdots\theta(x_{n-1} - x_n)$. By direct calculation, we can show that

$$[H,\ A(\lambda)] = 0, \qquad [H,\ B^{\dagger}(\lambda)] = \lambda^2 B^{\dagger}(\lambda) . \qquad (2.4a,b)$$

Equation (2.4a) indicates that $A(\lambda)$ is conserved. From (2.3a) we have

$$A(\lambda) = 1 + \sum_{\ell=1}^{\infty} A_{\ell}/(i\lambda)^{\ell} ,$$

$$A_1 = -\kappa I_1 , \qquad A_2 = -i\kappa I_2 + \kappa^2 I_1(I_1 - 1)/2 , \qquad (2.5)$$

$$A_3 = \kappa I_3 + i\kappa^2 I_2(I_1 - 1) - \kappa^3 I_1(I_1 - 1)(I_1 - 2)/6 ,$$

where

$$I_1 = N = \int dx \, \phi^\dagger \phi \quad , \qquad I_2 = P = -i \int dx \, \phi^\dagger \phi_x \quad ,$$

$$I_3 = H = \int dx (-\phi^\dagger \phi_{xx} + \kappa \phi^\dagger \phi^\dagger \phi \phi)$$

(2.6)

As seen in the next section, $[A(\lambda), A(\mu)] = 0$, which includes (2.4a) as a special case and shows that the conserved quantities are in involution; $[I_i, I_j] = 0$, $i,j = 1,2$, $\cdots$. On the other hand, equation (2.4b) indicates that the state obtained by operating $B^\dagger$'s on the vacuum $|0\rangle$, $B^\dagger(k_1) B^\dagger(k_2) \cdots B^\dagger(k_n)|0\rangle$, is the eigenstate of the Hamiltonian. This is the (unnormalized) Bethe state which consists of n elementary modes. We see that the quantum inverse scattering method gives an algebraic construction of the Bethe states [4,5].

3. Algebra of Scattering Data

We shall obtain the commutation relations among the scattering data [6,7]. We first fix the system size $V = 2L$ finite. Define a 2×2 matrix Jost function which satisfies the boundary conditions

$$G_L(x,\lambda)\Big|_{x=-L} = \begin{pmatrix} 1 & 0 \\ 0 & 1 \end{pmatrix}$$

(3.1)

$$G_L(x,\lambda)\Big|_{x=L} = \begin{pmatrix} A_L(\lambda) & \varepsilon B_L^\dagger(\lambda) \\ B_L(\lambda) & A_L^\dagger(\lambda) \end{pmatrix} \equiv T_L(\lambda) \ .$$

(3.2)

In the limit $L \to \infty$, we have the relation

$$G_L(x,\lambda) = \Psi(x,\lambda) V(L,\lambda) \ ,$$

(3.3)

where

$$\Psi(x,\lambda) = \begin{pmatrix} \psi_1(x,\lambda) & \bar{\psi}_1(x,\lambda) \\ \psi_2(x,\lambda) & \bar{\psi}_2(x,\lambda) \end{pmatrix} \ , \qquad V(x,\lambda) = \begin{pmatrix} e^{-i\lambda x/2} & 0 \\ 0 & e^{i\lambda x/2} \end{pmatrix} \ .$$

(3.4)

The matrix $T_L(\lambda)$ defined in (3.2) is called the transition matrix [8]. The scattering data matrix $T(\lambda)$ is related to the transition matrix $T_L(\lambda)$;

$$T(\lambda) = \begin{pmatrix} A(\lambda) & \varepsilon B^\dagger(\lambda) \\ B(\lambda) & A^\dagger(\lambda) \end{pmatrix} = \lim_{L\to\infty} V^{-1}(L,\lambda) T_L(\lambda) V^{-1}(L,\lambda) \ .$$

(3.5)

We consider a 4×4 matrix of operators which is the direct product of matrix Jost functions;

$$H_L^{\lambda\mu}(x) = G_L(x,\lambda) \otimes G_L(x,\mu) \ .$$

(3.6)

Using the auxiliary linear problem (2.1), we can show that

$$R H_L^{\lambda\mu}(x) = H_L^{\mu\lambda}(x) R \ ,$$

(3.7)

where

$$R = \begin{pmatrix} 1 & 0 & 0 & 0 \\ 0 & a & b & 0 \\ 0 & b & a & 0 \\ 0 & 0 & 0 & 1 \end{pmatrix} \ , \qquad \begin{aligned} a &= \frac{-i\kappa}{\lambda-\mu-i\kappa} \ , \\[2ex] b &= \frac{\lambda-\mu}{\lambda-\mu-i\kappa} \ . \end{aligned}$$

(3.8)

In particular, we have at x = L

$$R(T_L(\lambda)\otimes T_L(\mu)) = (T_L(\mu)\otimes T_L(\lambda))R \quad . \tag{3.9}$$

This is the Yang-Baxter relation [9,10] for the nonlinear Schrödinger model. Based on the Yang-Baxter relation, the lattice formulation of the quantum inverse scattering method has been developed by the Leningrad school [8,11,12]. The matrix elements in (3.9) give the commutation relations, some of which are

$$[A_L(\lambda), A_L(\mu)] = [B_L(\lambda), B_L(\mu)] = 0 \quad ,$$

$$[A_L(\lambda) + A_L^\dagger(\lambda), A_L(\mu) + A_L^\dagger(\mu)] = 0$$

$$[B_L(\lambda) + B_L^\dagger(\lambda), B_L(\mu) + B_L^\dagger(\mu)] = 0 \quad , \tag{3.10}$$

$$A_L^\dagger(\lambda)B_L^\dagger(\mu) = \frac{\lambda-\mu-i\kappa}{\lambda-\mu}B_L^\dagger(\mu)A_L^\dagger(\lambda) + \frac{i\kappa}{\lambda-\mu} B_L^\dagger(\lambda)A_L^\dagger(\mu) \quad ,$$

$$A_L(\lambda)B_L^\dagger(\mu) = \frac{\lambda-\mu+i\kappa}{\lambda-\mu}B_L^\dagger(\mu)A_L(\lambda) - \frac{i\kappa}{\lambda-\mu} B_L^\dagger(\lambda)A_L(\mu) \quad .$$

We shall consider the limit $L\to\infty$ (infinite volume limit) in (3.9). We introduce the expectation value of $H_L^{\lambda\mu}(x)$ between the vacuum state $|0\rangle$:

$$W_{\lambda\mu}(x+L) \equiv \langle 0|H_L^{\lambda\mu}(x)|0\rangle$$

$$= \begin{bmatrix} e^{-\frac{i}{2}(\lambda+\mu)(x+L)} & 0 & 0 & 0 \\ 0 & e^{-\frac{i}{2}(\lambda-\mu)(x+L)} & 0 & 0 \\ 0 & \frac{2\kappa}{\lambda-\mu}\sin\frac{1}{2}(\lambda-\mu)(x+L) & e^{\frac{i}{2}(\lambda-\mu)(x+L)} & 0 \\ 0 & 0 & 0 & e^{\frac{i}{2}(\lambda+\mu)(x+L)} \end{bmatrix} \tag{3.11}$$

It can be shown that $W_{\lambda\mu}(x)$ also satisfies the relation $RW_{\lambda\mu}(x) = W_{\mu\lambda}(x)R$. Multiplying both sides of (3.9) by $(W_{\mu\lambda}(L))^{-1}$ from the left and by $(W_{\lambda\mu}(L))^{-1}$ from the right, we obtain

$$RU_1(\lambda,\mu)(T(\lambda)\otimes T(\mu))U_2(\lambda,\mu) = U_1(\mu,\lambda)(T(\mu)\otimes T(\lambda))U_2(\mu,\lambda)R \quad , \tag{3.12}$$

where

$$U_1(\lambda,\mu) = (W_{\lambda\mu}(L))^{-1}\cdot(V(L,\lambda)\otimes V(L,\mu)) \quad ,$$

$$U_2(\lambda,\mu) = (V(L,\lambda)\otimes V(L,\lambda))\cdot(W_{\lambda\mu}(L))^{-1} \tag{3.13}$$

Finally, in the limit $L \to \infty$, we arrive at

$$R_1(\lambda,\mu)(T(\lambda)\otimes T(\mu)) = (T(\mu)\otimes T(\lambda))R_2(\lambda,\mu) \quad . \tag{3.14}$$

The middle parts of the 4×4 matrices, R_1 and R_2, are

$$R_1(\lambda,\mu) = \begin{pmatrix} -2\pi b\kappa\delta(\lambda-\mu) & b \\ c & 0 \end{pmatrix}, \quad R_2(\lambda,\mu) = \begin{pmatrix} 0 & b \\ c & -2\pi b\kappa\delta(\lambda-\mu) \end{pmatrix} , \tag{3.15}$$

with $c = (\lambda-\mu+i\kappa)/(\lambda-\mu)$. Equation (3.14) with (3.15) gives the commutation relations among scattering data:

$$[A(\lambda), A(\mu)] = [A(\lambda), A^{\dagger}(\mu)] = [B(\lambda), B(\mu)] = 0 \ ,$$

$$A(\lambda)B^{\dagger}(\mu) = \frac{\lambda-\mu+i\kappa}{\lambda-\mu+i0} B^{\dagger}(\mu)A(\lambda) \ ,$$

$$A^{\dagger}(\lambda)B^{\dagger}(\mu) = \frac{\lambda-\mu-i\kappa}{\lambda-\mu-i0} B^{\dagger}(\mu)A^{\dagger}(\lambda) \ ,$$

$$B(\lambda)B^{\dagger}(\mu) = \frac{(\lambda-\mu)^2+\kappa^2}{(\lambda-\mu)^2} B^{\dagger}(\mu)B(\lambda) + 2\pi\alpha^2 A(\lambda)A^{\dagger}(\mu)\delta(\lambda-\mu) \ .$$

(3.16)

For later reference, it is convenient to define the reflection coefficient operator as

$$R^{\dagger}(\lambda) = -i\varepsilon/\alpha \cdot B^{\dagger}(\lambda)(A^{\dagger}(\lambda))^{-1} \ , \tag{3.17}$$

which satisfies the commutation relations

$$[H, R^{\dagger}(\lambda)] = \lambda^2 R^{\dagger}(\lambda) \ , \tag{3.18}$$

$$R^{\dagger}(\lambda)R^{\dagger}(\mu) = S(\mu,\lambda)R^{\dagger}(\mu)R^{\dagger}(\lambda) \ , \tag{3.19a}$$

$$R(\lambda)R^{\dagger}(\mu) = S(\lambda,\mu)R^{\dagger}(\mu)R(\lambda) + 2\pi\delta(\lambda-\mu) \ , \quad \text{where} \tag{3.19b}$$

$$S(\lambda,\mu) = S(\lambda-\mu) = \frac{\lambda-\mu-i\kappa}{\lambda-\mu+i\kappa} \ . \tag{3.20}$$

Operators $R^{\dagger}(\lambda)$ and $R(\lambda)$ are the creation and annihilation operators of the elementary mode respectively. The state constructed by $R^{\dagger}$ operators, $R^{\dagger}(k_1)R^{\dagger}(k_2) \cdots R^{\dagger}(k_n)|0>$, is a properly normalized Bethe state.

4. Factorized S-matrix

We shall examine the scattering of the elementary modes for the repulsive case $\kappa > 0$. We proceed to consider two particle scattering. For $k_1 < k_2$, two particle in-state $|\psi^{(+)}(k_1,k_2)>$ and out-state $|\psi^{(-)}(k_1,k_2)>$ are given by

$$|\psi^{(+)}(k_1,k_2)> = R^{\dagger}(k_1)R^{\dagger}(k_2)|0> \ , \quad |\psi^{(-)}(k_1,k_2)> = R^{\dagger}(k_2)R^{\dagger}(k_1)|0>. \tag{4.1}$$

From (3.19a) we see that

$$|\psi^{(+)}(k_1,k_2)> = S(k_2,k_1)|\psi^{(-)}(k_1,k_2)> \ , \tag{4.2}$$

which means that $S(k_2,k_1)$ is the two-particle S matrix. Let us examine the scattering behavior of several particles [13]. The n particle in- and out-states are expressed as

$$|\psi^{(+)}(k_1, \cdots, k_n)> = A_n \sum_{P_\alpha} \theta(\alpha_1 < \cdots < \alpha_n)R^{\dagger}(\alpha_1) \cdots R^{\dagger}(\alpha_n)|0> \ ,$$

$$|\psi^{(-)}(k_1, \cdots, k_n)> = A_n \sum_{P_\alpha} \theta(\alpha_1 < \cdots < \alpha_n)R^{\dagger}(\alpha_n) \cdots R^{\dagger}(\alpha_1)|0> \ .$$

(4.3)

Here A_n is a normalization constant and the set $(\alpha_1, \alpha_2, \cdots, \alpha_n)$ denotes the set of momenta obtained by the permutation of a set $(k_1, k_2, \cdots, k_n)$. Using (3.19), we find that

$$S(k_1', \cdots, k_n'; k_1, \cdots, k_n) = <\psi^{(-)}(k_1', \cdots, k_n')|\psi^{(+)}(k_1, \cdots, k_n)>$$

$$= \prod_{j=2}^{n} \prod_{\ell=1}^{j-1} S(|k_j-k_\ell|)\delta_{sym}(k_1', \cdots, k_n'; k_1, \cdots, k_n) \ , \tag{4.4}$$

where δ_{sym} is the symmetrized δ-function [13]. The n-particle S matrix is written as a product of $n(n-1)/2$ two-particle S matrices. Thus the factorization of S matrix [14,15] is explicitly proved. The above analysis shows that R operators play the role of the formal generators $A_i(\lambda)$ in Zamolodchikov-Zamolodchikov's theory [15].

The physical meanings of the factorized S matrix are clear: (1) the number of particles does not change due to the collisions, (2) a set of momenta (k_1, k_2,$\cdots$, k_n) is conserved and only exchanges of momenta during the pair collisions occur, (3) the multiparticle scattering is essentially a sequence of independent pair collisions. It is interesting to note that the collision process among classical solitons has the same properties [16]. The factorization of S matrix is a common feature of completely integrable systems where the dynamics of the particles are severely restricted by an infinite number of conservation laws. Furthermore, Zamolodchikov [17] showed that the factorization equation for (1+1)-dimensional quantum field theory is closely related to the Yang-Baxter relation for 2-dimensional lattice statistical mechanics. More exactly, he found that the factorization equation for the Z_4 model is essentially the same as the Yang-Baxter relation for the 8-vertex model. Motivated by this discovery, many factorized S matrices and their equivalent solvable lattice and spin models were reported [18].

Related to the application of infinite volume algebra (3.19) we can comment on the periodic boundary condition. In the conventional Bethe ansatz method the periodic boundary condition is essential in investigating the ground state and excitations of a finite density system [19]. Under the periodic boundary condition, momenta k_1, $\cdots$, k_n are not arbitrary but they should satisfy

$$e^{2ik_jL} = \prod_{\ell \neq j} S(k_\ell - k_j) \ , \quad j = 1, 2, \cdots, n \ . \tag{4.5}$$

We examine how the condition (4.5) appears from the quantum inverse scattering method. When we use finite volume algebra (3.10), the condition (4.5) is obtained from the requirement that the state $B_L^\dagger(k_1)\cdots B_L^\dagger(k_n)|0\rangle$ is the eigenstate of the operator $A_L(\lambda) + A_L^\dagger(\lambda)$. We may consider the problem using infinite volume algebra. The n particle state is given by

$$|k_1, \cdots, k_n\rangle = R^\dagger(k_1)\cdots R^\dagger(k_n)|0\rangle \ . \tag{4.6}$$

Hypothetically we move k_j particle in (4.6) to the right passing $(n-j)$ particles, bring it back from $x = L$ to $x = -L$, and again move it to the original position passing $(j-1)$ particles. This operation should not change the state (4.6). Then we have

$$R^\dagger(k_1)\cdots R^\dagger(k_j)\cdots R^\dagger(k_n)|0\rangle$$

$$= \prod_{\ell=j+1}^{n} S(k_\ell - k_j)R^\dagger(k_1)\cdots (j)\cdots\cdots R^\dagger(k_n)R^\dagger(k_j)|0\rangle$$

$$= e^{-2ik_jL} \prod_{\ell=j+1}^{n} S(k_\ell - k_j)R^\dagger(k_j)R^\dagger(k_1)\cdots (j)\cdots\cdots R^\dagger(k_n)|0\rangle$$

$$= e^{-2ik_jL} \prod_{\ell \neq j}^{n} S(k_\ell - k_j)R^\dagger(k_1)\cdots R^\dagger(k_j)\cdots R^\dagger(k_n)|0\rangle \ . \tag{4.7}$$

The last equality gives the condition (4.5). The method mentioned above is the algebraic construction of the periodic boundary condition and it is also valid for particle-hole excitations.

5. Bound State Operators

For the attractive case $\kappa < 0$, the set of elementary modes is no longer complete. There appear bound states known as the "string" states. We shall construct operators which describe the bound states.

We introduce "Boost operator" [20,21]:

$$K = \int dx \, x\phi^\dagger(x)\phi(x) \quad . \tag{5.1}$$

Using the relations

$$U(a)\phi(x)U^{-1}(a) = e^{-ax}\phi(x) \quad , \quad U(a)\phi^\dagger(x)U^{-1}(a) = e^{ax}\phi^\dagger(x) \quad , \tag{5.2}$$

where $U(a) \equiv \exp(aK)$ and the linear auxiliary problem (2.1), we find that

$$U(a)A(\lambda)U^{-1}(a) = A(\lambda-ia) \quad , \qquad U(a)B(\lambda)U^{-1}(a) = B(\lambda-ia) \quad ,$$
$$U(a)A^\dagger(\lambda)U^{-1}(a) = A^\dagger(\lambda-ia) \quad , \quad U(a)B^\dagger(\lambda)U^{-1}(a) = B^\dagger(\lambda-ia) \quad . \tag{5.3}$$

Note that for complex λ, $[A(\lambda)]^\dagger = A^\dagger(\lambda^*)$ and $[B(\lambda)]^\dagger = B^\dagger(\lambda^*)$. The bound state occurs when momenta k_j make the "string";

$$k_j = p - i\kappa(n-2j+1)/2 \quad , \quad j = 1, 2, \cdots, n \quad . \tag{5.4}$$

The n-particle bound state operators are defined by

$$A_n(p) = U^{-1}(\tfrac{1}{2}\kappa(n-1))A(p-i\kappa(n-1))A(p-i\kappa(n-2))\cdots A(p)U(\tfrac{1}{2}\kappa(n-1))$$
$$= A(p - \tfrac{i}{2}\kappa(n-1))A(p - \tfrac{i}{2}\kappa(n-3))\cdots A(p + \tfrac{i}{2}\kappa(n-3))A(p + \tfrac{i}{2}\kappa(n-1)) \quad ,$$

$$B_n(p) = U^{-1}(\tfrac{1}{2}\kappa(n-1))B(p-i\kappa(n-1))B(p-i\kappa(n-2))\cdots B(p)U(\tfrac{1}{2}\kappa(n-1))$$
$$= B(p - \tfrac{i}{2}\kappa(n-1))B(p - \tfrac{i}{2}\kappa(n-3))\cdots B(p + \tfrac{i}{2}\kappa(n-3))B(p + \tfrac{i}{2}\kappa(n-1)) \quad , \tag{5.5}$$

$$A_n^\dagger(p) = U(\tfrac{1}{2}\kappa(n-1))A^\dagger(p)\cdots A^\dagger(p+i\kappa(n-2))A^\dagger(p+i\kappa(n-1))U^{-1}(\tfrac{1}{2}\kappa(n-1))$$
$$= A^\dagger(p + \tfrac{i}{2}\kappa(n-1))A^\dagger(p + \tfrac{i}{2}\kappa(n-3))\cdots A^\dagger(p - \tfrac{i}{2}\kappa(n-3))A^\dagger(p - \tfrac{i}{2}\kappa(n-1)) \quad ,$$

$$B_n^\dagger(p) = U(\tfrac{1}{2}\kappa(n-1))B^\dagger(p)\cdots B^\dagger(p+i\kappa(n-2))B^\dagger(p+i\kappa(n-1))U^{-1}(\tfrac{1}{2}\kappa(n-1))$$
$$= B^\dagger(p + \tfrac{i}{2}\kappa(n-1))B^\dagger(p + \tfrac{i}{2}\kappa(n-3))\cdots B^\dagger(p - \tfrac{i}{2}\kappa(n-3))B^\dagger(p - \tfrac{i}{2}\kappa(n-1)) \quad .$$

The second lines are formal expressions. Using them and algebra (3.16) with care, we obtain

$$[A_n(p), A_{n'}(p')] = [A_n(p), A_{n'}^\dagger(p')] = [B_n(p), B_{n'}(p')] = 0 \quad ,$$

$$A_{n'}(p')B_n^\dagger(p) = \prod_{j=1}^{\bar{n}} \frac{p'-p+i\kappa(2j+|n'-n|)/2}{p'-p+i\kappa(2j-n-n')/2+i0} \, B_n^\dagger(p)A_{n'}(p') \quad ,$$

$$A_{n'}^\dagger(p')B_n^\dagger(p) = \prod_{j=1}^{\bar{n}} \frac{p'-p-i\kappa(2j+|n'-n|)/2}{p'-p+i\kappa(n+n'-2j)/2-i0} \, B_n^\dagger(p)A_{n'}^\dagger(p') \quad ,$$

$$B_n(p)B_{n'}^\dagger(p') = 2\pi n\delta_{nn'}\delta(p-p')A_n(p)A_{n'}^\dagger(p')\alpha^2$$
$$+ \frac{(p-p')^2 + \kappa^2(n+n')^2/4}{(p-p'+i0)^2 + \kappa^2(n-n')^2/4} \, B_{n'}^\dagger(p')B_n(p) \tag{5.6}$$

where $\bar{n} = \min(n,n')$. Extending the reflection coefficient operator (3.17), we introduce

$$R_n^\dagger(\xi) \equiv i^n/\alpha \cdot B_n^\dagger(\xi)(A_n^\dagger(\xi))^{-1} \quad ,$$
$$R_n(\xi) \equiv (-i)^n/\alpha \cdot (A_n(\xi))^{-1}B_n(\xi) \quad . \tag{5.7}$$

These operators are the creation and annihilation operators of the n-particle bound state. From (5.6), we get the commutation relations

$$A(\lambda)R_n^\dagger(p) = \frac{\lambda-p+i\kappa(n+1)/2}{\lambda-p-i\kappa(n-1)/2} \, R_n^\dagger(p)A(\lambda) \quad , \tag{5.8}$$

$$R_n(\xi)R_{n'}^\dagger(\xi') = 2\pi n\delta_{nn'}\delta(\xi-\xi') + S_{nn'}(\xi-\xi')R_{n'}^\dagger(\xi')R_n(\xi) \quad , \tag{5.9a}$$

$$R_n^\dagger(\xi)R_{n'}^\dagger(\xi') = S_{nn'}(\xi'-\xi)R_{n'}^\dagger(\xi')R_n^\dagger(\xi) \quad . \tag{5.9b}$$

where setting $\bar{n} = \min(n,n')$

$$S_{nn'}(\lambda) = \prod_{j=1}^{\bar{n}} \frac{\lambda-i\kappa(2j+|n-n'|)/2}{\lambda+i\kappa(2j+|n-n'|)/2} \cdot \frac{\lambda+i\kappa(2j-n-n')/2}{\lambda-i\kappa(2j-n-n')/2} \tag{5.10}$$

Combining (5.8) with (2.5), we obtain

$$NR_n^\dagger(p)|0\rangle = nR_n^\dagger(p)|0\rangle \quad , \qquad PR_n^\dagger(p) = npR_n^\dagger(p)|0\rangle \quad ,$$

$$HR_n^\dagger(p)|0\rangle = (np^2-\kappa^2 n(n^2-1)/12)R_n^\dagger(p)|0\rangle \quad . \tag{5.11}$$

This confirms that $R_n^\dagger(p)$ creates the bound state of n elementary modes with total momentum np and the binding energy $-\kappa^2 n(n^2-1)/12$. Equation (5.9b) shows that $S_{nn'}(\lambda)$ is the S matrix for the scattering of n'-particle bound state on n-particle bound state. We note that all the formulas in this section reduce to the ones in §3 if we set $n = n' = 1$, $A_1(\xi) = A(\xi)$, $B_1(\xi) = B(\xi)$ and $R_1(\xi) = R(\xi)$.

6. Quantum Gelfand-Levitan equation

For the attractive case $\kappa < 0$, we shall derive the Gelfand-Levitan equation [20, 22], which gives the basic tool for reconstructing the field variable $\phi(x)$ from scattering data.

We define an operator function g by

$$g(x,\lambda) \equiv \bar{\chi}(x,\lambda) - i\varepsilon\alpha\chi(x,\lambda)R(\lambda) \quad , \tag{6.1}$$

$$= \begin{pmatrix} \tilde{A}(\lambda)e^{-\frac{i}{2}\lambda x} \\ 0 \end{pmatrix} \qquad \text{as} \quad x \to -\infty \quad , \tag{6.2}$$

where $\tilde{A}(\lambda) = A^\dagger(\lambda) + i\alpha B^\dagger(\lambda)R(\lambda)$. Making use of the algebra (5.6), we have for $m \geq 1$

$$\tilde{A}(\lambda)B_m^\dagger(\mu) = \frac{\lambda-\mu-i\kappa(m+1)/2}{\lambda-\mu+i\kappa(m-1)/2-i0} B_m^\dagger(\mu)\tilde{A}(\lambda). \tag{6.3}$$

Equation (6.3) implies that the operator $\tilde{A}(\lambda)$ has discontinuities at $\mathrm{Im}\lambda = -\kappa m/2$, $m = 1,2,\cdots$. The discontinuity at $\lambda = \zeta-i\kappa m/2$ is

$$\tilde{A}(\xi-i\kappa m/2 + i0) - \tilde{A}(\xi-i\kappa m/2 - i0)$$

$$= B_{m+1}^\dagger(\xi)(A_m^\dagger(\xi+i\kappa/2))^{-1}A_{m+1}(\xi)^{-1}B_{m+1}(\xi)$$

$$= i\alpha^2 B^\dagger(\xi-i\kappa m/2)R_m^\dagger(\xi+i\kappa/2)R_{m+1}(\xi) \quad . \tag{6.4}$$

Recalling the asymptotic behavior of $\chi(x,\lambda)$, we are led to

$$g(x, \xi-i\kappa m/2 + i0) - g(x, \xi-i\kappa m/2 - i0)$$

$$= -i\kappa\chi(x, \xi-i\kappa m/2)R_m^\dagger(\xi+i\kappa/2)R_{m+1}(\xi) \ . \tag{6.5}$$

We define a piecewise analytic function;

$$\Phi(x,\lambda) = \begin{cases} g(x,\lambda)e^{\frac{i}{2}\lambda x} \ , & \mathrm{Im}\lambda > 0, \quad \mathrm{Im}\lambda \neq -\kappa/2, \ -\kappa, \ \cdots, \\ \bar\chi(x,\lambda)e^{\frac{i}{2}\lambda x} \ , & \mathrm{Im}\lambda < 0 \ . \end{cases} \tag{6.6}$$

Noting (6.5) we have an integral representation for $\Phi(x,\lambda)$;

$$\Phi(x,\lambda) = \bar\chi(x,\lambda)e^{\frac{i}{2}\lambda x}\Big|_{|\lambda|=\infty} + \frac{1}{2\pi i}\sum_{m=0}^\infty \int_{-\infty}^\infty \frac{d\xi'}{\xi'-i\kappa m/2-\lambda}$$

$$\times \ (-i\kappa)\chi(x, \xi'-i\kappa m/2)R_m^\dagger(\xi'+i\kappa/2)R_{m+1}(\xi')e^{\frac{i}{2}(\xi'-i\kappa m/2)x} \tag{6.7}$$

Setting $\lambda = \xi-i0$, we arrive at the Gelfand-Levitan equation

$$\bar\chi(x, \xi-i0)e^{\frac{i}{2}\xi x} = \begin{pmatrix} 1 \\ 0 \end{pmatrix} - \frac{\kappa}{2\pi}\sum_{m=0}^\infty \int_{-\infty}^\infty \frac{d\xi'}{\xi'-\xi-i\kappa m/2 + i0}$$

$$\times \ \chi(x, \xi'-i\kappa m/2)R_m^\dagger(\xi'+i\kappa/2)R_{m+1}(\xi')e^{\frac{i}{2}(\xi'-i\kappa m/2)x} \ , \tag{6.8}$$

where $R_o^\dagger(\lambda) \equiv 1/\alpha$. This equation enables the Jost function $\bar\chi(x,\xi)$ to be determined in terms of the bound state operators. The field $\phi(x)$ is related to the Jost function as

$$\chi_1(x,\lambda)e^{-\frac{i}{2}\lambda x} = \frac{\alpha}{\lambda}\phi^\dagger(x) + 0(\frac{1}{\lambda^2}) \ ,$$

$$\chi_2(x,\lambda)e^{-\frac{i}{2}\lambda x} = 1 + i\frac{\kappa}{\lambda}\int_x^\infty dy\phi^\dagger(y)\phi(y) + 0(\frac{1}{\lambda^2}) \ . \tag{6.9}$$

The equations (6.8) and (6.9) with (2.2) complete the inverse problem, that is, the reconstruction of the field $\phi(x)$ from the scattering data. If we take only an $m = 0$ term in (6.8), we recover the Gelfand-Levitan equation for $\kappa > 0$ case.

7. Classical Soliton

Compared to the classical theory, the potential power of the Gelfand-Levitan equation has not been fully examined. Thacker [23] used it to evaluate the two-point equal time correlation function. In this section I want to show you an example of the applications [24]. Using the Gelfand-Levitan equation, we shall evaluate the matrix elements of $\phi(x)$ and $\phi^\dagger(x)\phi(x)$ to discuss the classical limit of the theory. We define n-particle bound state by

$$|n, p> \ = \ 1/n\cdot R_n^\dagger(p)|0> \ , \tag{7.1}$$

which is normalized as $<n',p'|n,p> = (2\pi/n)\cdot\delta_{nn'}\delta(p-p')$. For complex λ, the Gelfand-Levitan equation reads as

$$\chi_1^\dagger(x,\lambda)e^{\frac{i}{2}\lambda^* x} = \frac{\kappa}{2\pi}\sum_{m=0}^\infty \int \frac{d\xi'}{\xi'-\lambda^*-i\kappa m/2} \chi_2(x, \xi'-i\kappa m/2)R_m^\dagger(\xi'+i\kappa/2)$$

$$\cdot \ R_{m+1}(\xi')e^{\frac{i}{2}(\xi'-i\kappa m/2)x} \ , \tag{7.2a}$$

$$\chi_2(x,\lambda)e^{-\frac{i}{2}\lambda x} = 1 - \frac{\kappa}{2\pi}\sum_{m=0}^{\infty}\int\frac{d\xi'}{\xi'-\lambda+i\kappa m/2}\,R_{m+1}(\xi')R_m(\xi'-i\kappa/2)$$

$$\cdot\,\chi_1^{\dagger}(x,\,\xi'-i\kappa m/2)e^{-\frac{i}{2}(\xi'+i\kappa m/2)x}\quad. \tag{7.2b}$$

Making use of (7.2) and (5.9), we can show that

$$<0|R_{n-1}(p'-i\kappa/2)\chi_1^{\dagger}(x,\,p'-i\kappa(n-1)/2)R_n^{\dagger}(p+i\kappa/2)|0>$$

$$= \frac{\kappa^2 n(n+1)}{(p-p'-i\kappa(2n-3)/2)(p-p'-i\kappa(2n-5)/2)}\,e^{\frac{i}{2}(p-p'-i\kappa(2n-3)/2)x}e^{\frac{i}{2}(p-p'-i\kappa(2n-5)/2)x}$$

$$\times\,<0|R_{n-2}(p'-i\kappa)\chi_1^{\dagger}(x,\,p'-i\kappa(n-3)/2)R_{n-1}^{\dagger}(p+i\kappa)|0>\quad. \tag{7.3}$$

This recursion formula is quite useful in the following analysis.

First we calculate the matrix element of $\phi(x)$. From (7.2), we have

$$<n,\,p'|\chi_1^{\dagger}(x,\lambda)e^{\frac{i}{2}\lambda^*x}|n+1,\,p>$$

$$= \frac{\kappa^2}{(p-\lambda^*-i\kappa n/2)(p-p'-i\kappa(2n-1)/2)}\,e^{\frac{i}{2}(2p-p'-i\kappa(3n-1)/2)x}$$

$$\cdot\,<0|R_{n-1}(p'-i\kappa/2)\chi_1^{\dagger}(x,\,p'-i\kappa(n-1)/2)R_n^{\dagger}(p+i\kappa/2)|0>\quad. \tag{7.4}$$

Using the formula (7.3) repeatedly, we obtain

$$<0|R_{n-1}(p'-i\kappa/2)\chi_1^{\dagger}(x,\,p'-i\kappa(n-1)/2)R_n^{\dagger}(p+i\kappa/2)|0>$$

$$= -\kappa^{2(n-1)}n!(n-1)!\,e^{i(n-1)(p-p')x}e^{\frac{i}{2}(2p-p')x}e^{-\frac{1}{4}\kappa(3n-1)x}$$

$$\frac{1}{p-p'+i\kappa(2n-1)/2}\prod_{j=1}^{n-1}\frac{1}{[p-p'-i\kappa(2n-4j+1)/2][p-p'-i\kappa(2n-4j-1)/2]}\quad. \tag{7.5}$$

Here $R_0(\lambda) = R_0^{\dagger}(\lambda) = 1/\alpha$, and $<0|\chi_1^{\dagger}(x,\lambda)R^{\dagger}(\mu)|0> = -\alpha/(\mu-\lambda^*)\exp[i(2\mu-\lambda^*)x/2]$ have been used. Substituting (7.5) into (7.4), we have

$$<n,\,p'|\chi_1^{\dagger}(x,\lambda)\,e^{\frac{i}{2}\lambda^*x}|n+1,\,p>$$

$$= -\frac{\kappa^{2n}n!(n-1)!}{p-\lambda^*-i\kappa n/2}e^{i((n+1)p-np')x}\prod_{j=1}^{2n}\frac{1}{(p-p'+i\kappa(2n-2j+1)/2)} \tag{7.6}$$

Comparing this with (6.10), we arrive at

$$<n,p'|\phi(x)|n+1,p> = \alpha^{4n-1}n!(n-1)!\,e^{i((n+1)p-np')x}\cdot\prod_{j=1}^{n}\frac{1}{[(p-p')^2+\kappa^2(2n-2j+1)^2/4]} \tag{7.7}$$

In a similar way, we can evaluate the matrix element of $\phi^{\dagger}(x)\phi(x)$;

$$<n,\,p'|\chi_2(x,\lambda)e^{-\frac{1}{2}\lambda x}|n,\,p>$$

$$= \frac{2\pi}{n}\delta(p-p') - \frac{\kappa}{n}\frac{1}{p'-\lambda+i\kappa(n-1)/2}e^{-\frac{i}{2}(p'+\frac{i}{2}\kappa(n-1))x}$$

$$\times\,<0|R_{n-1}(p'-i\kappa/2)\chi_1^{\dagger}(x,\,p'-i\kappa(n-1)/2)R_n^{\dagger}(p)|0>$$

$$= \frac{2\pi}{n}\delta(p-p') + \frac{\kappa^{2n-1}(n-1)!(n-1)!e^{in(p-p')x}}{(p'-\lambda+i\kappa(n-1)/2)(p-p')} \prod_{\ell=1}^{n-1} \frac{1}{[(p-p')^2+\kappa^2\ell^2]} \quad . \tag{7.8}$$

Comparison of (7.8) and (6.9) leads to

$$<n,p'|\phi^\dagger(x)\phi(x)|n,p> = -\kappa^{2(n-1)}n!(n-1)!e^{in(p-p')x} \prod_{\ell=1}^{n-1} \frac{1}{[(p-p')^2+\kappa^2\ell^2]} \quad . \tag{7.9}$$

In [25], the Bethe ansatz wave function in coordinate representation has been used for the same problem. We see that the Gelfand-Levitan equation simplifies the calculations and gives the results in a compact form.

Let us discuss the relationship between the matrix element and the classical soliton. In the limit $n \to \infty$, the expressions (7.7) and (7.9) reduce to

$$\lim_{n\to\infty} <n,p'|\phi(x)|n+1, p> = \frac{\pi}{\alpha} e^{i((n+1)p-np')x} \operatorname{sech}(\tfrac{\pi}{\kappa}(p-p')) \quad , \tag{7.10}$$

$$\lim_{n\to\infty} <n,p'|\phi^\dagger(x)\phi(x)|n, p> = \frac{n\pi}{\kappa} e^{in(p-p')x} \frac{p-p'}{\sinh(\pi(p-p')/\kappa)} \tag{7.11}$$

where the formulas of infinite products

$$\cosh x = \prod_{\ell=1}^{\infty} (1+4x^2/\pi^2(2\ell-1)^2) \quad , \qquad \sinh x = x \prod_{\ell=1}^{\infty} (1+x^2/\pi^2\ell^2) \quad , \tag{7.12}$$

have been used. The time dependence of the theory will be considered in the Schrödinger picture;

$$|n,p,t> = e^{-iHt}|n,p>$$

$$= \exp[-i(np^2 - \tfrac{1}{12}n(n^2-1)\kappa^2)t]|n,p> \quad . \tag{7.13}$$

We introduce a wave packet

$$|n,X,t> = \int \frac{dP}{2\pi} e^{-iPX} e^{i(P^2/n)t}|n, P/n, t> \quad . \tag{7.14}$$

The wave packet $|n,X,t>$ has its center of mass located at X for all t. It is to be noted that this wave packet is not an eigenstate. Using (7.14), (7.13) and (7.10), we obtain

$$\lim_{n\to\infty} <n,X',t|\phi(x)|n+1, X, t>$$

$$= \tfrac{1}{2}\alpha(n+1)e^{i\kappa^2 n(n+1)t/4} \operatorname{sech}[\tfrac{1}{2}\kappa(n+1)(x-X)]\delta(X'-X + \tfrac{1}{n}x) \quad . \tag{7.15}$$

Setting $4\eta = |\kappa|(n+1)$, we arrive at

$$\lim_{n\to\infty} <n,X',t|\phi(x)|n+1, X, t> = \phi^S(x,t)|_{v=0, \, x_0=X} \; \delta(X'-X + \tfrac{x}{n}) \quad , \tag{7.16}$$

where $\phi^S(x,t)$ is the one-soliton solution;

$$\phi^S(x,t) = \frac{2\eta}{\alpha} \exp[4i\eta^2 t - \tfrac{i}{4}v^2 t + \tfrac{i}{2}vx + i\theta_0]\cdot\operatorname{sech}(2\eta(x-x_0-vt)) \quad . \tag{7.17}$$

Thus, we have shown that the classical soliton solution at rest is related to the matrix element of $\phi(x)$ in the limit $n \to \infty$. Introducing the Galilean transformed state $|n,X,t>_v$ by $|n,X,t>_v \equiv G(v)|n,X,t>$, where $G(v)$ is the generator of the Galilean transformation [25], we obtain for the moving soliton

$$\lim_{n\to\infty} {}_v<n,X',t|\phi(x)|n+1, X, t>_v = \phi^S(x,t)|_{x_0=X} \; \delta(X'-X + (x-vt)/n) \quad . \tag{7.18}$$

Similarly, we obtain

$$\lim_{n\to\infty} {}_V\langle n,X',t|\phi^\dagger(x)\phi(x)|n,X,t\rangle_V = \phi^{*S}(x,t)\phi^S(x,t)\big|_{x_0=X}\,\delta(X'-X)\ . \qquad (7.19)$$

We conclude that the moving soliton as well as the soliton at rest appear as the matrix elements of the field operator in the limit $n \to \infty$, where n is the particle number making the bound state.

8. Concluding Remarks

In this lecture the quantum inverse scattering method was introduced and discussed. This new development of the soliton theory is extremely useful in showing the existence of an infinite number of conservation laws, the algebraic construction of the Bethe state, the explicit verification of the factorized S matrix and so on. These results characterize the completely integrable quantum system. Furthermore, scattering method relates the (1+1)-dimensional field theory to 2-dimensional lattice statistical mechanics and gives a unified viewpoint on the solvable models in physics.

Recently, the Kondo problem [26] and Anderson model [27] have been solved by the Bethe ansatz method. Since the quantum inverse scattering method includes the Bethe ansatz method, it is a simple conjecture that those problems and related interesting models in solid state physics should be investigated by the quantum inverse scattering method.

References

1.   V.E. Zakharov and A.B. Shabat: Sov. Phys. JETP 34, 62 (1972).
2.   I.B. McGuire : J. Math. Phys. 6, 432 (1965).
3.   E.H. Lieb and W. Liniger: Phys. Rev. 130, 1605 (1963).
4.   E.K. Sklyanin and L.D. Faddeev: Sov. Phys. Dokl. 23, 902 (1978).
5.   H.B. Thacker and D. Wilkinson: Phys. Rev. D19, 3660 (1979).
6.   E.K. Sklyanin: Sov. Phys. Dokl. 24, 107 (1979).
7.   H.B. Thacker: Rev. Mod. Phys. 53, 253 (1981).
8.   L.D. Faddeev: Sov. Sci. Rev. Math. Phys. C1, 107 (1980).
9.   C.N. Yang: Phys. Rev. Lett. 19, 1312 (1967).
10. B.J. Baxter: Ann. Phys. 70, 193 (1972).
11. L.A. Takhatajan and L.D. Faddeev: Russian Math. Survey 34, 11 (1979).
12. P.P. Kulish and E.K. Sklyanin: Lecture Notes in Physics 151 (Springer-Verlag, 1982) p.61.
13. K. Sogo, M. Uchinami, A. Nakamura and M. Wadati: Prog. Theor. Phys. 66, 1 (1981).
14. K. Karowski, H.J. Thun, T.T. Truong and P.H. Weisz: Phys. Lett. 67B, 321 (1977).
15. A.B. Zamolodchikov and A.B. Zamolodchikov: Ann. Phys. (USA) 120, 523 (1979).
16. M. Wadati and M. Toda: J. Phys. Soc. Japan 32, 1403 (1972).
17. A.B. Zamolodchikov: Commun. Math. Phys. 69, 165 (1979).
18. A.B. Zamolodchikov and V.A. Fateev: Sov. J. Nucl. Phys. 32, 293 (1980).
     J.H.H. Perk and C.L. Schultz: Phys. Letters 84A, 407 (1981).
     K. Sogo and M. Wadati: Prog. Theor. Phys. 68 , 85 (1982).
     K. Sogo, M. Uchinami, Y. Akutsu and M. Wadati: Prog. Theor. Phys. 68, 508 (1982).
     K. Sogo, Y. Akutsu and T. Abe: Prog. Theor. Phys. 70, 730, 739 (1983).
19. C.N. Yang and C.P. Yang: J. Math. Phys. 10, 1115 (1969).
20. M. Göckeler: Z. Phys. C7, 263 (1981), 11, 125 (1981).
21. K. Sogo and M. Wadati: Prog. Theor. Phys. 69, 431 (1983).
22. F.A. Smirnov: Dokl. Akad. Nauk SSSR 262, 78 (1982).
23. H.B. Thacker: Lecture Notes in Physics 151 (Springer-Verlag, 1982) p.1.

24. M. Wadati: preprint.
25. M. Wadati and M. Sakagami: J. Phys. Soc. Japan 53, 1983 (1984).
26. P.B. Wiegman: JETP Lett. 31, 364 (1980).
     N. Andrei and J.H. Lowenstein: Phys. Rev. Lett. 46, 356 (1981).
     N. Andrei, K. Furuya and J.H. Lowenstein: Rev. Mod. Phys. 55, 331 (1983).
27. P.B. Wiegman: Phys. Lett. 80A, 163 (1980).
     A. Okiji and N. Kawakami: Solid State Commun. 43, 365 (1982).

Yang-Baxter Algebras and Integrable Models in Field Theory and Statistical Mechanics

H.J. de Vega

LPTHE, Tour 16, 1e-étage, Univ. P. et M. Curie, 4, Place Jussieu
F-75230 Paris, France

The investigation of two-dimensional classical and quantum theories has shown in recent years that the Yang-Baxter equations and the Yang-Baxter algebras are the basic concepts of integrability. In statistical models and field theories the commutativity of transfer matrices $t(\lambda)$ at different values of the spectral parameter follows directly from the Yang-Baxter algebra. The expansion in powers of λ of log $t(\lambda)$ (or $t(\lambda)$) provides an infinite number of commuting operators including the Hamiltonian. So, we can say that the theory is integrable since there are as many commuting operators as degrees of freedom (infinity). More precisely, one associates in many theories a local transition matrix $L_n(\lambda)$ and the monodromy operator $T(\lambda)=\overleftarrow{\Pi}_n L_n(\lambda)$, the trace of which is the transfer matrix $t(\lambda)$. In an integrable theory $T(\lambda)$ verifies the Yang-Baxter algebra.

$$R(\lambda,\mu)[T(\lambda)\otimes T(\mu)] = [T(\mu)\otimes T(\lambda)] \, R(\lambda,\mu) \tag{1}$$

where the model dependent numerical matrix $R(\lambda,\mu)$ satisfies the consistency condition (Yang-Baxter equation)

$$R_{12}(\lambda,\mu) \, R_{13}(\lambda,\rho) \, R_{23}(\mu,\rho) = R_{2,3}(\mu,\rho) \, R_{13}(\lambda,\rho) \, R_{12}(\lambda,\mu) \; . \tag{2}$$

Moreover, the off-diagonal elements of $T(\lambda)$ provide creation and annihilation operators for the Bethe eigenstates of $t(\lambda)$[1]. These integrable quantum models have classical counterparts which are solvable by the classical inverse scattering method. They possess an associated system of linear equations (Lax pair) and its monodromy matrix $T(\lambda)$ which is the classical analogue of the quantum monodromy operator. Poisson brackets of the elements of $T(\lambda)$ obey a classical Yang-Baxter algebra.

$$\{T(\lambda)\otimes T(\mu)\} = [r(\lambda,\mu),T(\lambda)\otimes T(\mu)] \tag{3}$$

where the model dependent numerical matrix $r(\lambda,\mu)$ satisfies the classical Yang-Baxter equation

$$[r_{12}(\lambda,\mu),r_{13}(\mu,\rho)] + [r_{13}(\mu,\rho),r_{23}(\rho,\lambda)] + [r_{23}(\rho,\lambda),r_{12}(\lambda,\mu)] = 0 \; . \tag{4}$$

Usually, $R(\lambda,\mu)$ and $r(\lambda,\mu)$ are related through

$$R(\lambda,\mu) \underset{\hbar\to 0}{=} P[1 - i\hbar \, r(\lambda,\mu) + O(\hbar^2)] \qquad (P = \text{permutation matrix}) \; . \tag{5}$$

In the infinite volume limit (with suitable normalization of $T(\lambda)$ and $r(\lambda,\mu)$) the trace of $T(\lambda)$ provides infinitely many commuting local conserved charges (action variables) including the Hamiltonian, whereas the off-diagonal elements furnish angle variables.

In a remarkable class of field theories the whole monodromy operator (and not) only its trace) commutes with the Hamiltonian and generates an infinite sequence of non-local conserved charges. This class contains two-dimensional field theories with non-abelian internal symmetries which classically are scale invariant. Examples are the non-linear sigma models in symmetric spaces, the Gross-Neveu models, the fermionic chiral invariant models of ref[2] and the non-linear sigma model including the Wess-Zumino functional in the action. In all these theories the equations of motion read

$$\partial_\mu A^\mu(x) = 0 . \tag{6}$$

Here, the current $A_\mu(x)$ takes values in a Lie algebra and it is a local function of the fundamental fields. It obeys the flatness condition

$$\partial_0 A_1 - \partial_1 A_0 + [A_0,A_1] = 0 . \tag{7}$$

Eqs.(6)-(7) admit an associated matrix linear system[3-5]

$$(\partial_\mu + L_\mu)\Phi = 0 , \qquad\qquad \mu = 0,1$$

$$L_\mu(x,\lambda) = \frac{\lambda}{\lambda^2-1} [\lambda A_\mu(x) - \varepsilon_{\mu\nu}A^\nu(x)] , \qquad \varepsilon_{01} = +1 \tag{8}$$

where λ is a spectral parameter. Eqs.(6)-(7) ensure the compatibility of eqs.(8), i.e.

$$[\partial_0 + L_0, \partial_1 + L_1] = 0 . \tag{9}$$

Assuming the finite energy boundary conditions

$$\lim_{|x^1|\to\infty} |x|^{1+\varepsilon} A_\mu(x) = 0 , \qquad\qquad \varepsilon>0 \tag{10}$$

we can introduce a matrix solution $\Phi(x,\lambda)$ of eq.(8) such that

$$\lim_{x^1\to+\infty} \Phi_{ab}(x) = \delta_{ab} . \tag{11}$$

The monodromy matrix is defined as

$$T_{ab}(\lambda) = \lim_{x^1\to-\infty} \Phi_{ab}(x) . \tag{12}$$

Expansion in powers of λ provides an infinite number of non-local conserved charges

$$T(\lambda) = 1 + \sum_{n=1}^{\infty} \lambda^n T^{(n)} \tag{13}$$

$$\log T(\lambda) = \sum_{n=0}^{\infty} \lambda^{n+1} Q^{(n)} \quad \text{where} \tag{14}$$

$$Q^{(0)} = - \int_{-\infty}^{+\infty} A_0(x)dx \tag{15}$$

$$Q^{(1)} = - \frac{1}{2} \int_{-\infty}^{+\infty} dx'dy' \ \varepsilon(x'-y')A_0(x)A_0(y) - \int_{-\infty}^{+\infty} dx'A_0(x) . \tag{16}$$

For later use, we note the transformation properties of $L_\mu(x,\lambda)$ and $T(\lambda)$ under P (parity) and τ (time reversal).

$$
P : \quad
\begin{aligned}
A_\mu(t,x) &\longrightarrow A^\mu(t,-x) \\
T(\lambda) &\longrightarrow T(-\lambda)^{-1}
\end{aligned}
\qquad
\tau : \quad
\begin{aligned}
A_\mu(t,x) &\longrightarrow -A^\mu(-t,x) \\
T(\lambda) &\longrightarrow T(-\lambda) .
\end{aligned}
\tag{17}
$$

The canonical algebra of monodromy matrices (and of non-local charges) has been recently determined for several fermionic theories[2]. In all these cases the

Poisson bracket of two T's read

$$\{T(\lambda)\otimes T(\lambda)\} = \frac{g}{\lambda^{-1}-\mu^{-1}} [\pi, T(\lambda)\otimes T(\mu)] \tag{18}$$

where g is the coupling constant and π a numerical matrix depending on the model. Here we use the tensor product notations

$$\begin{aligned}
(A\otimes B)_{ab,cd} &= A_{ac} B_{bd} \\
\{A\otimes B\}_{ab,cd} &= \{A_{ac}, B_{bd}\} \, .
\end{aligned} \tag{19}$$

In any classical field theory where eqs.(6-10) hold, the classical monodromy matrix fulfills a simple factorization property.

Let us consider a field configuration formed by two separated lumps

$$\begin{aligned}
A_\mu(x) &= A_\mu^{(1)}(x) \, , & x' &\leq A \\
A_\mu(x) &= 0 \, , & A &\leq x' \leq B \\
A_\mu(x) &= A_\mu^{(2)}(x) \, , & B &\leq x' \, .
\end{aligned} \tag{20}$$

It follows from eq.(8) that

$$T_{ab}(\lambda,A_\mu) = T_{ac}(\lambda,A_\mu^{(1)}) \, T_{cb}(\lambda,A_\mu^{(1)}) \, . \tag{21}$$

This relation admits a very useful quantum generalization that will be discussed below.

The existence of a conserved $T(\lambda)$ in the quantum theory involves a non-trivial renormalization problem since $T(\lambda)$ is built of products of an arbitrary number of fields at different and equal points.

Assuming the existence of $T(\lambda)$ as a time independent operator in the quantum theory, we have developed a method[6,7] providing an explicit construction for it. Namely, our procedure gives a closed form expression for all matrix elements of $T(\lambda)$ in asymptotic states. The explicit solution obtained in this way indicates that the non-trivial renormalization problem of $T(\lambda)$ is solvable.

The quantum $T(\lambda)$ is constructed from the following general properties:
(a) $T(\lambda)$ exists as a quantum operator and is conserved.
(b) $T(\lambda)$ fulfills a quantum factorization principle.
(c) P, τ and internal symmetries hold in the quantum theory.
Let us denote by $|\theta,\alpha\rangle$ the one particle asymptotic states of the theory where θ stands for the rapidity and α for the set of quantum numbers. Here we assume all physical particles to be massive.

The quantum factorization principle relates the action of $T_{ab}(\lambda)$ on K-particle asymptotic states to a product of $T_{ab}(\lambda)$ applied on one-particle states in close analogy with the classical property (21)[8]

$$T_{ab}(\lambda)|\theta_1\alpha_1,\cdots,\theta_K\alpha_K\rangle_{out} = \sum_{a_1\cdots a_{K-1}} T_{aa_1}(\lambda)|\theta_1\alpha_1\rangle$$
$$T_{a_1a_2}(\lambda)|\theta_2\alpha_2\rangle \cdots T_{a_{K-1}b}|\theta_K\alpha_K\rangle \quad \text{and} \tag{22}$$

$$T_{ab}(\lambda)|\theta_1\alpha_1, \theta_2\alpha_2, \cdots, \theta_K\alpha_K\rangle_{in} =$$

$$= \sum_{a_1\cdots a_{K-1}} T_{a_1b}(\lambda)|\theta_1\alpha_1\rangle \, T_{a_2b_1}(\lambda)|\theta_2\alpha_2\rangle \cdots T_{a\,a_{K-1}}(\lambda)|\theta_K\alpha_K\rangle \tag{23}$$

where $\theta_i > \theta_j$ for $i>j$. Using property (a) the one-particle matrix elements read

$$<\theta\alpha|T_{ab}(\lambda)|\theta'\beta> = \delta(\theta-\theta') \, T_{a\alpha,b\beta}(\lambda,\theta) \tag{24}$$

where $T_{a\alpha,b\beta}(\lambda,\theta)$ is the same for in and out states. It also follows using eqs. (23)-(24) that all matrix elements of $T_{ab}(\lambda)$ between states with different numbers of particles vanish.

The asymptotic states in the theory being connected as usual by the S-matrix through

$$|in> = S|out>$$

we have the identity

$$<in|S \, T_{ab}(\lambda)|in> = <out|T_{ab}(\lambda) \, S|out> \; . \tag{25}$$

With the help of eqs.(23)-(24) this gives for two-particle states

$$\sum_{c\delta_1\delta_2} S_{\gamma_1\gamma_2,\delta_1\delta_2}(\theta_2-\theta_1) \, T_{a\delta_2,c\gamma_2'}(\lambda,\theta_2) \, T_{c\delta_1,b\gamma_1'}(\lambda,\theta_1) \tag{26}$$

$$= \sum_{c\delta_1\delta_2} T_{a\gamma_1,c\delta_1}(\lambda,\theta_1) \, T_{c\gamma_2b\delta_2}(\lambda,\theta_2) \, S_{\delta_1\delta_2,\gamma_1'\gamma_2'}(\theta_2-\theta_1)$$

where $S_{\alpha_1\alpha_2,\alpha_1'\alpha_2'}(\theta_1-\theta_2)$ stands for the two-particle S-matrix.

Following (c), $T_{ab}(\lambda)$ is restricted by P and τ invariance. So, a quantum unitary operator P and an antiunitary operator τ must exist such that (cf.eq. (17))

$$PT(\lambda)P^{-1} = T(-\lambda)^{-1} \; , \qquad \tau T(\lambda)\tau^{-1} = T(-\lambda) \; . \tag{27}$$

This gives on one-particle states

$$T_{a\alpha,c\beta}(+\lambda,+\theta) \, T_{c\beta,b\gamma}(-\lambda,-\theta) = \delta_{ab}\delta_{\alpha\gamma} \tag{28}$$

$$T_{a\alpha,c\beta}(\lambda,\theta) \, T^*_{c\beta,b\gamma}(\lambda,\theta) = \delta_{ab}\delta_{\alpha\gamma} \; .$$

Since $|a\theta>$ are possible particle states in the theory (they usually correspond to the fundamental fields) we can consider the S-matrix $S_{a\alpha,b\beta}(\theta)$. It verifies the factorization equation[9] (it has the form of eq.(2)), the unitarity relation and the real analiticity equations

$$S_{a\alpha,c\gamma}(\theta) \, S^*_{c\gamma,b\beta}(\theta^*) = \delta_{ab}\delta_{\alpha\beta}$$

$$S^*_{a\alpha,b\beta}(\theta^*) = S_{a\alpha,b\beta}(-\theta) \; . \tag{29}$$

This provides an explicit solution of eqs.(26) and (28)

$$[T_{ab}(\lambda,\theta)]_{\alpha\beta} = S_{a\alpha,b\beta}(\theta+\gamma(\lambda)) \, e^{i\phi(\lambda,\theta)} \tag{30}$$

where $\gamma(\lambda)$ and $\phi(\lambda,\theta)$ are real functions. Eqs.(28)-(29) imply $\gamma(-\lambda)=-\gamma(\lambda)$ and $\phi(-\lambda,-\theta)=-\phi(\lambda,\theta)$.

The classical monodromy matrix is invariant under Lorentz transformations.

$$x \pm t \to x' \pm t' = e^{\pm\varepsilon}(x \pm t) \; . \tag{31}$$

However, since the rapidity transforms as $\theta' \to \theta+\varepsilon$ the quantum spectral parameter $\gamma(\lambda)$ carries a transformation of the Lorentz group

84

$$\lambda \to \lambda'(\varepsilon) , \qquad \gamma(\lambda) \to \gamma' = \gamma(\lambda) + \varepsilon . \tag{32}$$

All matrix elements of $T_{ab}(\lambda)$ can now be computed using the one-particle matrix elements (30) and the factorization principle (22), (23)

$$_{out}\langle\theta'_1\alpha'_1,\cdots,\theta'_K\alpha'_K|T_{ab}(\lambda)|\theta_1\alpha_1,\cdots,\theta_L\alpha_L\rangle_{out}$$

$$= \delta_{KL} \prod_{i=1}^{n} \delta(\theta_i-\theta'_i) \sum_{a_1\cdots a_{K-1}} S_{a\alpha'_1,a_1\alpha_1}(\theta_1+\gamma(\lambda))\cdots S_{a_{K-1}\alpha'_K,b\alpha_K}(\theta_K+\gamma(\lambda))$$

$$\times \exp\left[i\sum_{j=1}^{K} \phi(\lambda,\theta_j)\right] . \tag{33}$$

The algebra of quantum monodromy matrices follows from eq.(33) and the Yang-Baxter eq.(2) for $S_{a\alpha,b\beta}(\theta)$. It can be written in a compact way

$$R(\lambda,\mu)[T(\lambda)\otimes T(\mu)] = [T(\mu)\otimes T(\lambda)] R(\lambda,\mu) \quad \text{where} \tag{34}$$

$$R(\lambda,\mu) = PS(\gamma(\lambda) - \gamma(\mu))$$

$$P_{ab,cd} = \delta_{ad}\delta_{bc} . \tag{35}$$

Since eq.(34) takes the same form in all K-subspaces, it holds as an operatorial equation in the whole Fock space.

So, we find that the quantum R matrix equals P times the two-body S-matrix for particle states $|a\theta,b\theta'\rangle$ at $\theta=\gamma(\lambda)$, $\theta'=\gamma(\mu)$. It must be noted that our derivation of $R(\lambda,\mu)$ does not involve the local operator $L_n(\lambda)[1]$. This is particularly interesting for cases like the non-linear σ model where the construction of a quantum $L_n(\lambda)$ is particularly delicate.

The matrix elements of $T_{ab}(\lambda)$ [eqs.(30)-(33)] has been checked with the renormalized expression of $T(\lambda)$ for several field theories[6-7]

$$\log T(\lambda) = -\lambda \int_{-\infty}^{+\infty} dx\, A_0(x) - \lambda^2 \left[\frac{1}{2} \iint_{|x-y|>\delta} dxdy\, \varepsilon(x-y)A_0(x)A_0(y)\right.$$

$$\left. + Z(\delta)\int_{-\infty}^{+\infty} A_1(y)dy\right] + O(\lambda^3) , \qquad \delta\to 0^+ . \tag{36}$$

The renormalization constant $Z(\delta)$ diverges logarithmically as $\delta\to 0^+$. Its value determines $\gamma(\lambda)$ for small λ as

$$\gamma(\lambda) = \frac{c}{\lambda} + O(\lambda) . \tag{37}$$

The constant c depends on the model chosen. For the O(N) non-linear sigma model $c=\pi/(N-2)$, for the SU(N) chiral Gross-Neveu model $c=\pi/(2gN)$ and $c=\pi/[8g(N-1)]$ for the SU(N) Gross-Neveu model. Here g stands for the coupling constant. The phase $\phi(\lambda\theta)$ in eq. (30) turns out to be consistent with zero provided one takes for S the so-called "physical S-matrix"[7]. That is the S-matrix where the asymptotic particles obey exotic statistics.

The explicit algebras of $T(\lambda)$ follows by insertion of the known S-matrices in eqs.(34)-(35). One finds, for example, for the SU(N) chiral Gross-Neveu model

$$[T_{ac}(\lambda),T_{bd}(\mu)] = \frac{2\pi i}{N} \frac{T_{bc}(\lambda)T_{ab}(\mu) - T_{bc}(\mu)T_{ad}(\lambda)}{\gamma(\lambda) - \gamma(\mu)} . \tag{38}$$

In the classical limit this reproduces the classical Poisson bracket algebra[2].

Slightly more involved expressions follow for the non-linear sigma model and Gross-Neveu model[6,7].

Bilinear algebras of the Yang-Baxter type (1) are characteristic of integrable field theories and integrable statistical models. Eqs.(34)-(35) are a new realization of these algebras in the Fock space of a class of quantum field theories. At each section with fixed number N of particles the representation of the YB algebra turns out to be isomorphic with a statistical model in a lattice with N sites on a side. In such cases

$$T_{ab}(\theta) = \sum_{a_1 \cdots a_{N-1}=1}^{q} t_{aa_1}(\theta) \otimes t_{a_1 a_2}(\theta) \otimes \cdots \otimes t_{a_{N-1}b}(\theta) \tag{39}$$

where $[t_{ab}(\theta)]_{\alpha\beta} = R_{\alpha a, b\beta}(\theta)$ and q is the number of possible states per link.

It is interesting to study the symmetries of the Yang-Baxter algebra (1). It is invariant under transformations of the monodromy operator

$$T(\theta) \longrightarrow g\, T(\theta) \tag{40}$$

provided

$$[R(\theta), g \otimes g] = 0 , \qquad \forall\, \theta . \tag{41}$$

Then the operator $t_g(\theta) = \mathrm{Tr}[gT(\theta)]$ gives a family of commuting transfer matrices[10].

$$[t_g(\theta), t_g(\theta')] = 0 . \tag{42}$$

There is another symmetry that leaves eq.(1) invariant. One can add an arbitrary parameter μ_i depending on the site to the argument θ in each $t_{a_{i-1}a_i}(\theta)$ of eq.(39). That is, the transformation

$$t_{a_{i-1}a_i}(\theta) \longrightarrow t_{a_{i-1}a_i}(\theta+\mu_i)$$

leaves eq.(1) invariant[11] [Eq.(33) provides an explicit example of an inhomogeneous YB algebra]. Eq.(42) can be generalized as

$$[t_g(\theta,\{\mu\}), t_g(\theta',\{\mu\})] = 0 . \tag{43}$$

These operators $t_g(\theta,\{\mu\})$ generate one-dimensional Hamiltonians with inhomogeneous parameters $\{\mu_k\}$ provided one adequately chooses g and the expansion point in the spectral variable θ. Expansion around $\theta=\infty$ provides non-local hermitian Hamiltonians. They couple all pairs of sites in a line. The higher orders in the expansion around $\theta=\infty$ give operators that couple triples, quartets and more sites[10].

Let us briefly discuss the six-vertex model and its descendants. It is defined by

$$t(\theta)_{11} = \begin{pmatrix} \mathrm{sh}(\theta+i\eta) & 0 \\ & \\ 0 & \mathrm{sh}\theta \end{pmatrix} , \qquad t(\theta)_{22} = \begin{pmatrix} \mathrm{sh}\theta & 0 \\ & \\ 0 & \mathrm{sh}(\theta+i\eta) \end{pmatrix}$$

$$\tag{44}$$

$$t(\theta)_{12} = i\sigma_- \,\mathrm{sh}\eta , \qquad t(\theta)_{21} = i\sigma_+ \,\mathrm{sh}\eta , \qquad \sigma_- = \sigma_+^T = \begin{pmatrix} 0 & 1 \\ 0 & 0 \end{pmatrix} .$$

The corresponding YB algebra is invariant under the group $D \otimes Z_2$ where

$$\begin{pmatrix} \sigma & 0 \\ 0 & \sigma^{-1} \end{pmatrix} \in D , \quad \sigma \in R \quad \text{and} \quad \left\{ \begin{pmatrix} 0 & 1 \\ 1 & 0 \end{pmatrix}, I_2 \right\} = Z_2 .$$

Expansion around $\theta=0$ of $\log\tau_I(\theta)$ generates what is known as the XXZ Heisenberg hamiltonian. Another commuting family here is given by

$$\tilde{\tau}_\beta(\theta,\{\mu\}) = \beta\, e^{-inS_3}\, T_{11}(\theta) + \beta^{-1}\, e^{inS_3}\, T_{22}(\theta) \tag{45}$$

where S_3 is the third component of the total spin. Expansion around $\theta=\infty$ for $\beta=i$ gives

$$\tilde{\tau}_\beta(\theta,\{\mu\}) = \frac{i}{2}\left(\frac{e^{\theta+in/2}}{2}\right)^{N-2} sinn\; H[1+0(e^{-2\theta N})] \tag{46}$$

where H is the hermitian operator

$$H\{\mu_\ell\} = \frac{sinn}{4} \sum_{1\leq a\neq b\leq N} sign(a-b)\, e^{-\mu_a-\mu_b}(\vec{\sigma}_a\wedge\vec{\sigma}_b)_3 + \sum_{a=1}^{N} e^{-2\mu_a}(\vec{\sigma}_a)_3 \;. \tag{47}$$

Here the $\vec{\sigma}_a$ are Pauli matrices. In the isotropic limit $(n\to 0)$, one gets the Hamiltonian

$$h\{\nu_\ell\} = \frac{1}{4} \sum_{1\leq a\neq b\leq N} sign(a-b)(\vec{\sigma}_a\wedge\vec{\sigma}_b)_3 - \sum_{a} \nu_a(\vec{\sigma}_a)_3 \;. \tag{48}$$

These Hamiltonians correspond to a long-range Dzialozhinski-Moriya interaction over a spin chain plus external magnetic fields. Both the D-M coupling and the magnetic field point in the third axis direction.

The $\{\mu_h\}$ and $\{\nu_\ell\}$ are arbitrary numbers in eq.(47)-(48). So one can say that they are given by some probability distribution $\rho(\mu)$ [or $\rho(\nu)$]. In other words, we have an integrable model with disorder.

The presence of the $\{\mu_h\}$ or $\{\nu_\ell\}$ does not affect the algebraic structure of the underlying Bethe Ansatz. In this way eigenvalues and eigenvectors of the transfer matrix $t_\beta(\theta,\{\mu\})$ can be obtained for arbitrary distributions $\rho(\mu)$ without special effort. As a concrete application we have derived the spectrum of states of $h\{\nu\}$ around the ferromagnetic vacuum[10]. One finds for an n-string state the dispersion relation

$$q_n(E) = 2 \int_{-\infty}^{+\infty} d\nu\rho(\nu)\; arctg(\frac{E-2\nu n}{n^2}) \;, \qquad n = 1,2,3,\cdots,\frac{N}{2} \;.$$

Here E stands for the energy, q for the momentum $(q=\frac{2\pi}{N}(m+\frac{n}{2}),\; m=-\frac{N}{2},\cdots,\frac{N}{2}-1)$ and $\rho(\nu)$ for the density of inhomogeneous parameters

$$\rho(\nu) \equiv \frac{1}{N} \sum_{e=1}^{N} \delta(\nu-\nu_e) \;.$$

The lower energy eigenstates of $h\{\nu\}$ for N large but finite are found from the longest string with momentum closest to the Brillovin zone bord[10].

It can be pointed out that the structure of operator (47)-(48) in the homogeneous case $(\mu_\ell=\nu_h=0)$ is like the first non-local charge in sigma models [eq.(16)].

We would like to conclude this lecture with some comments about Kac-Moody and Yang-Baxter algebras and their connections.

The equal-time commutator of currents $A_\mu(x)$ can be computed perturbatively in fermionic models like those of ref[2]. A c-number term appears in this commutator. It is the so-called Goto-Imamura-Schwinger term and it is finite in two dimensions[2,12].

$$[A_\mu(t,x) \otimes A_\nu(t,y)] = -4ig\delta(x-y)[1 \otimes A_{|\mu-\nu|}(x),\Pi] + \frac{8g^2 i}{\pi} \delta_{|\mu-\nu|,1} \Pi\delta'(x-y) \quad . \quad (49)$$

It is convenient to go to light cone coordinates $A_\pm = A^\mp = (1/\sqrt{2})(A_0 \pm A_1)$ and define a rescaled current[13] $J_\mu = -(1/g\sqrt{8})A_\mu$.   One gets

$$[J_-(x) \otimes J_-(y)] = 2i\delta(x-y)[1 \otimes J_-(x),\pi] - \frac{i}{\pi}\delta'(x-y)\Pi \quad . \tag{50}$$

This is precisely a Kac-Moody algebra. More generally, in a Kac-Moody algebra the c-number term in eq.(50) is multiplied by a number k, but well defined unitary representation only exists for integer values of k. This suggests that higher order perturbative contributions do not change the last term in eq.(50).

A link with the Yang-Baxter algebra follows from eqs.(8-11) since we can write at the classical level

$$T(\lambda) = P \exp\left[\frac{g\sqrt{8}}{1-\lambda^{-1}} \int_{-\infty}^{+\infty} J_-(x)dx^- \right] \quad . \tag{51}$$

Here P stands for "ordered exponential". At the quantum level the renormalization prescription R must be added to P.

Eq.(51) and its quantum counterpart provide an explicit link between Kac-Moody and Yang-Baxter algebra generators.

I would like to thank H. Eichenherr and J.M. Maillet in collaboration with whom a large part of this work was done.

References

1.   See, for reviews
     L.D. Faddeev, Les Houches lectures 1982, Saclay preprint T-82-76.
     P.P. Kulish and E.K. Sklyanin, in Tvärminne Lectures, Springer Lectures in Physics vol.151 (1982).
     H. Thacker, Revs. Mods. Physics 53, 253 (1981).
2.   H.J. de Vega, H. Eichenherr and J.M. Maillet, Phys. Lett. 132B, 337 (1983).
3.   A.V. Mikhailov and V.E. Zakharov, JETP 47, 1017 (1978).
     H. Eichenherr and M. Forger, Nucl. Phys. B155, 381 (1979).
4.   M. Lüscher and K. Pohlmeyer, Nucl. Phys. B137, 46 (1978).
5.   H.J. de Vega, Phys. Lett. 87B, 233 (1979).
6.   H.J. de Vega, H. Eichenherr and J.M. Maillet, Comm. Math. Phys. 92, 507 (1984).
7.   H.J. de Vega, H. Eichenherr and J.M. Maillet, Nucl. Phys. B (to appear).
8.   Al.B. Zamolodchikov, Dubna preprint E2, 11485 (1978).
9.   A.B. Zamolodchikov and Al.B. Zamolodchikov, Ann. Phys. 120, 253 (1979).
10. H.J. de Vega, Nucl. Phys. B (to appear); LPTHE Paris preprint 84/2.
11. R.J. Baxter, Studies in Appl. Math. L51 (1971).
12. See for example, H.J. de Vega and H.O. Girotti, Nucl. Phys. B79, 77 (1974).
13. E. Witten, Comm. Math. Phys. 92, 455 (1984).

Solitons in the Quantum Toda Lattice

Franz G. Mertens and Manfred Hader

Physikalisches Institut, Universität Bayreuth
D-8580 Bayreuth, Fed. Rep. of Germany

1. Introduction

One-dimensional quantum systems with continuous variables were
treated exactly by Bethe's ansatz beginning with the work of LIEB
and LINIGER [1] and YANG and YANG [2] on the Bose gas with repulsive
delta-function interactions. SUTHERLAND [3] generalized the method
to other integrable systems, where Bethe's ansatz does not give the
exact but only the asymptotic wave functions; however, this is sufficient to obtain the ground state energy and the excitation spectrum.

Recently this method has been applied to the quantum Toda lattice
[5]; analytical results have been obtained for two cases: In the
strongly anharmonic (weak coupling) regime the system behaves effectively like a gas of hard spheres with constant attraction where
the radius of the spheres is the scattering length. In the classical
(strong coupling) limit SUTHERLAND's result [4] has been rederived:
the particle and hole excitations of the quantum problem reduce to
solitons and phonons, respectively.

In this paper we consider the case of an arbitrary coupling constant. In particular we test whether the classical Toda solitons sur-
vive the quantization: this had been conjectured because of thermodynamic results obtained by a path-integral method [6].

2. Ground State

We write the Toda potential in the form

$$V(r) = \frac{m\omega^2}{\gamma^2} [e^{-\gamma(r-r_o)} + \gamma(r - r_o)] \tag{2.1}$$

After an appropriate scaling of all lengths by $1/\gamma$ and energies by
$\hbar^2\gamma^2/2m$ the Hamiltonian reads

$$H = \sum_n^N \{-\frac{\partial^2}{\partial x_n^2} + V_o + 2C^2 \exp [-(x_{n+1} - x_n - r_o)]\} \tag{2.2}$$

Here $C = m\omega/\hbar\gamma^2$ is the coupling constant; $1/C$ is a measure for the
anharmonicity. The term $V_0 = 2C^2(a - r_0)$ results from the linear
part of the potential applying periodic boundary conditions to a
chain of length $L = Na$.

The method of LIEB and LINIGER [1] leads to a formal description
of the system as an ideal Fermi gas. The ground state energy is

$$E_o/N = V_o + a \int_{-k_F}^{k_F} k^2 \rho(k)\, dk \tag{2.3}$$

The Fermi momentum k_F is fixed by normalization of ρ to $1/a$. The density of states $\rho(k)$ is determined by

$$2\pi\,\rho(k) = 1 - \int_{-k_F}^{k_F} dk'\ \delta'(k - k')\,\rho(k') \qquad (2.4)$$

Here the kernel is the derivative $\delta'(k) = r_0 + 2\ln C - 2\mathrm{Re}\{\psi(1 + ik)\}$ of the two-particle phase shift, ψ is the digamma function.

The integral equation for ρ is solved numerically for $|k| \leq k_F$ by replacing it by a system of linear equations (an iteration procedure converges only for $C < 1$). The result is inserted into the integrand and gives ρ for $|k| > k_F$. Figure 1 shows that ρ is qualitatively very different from the corresponding classical result of SUTHERLAND [4]. The ground state energy as a function of the coupling constant C is shown for two different boundary conditions. By the comparison

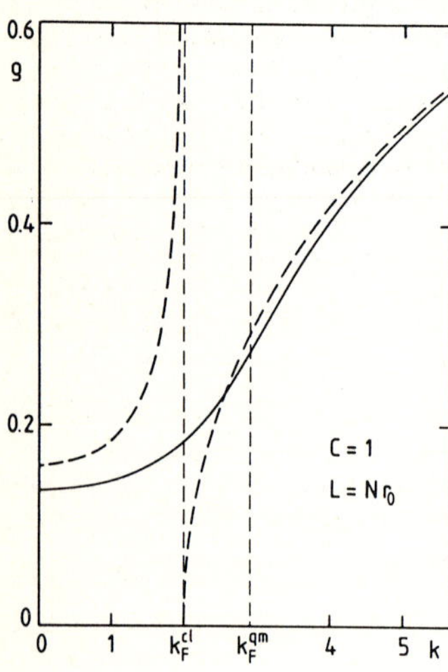

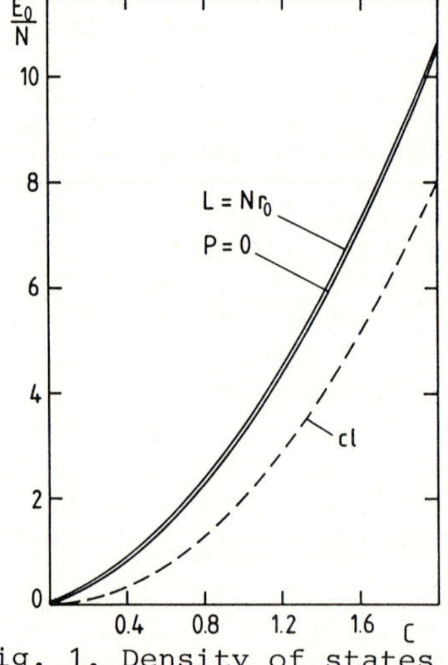

Fig. 1. Density of states ρ and ground state energy E_0 (for constant length L or zero pressure P) compared to the classical results (dashed lines)

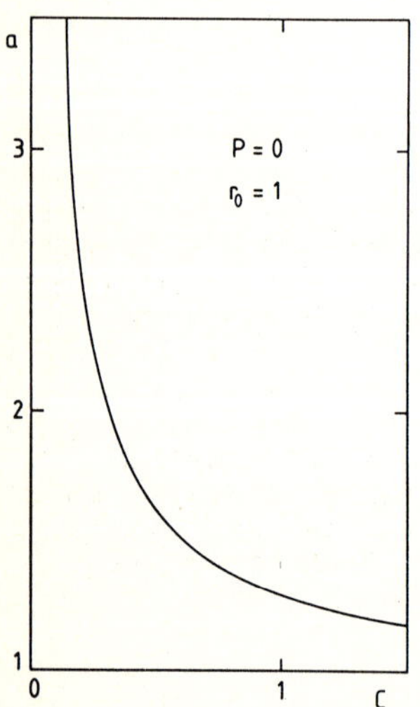

Fig. 2. Lattice parameter a as a function of the coupling constant C for zero pressure P

with the classical result it is seen that there are large quantum
fluctuations in the ground state for $C \lesssim 10$. They cause a considerable expansion of the lattice in the case of zero pressure (Fig. 2).

3. Solitons and Phonons

As our system is formally described by a Fermi gas the excitations
have a particle-hole character. The momentum $p(k)$ is obtained by the
integration of ρ from k to k_F (holes) or from k_F to k (particles).
The excitation energies are $|\varepsilon(k)|$ where the YANG and YANG [2] in-
tegral equation for ε in our case has the form [5]

$$\varepsilon(k) = k^2 - \mu - 2C^2 r_o - \int_{-k_F}^{k_F} \frac{dk'}{2\pi} \; \delta'(k - k') \; \varepsilon(k') \qquad (3.1)$$

The chemical potential μ is determined by $\varepsilon(k_F) = 0$. The elimination
of k from $p(k)$ and $|\varepsilon(k)|$ finally leads to the dispersion relations
$\varepsilon_1(p)$ and $\varepsilon_2(p)$ for the particle and hole excitations, respectively.

Equation (3.1) is solved numerically in the same way as (2.4).
For $C \gtrsim 5$ the dispersion curves are indistinguishable (in the gauge
of Fig. 3) from those of the classical solitons and phonons, which
are given [4,5] by

$$\varepsilon_1(\alpha) = 2C^2\{\sinh(2\alpha) - 2\alpha\}$$
$$p(\alpha) = 4C/a\{\alpha \cosh \alpha - \sinh \alpha\} \qquad (3.2)$$

and

$$\varepsilon_2(p) = 4C|\sin(pa/2)| \qquad (3.3)$$

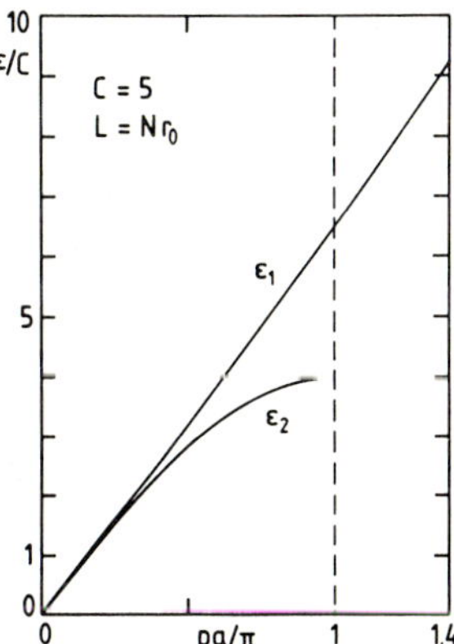

Fig. 3. For small anharmonicity the dispersion of the excitations is identical to that
of classical solitons (ε_1) and phonons (ε_2)

On the other hand the ground state certainly has to be described
quantum mechanically, as can be seen from the density of states and
the ground state energy (section 2). E.g. for $C = 5$ the zero point
motion gives a 13% contribution to the energy and expands the lattice by 5.5% in the case of zero pressure.

4. Conclusion and Discussion

Despite the quantum fluctuations in the ground state the upper
branch of the excitation spectrum can be identified with classical
solitons for $C \gtrsim 5$, which means not too large an anharmonicity. For
the lower branch we get the familiar result that the phonon dispersion is nearly the same classically or quantum mechanically for a
small anharmonicity.

For $C = 1$ the differences to the classical dispersion are still
rather small (Fig. 4), with increasing anharmonicity they become
larger and larger. The question is how these quantum excitations can
be interpreted. The lower branch always describes phonons [5]. For
the upper branch we are presently investigating whether these excitations have soliton properties. This is done by a variational
calculation of the wave functions of the ground state and the excitations by means of a quantum transfer-integral method [7]. For the
ground state energy this gives the same result as the Bethe ansatz
(within the numerical error). For the phonon dispersion the error is
in the order of 5%. Therefore it can be expected that the upper
branch is also well represented by this method and can be inspected
for soliton properties.

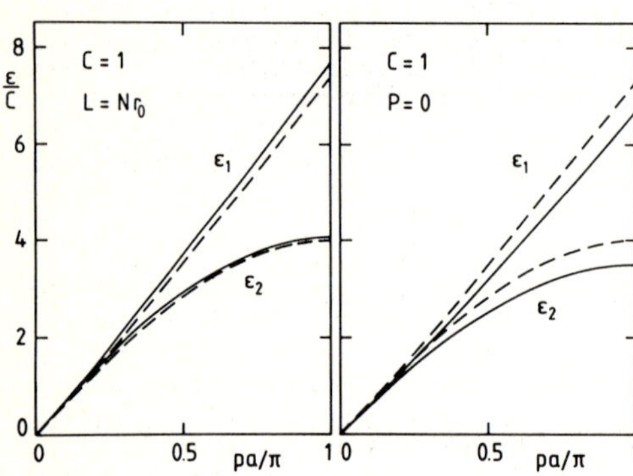

Fig. 4. Dispersion of the excitations compared to the classical solitons and phonons (dashed
lines)

References

1. E.H. Lieb, W. Liniger: Phys. Rev. 130, 1605 (1963)
   E.H. Lieb: Phys. Rev. 130, 1616 (1963)
2. C.N. Yang, C.P. Yang: J. Math. Phys. 10, 1115 (1969)
3. B. Sutherland: J. Math. Phys. 12, 246, 251 (1971)
4. B. Sutherland: Rocky Mount. J. of Math. 8, 413 (1978)
5. F.G. Mertens: Z. Physik B 55, 353 (1984)
6. F.G. Mertens, H. Büttner: EPS Conf., Antwerpen (1980),
       Europhys. Conf. Abstr. 3G, 46 (1980)
7. F.G. Mertens, W. Biem: J. Low Temp. Phys. 24, 379 (1975)

Recent Progress in Quantum Inverse Scattering Method – Baxter Hierarchy and Its Applications

K. Sogo

Department of Physics, Fronczak Hall, State University of New York at Buffalo
Buffalo, NY 14260, USA

The Baxter model is generalized to the higher dimensional representations of SU(2) group. The factorized S-matrix, q-state vertex model and XXZ Heisenberg model of spin S are treated from the unified point of view. Some features of SU(2) Thirring model are also discussed.

1. Introduction

Recent developments of the quantum inverse scattering method (QISM) have revealed that close links exist between some problems in the theoretical physics. Actually at the basis of the exact theories such as the quantum theory of solitons, exactly solvable models in statistical mechanics and the factorized S-matrix theory in the field theory there lie the same mathematical equations, that is, the factorization equations or the Yang-Baxter relation. Indeed if the Yang-Baxter relation holds, we can solve the problem exactly by the procedure of QISM (i.e. Bethe ansatz). In other words the factorization equations or the Yang-Baxter relation manifest the complete integrability of the system.

In this note we review the recent progress in QISM, focusing on the Baxter hierarchy and its applications. The factorization equations are classified by the specific group of symmetry which the S-matrix has and to which, therefore, the system itself belongs. Here we consider a large family of completely integrable systems which possess the XXZ-charge conservation symmetry. Mathematically this symmetry gives a sequence of higher dimensional representations of the SU(2) group, which we call "Baxter hierarchy." This family produces the exact theory of some physically important problems such as a) the two-dimensional q-state model, b) the Heisenberg XXZ chain of arbitrary spin S and c) the anisotropic SU(2) Thirring Model. It should be emphasized that to these different problems is applied the same Bethe ansatz!
In section 2 the inductive method to generate the factorized S-matrix is formulated and the Baxter hierarchy is introduced. Problem a), that is, the phase transition of the q-state model is discussed in section 3. Section 4 is devoted to problem b); the ground state and spin wave excitations are found for the Heisenberg XXZ chain of spin S. Problem c) is discussed in section 5. Section 6 is conclusions.

Because of the limitation of length, most of the formulas are given without derivations. The details are to be found in the author's papers referred. A concise introduction to the QISM and the Yang-Baxter relation is given in the articles by WADATI [1] and DE VEGA [2] of this volume.

2. Factorized S-Matrix and Baxter Hierarchy

We consider a system with q quantum states. The XXZ symmetric S-matrix is defined by the property:

$$S_{jl}^{ik}(\theta) = 0 \qquad \text{unless} \qquad i + j = k + l, \tag{1}$$

which we call the charge conservation symmetry. The index takes q = 2S+1 values of S, S-1, ..., -S (S = integer or half-integer). The S-matrix is assumed to also have the usual C, P. T and crossing symmetry. The factorization equations are given by

$$S^{i\alpha}_{j\beta}(12)\ S^{\alpha p}_{k\gamma}(13)\ S^{\beta q}_{\gamma r}(23) = S^{j\beta}_{k\gamma}(23)\ S^{i\alpha}_{\gamma r}(13)\ S^{\alpha p}_{\beta q}(12),\tag{2}$$

where $12 = \theta_1 - \theta_2$ and repeated indices are to be summed from S to -S.

The solution of the functional equations (2) under the condition (1) is derived in [3] by the inductive method. The inductive method is a procedure to generate a new factorized S-matrix by the use of the mixed factorization equation, which is shown graphically in Fig. 1.

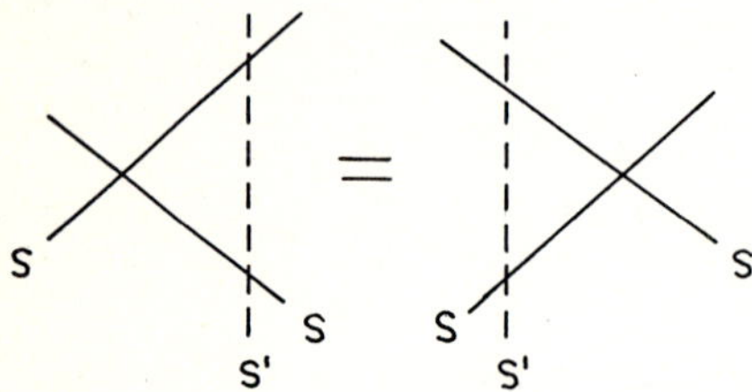

Fig. 1 Inductive method

Choosing S, S' = 1/2, 1, ... in Fig. 1, we obtain a sequence of factorized S-matrices from an initial S-matrix of S = 1/2 which is Baxter's one. We call this series of S-matrix the Baxter hierarchy. The hierarchy is shown schematically in Fig. 2. The vertical series of the XYZ symmetry is derived by changing the condition (1) to

$$S^{ik}_{jl}(\theta) = 0 \qquad \text{unless} \qquad i + j = k + l \pmod 2.\tag{3}$$

For this case the solutions are expressed in terms of the elliptic functions. The cases of XXZ and XXX are special cases of XYZ symmetry.

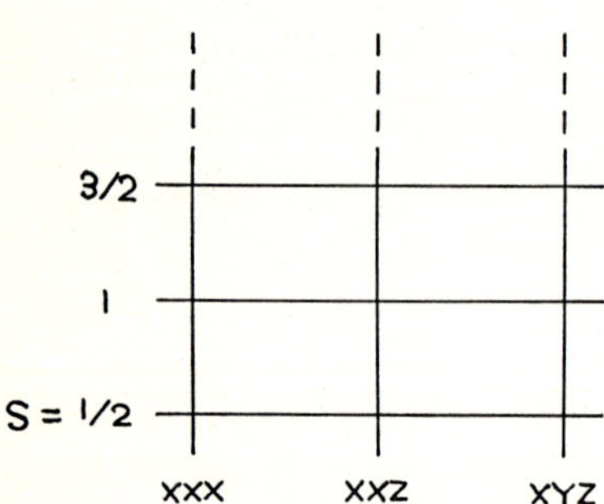

Fig. 2 Baxter hierarchy

3. Exactly Solvable q-State Model

We interprete the S-matrix elements as the Boltzmann weights of two-dimensional q-state vertex model;

$$S^{ik}_{jl}(\theta) = \text{weight of } i \underset{j}{\overset{l}{+}} k .\tag{4}$$

Then the factorization equation (2) becomes the Yang-Baxter relation,

$$R(\theta-\theta')(L_n(\theta)\otimes L_n(\theta')) = (L_n(\theta')\otimes L_n(\theta))R(\theta-\theta').\tag{5}$$

The partition function of this vertex model is computed in [4] by the inversion method. The inversion method is a procedure to calculate the largest (nondegenerate) eigenvalue of the transfer matrix by solving a couple of functional equations. In our case they are given by (for the high temperature region),

$$\kappa(u)\kappa(-u) = \rho_0^2 \prod_{p=1}^{2S} \frac{sh(\lambda p+u)sh(\lambda p-u)}{sh^2(\lambda p)} ,$$ (6.a)

$$\kappa(u) = \kappa(\lambda-u) ,$$ (6.b)

where $\kappa(u)$ is the partition function per site and λ is a parameter related to the temperature. The solution of (6) is given by

$$\ln\kappa(u) = \ln \rho_0 + \sum_{n=1}^{\infty} \sum_{p=1}^{2S} \frac{2e^{-2p\lambda n}sh(nu)sh[n(\lambda-u)]}{n\; ch(n\lambda)} .$$ (7)

This result can be also derived by the Bethe ansatz given in the next section. Near the critical point $\lambda = 0$, the free energy behaves as $\ln \kappa \sim exp(-const./\lambda)$, that is, an essential singularity.

Our q-state model is related to the solid on solid model of crystal growth and the classical XY model in two dimensions. The essential singularity obtained above implies that the system undergoes the Kosterlitz-Thouless transition. These problems are discussed in [4].

4. Heisenberg XXZ Chain of Arbitrary Spin S

The transfer matrix $T_N(\lambda)$ of our q = 2S+1 state vertex model can be diagonalized by the Bethe ansatz. This is performed by the generalized QISM where the mixed factorization equation is used effectively [5]. The eigenstate $\Psi(\lambda_1..., \lambda_M)$ for the transfer matrix $T_N(\lambda)$ is characterized by M complex numbers λ_j, and the eigenvalue for this state is given by

$$t_N(\lambda) = \sum_{m=-S}^{S} \alpha_m(\lambda)^N \prod_{j=1}^{M} C_m(\lambda-\lambda_j), \quad \text{where}$$ (8)

$$\alpha_m(\lambda) = \prod_{l=1}^{2S} \sin[\lambda+(m+l-S)\gamma],$$ (9)

$$C_m(\lambda) = \frac{\sin(\lambda-S\gamma)\sin[\lambda+(S+1)\gamma]}{\sin(\lambda+S\gamma)\sin[\lambda+(m+1)\gamma]} ,$$ (10)

and the λ_j's are determined by the equations

$$\frac{\sin(\lambda_j-S\gamma)}{\sin(\lambda_j+S\gamma)}^N = \prod_{\substack{k=1 \\ (k\neq j)}}^{M} \frac{\sin(\lambda_j-\lambda_k-\gamma)}{\sin(\lambda_j-\lambda_k+\gamma)} .$$ (11)

Since the transfer matrix $T_N(\lambda)$ is related to the Heisenberg XXZ model by the formula.

$$H_S = \frac{\sin(2S\gamma)}{4S} \frac{d}{d\lambda} \ln T_N(\lambda)\big|_{\lambda=0}$$

$$= \frac{1}{4S^2} \sum_{n=1}^{N} \{S_n^x S_{n-1}^x + S_n^y S_{n-1}^y + cos(2S\gamma)S_n^z S_{n-1}^z$$

+ higher terms} , (12)

we now have all the eigenstate and eigenvalue of H_c by using (8) $\sim$ (11). The ground state energy and low-lying excitations (spin wave modes) are calculated in [6] for both cases of the Ising-like anisotropy and the XY-like anisotropy. The ground state is singlet for both cases. The spin wave dispersion for the XY-like case is gapless and is given by

$$\varepsilon(k) = \frac{\pi\sin\Xi}{2\Xi} \; |\sin k| \; , \tag{13}$$

where we put $\Xi = 2S\gamma$.

The spin wave dispersion for the Ising-like case is

$$\varepsilon(k) = \frac{K_1 sh\Phi}{2\pi S} \; (1-k_1^2\cos^2 k)^{1/2} \; , \tag{14}$$

where the modulus k_1 is defined by $K_1'/K_1 = \Phi/2\pi S$ and we put $\gamma = i\Phi/2S$ (pure imaginary). Hence the spin wave mode for the Ising-like case has a gap

$$G = (1-k_1^2)^{1/2} \; K_1 sh\Phi/(2\pi S) \; , \tag{15}$$

which behaves for small Φ/S as $\sim 2\pi \exp(-\pi^2 S/\Phi)$.

Since the higher order terms of (12) will be irrelevant in the universality argument, our results show the exact behaviors of the usual Heisenberg model. This conclusion negates the conjecture by HALDANE [7] on the excitations in the Heisenberg model. His conjecture says that the existence of a gap depends on whether the magnitude S of spin is integer or half integer. Our results, however, show that it depends not on the parity of 2S, but on the anisotropy: XY-like or Ising-like. This conclusion will also help to clarify a controversial situation in the numerical calculations [8], [9].

5. SU(2) Thirring Model

So far we have reviewed some facets (vertex model, spin system) of the Baxter heirarchy. Here we give another application of the Baxter hierarchy: anisotropic SU(2) Thirring model. We consider a one-dimensional fermion system with spin S. The Hamiltonian is given by

$$H = \quad dx \; \{-i(\psi_1^\dagger \frac{d}{dx} \psi_1 - \psi_2^\dagger \frac{d}{dx} \psi_2)$$

$$+ 4V(\alpha\beta,\alpha'\beta')\psi_{1\alpha}^* \psi_{2\beta}^* \psi_{1\alpha'} \psi_{2\beta'} \tag{16}$$

with

$$V(\alpha\beta,\alpha'\beta') = S^{-2} \cdot \sum_{j=0}^{3} v_j(\tau^j)_{\alpha\alpha'} \cdot (\tau^j)_{\beta\beta'} \tag{17}$$

where τ^j's are generators of SU(2); $\tau^0 = S\cdot 1$ and $[\tau^1, \tau^2] = i\tau^3$ (cyclic). The case of $S = 1/2$ is the Bychkov-Gorkov-Dzyaloshinskii model.

The Hamiltonian (16) can be diagonalized by the Bethe ansatz,

$$\Psi = \sum_p \xi_p(Q)\exp(i \sum_{j=1}^{N} k_{Pj}x_{Qj}), \tag{18}$$

where $N = N_1 + N_2$ and P, Q are the permutations of $(1, \ldots, N)$. Applying (16) to (18), it is shown that the transition matrix $R \equiv \xi_p(Q)/\xi_p(Q')$ is given by $R = (1-iv)/(1+iv)$. And moreover the matrix R is nothing but the S-matrix of our Baxter hierarchy:

$$R^{\alpha\alpha'}_{\beta\beta'} = S^{\alpha\alpha'}_{\beta\beta'}(\theta=1) \tag{19}$$

The factorization equation (2) is a sufficient condition for the applicability of
the Bethe ansatz.

Thus we can solve the problem exactly by the ordinary technique of Bethe ansatz.
One of the interesting features is the excitation spectrum. There are two types of
excitations, 1) massless density fluctuations and 2) massive spin excitations. As
in the case of Heisenberg model discussed in the section 4, the spin excitation
becomes massless either in the isotropic limit ($v_1=v_2=v_3$) or in the S$\rightarrow\infty$ limit. Due
to the regularization, therefore, there are two possibilities for the dynamic mass
generation:

1) the finite S and isotropic model,

2) the anisotropic and S$\rightarrow\infty$ model.

Case 1) is an extension of the result by McCOY-WU [10], who discussed the regularization of the spinless Thirring model, to the arbitrary spin. Case 2) is a new one,
and is possibly related to the mass generation of the nonlinear σ model. There may
be the same mechanisms also in the nonabelian gauge theory, which offer a replacement
of the Higgs field.

6. Conclusions

To summarize, we may conclude as follows: Firstly, we add a new parameter S to
Baxter's eight vertex model by the inductive method. We find that the inductive
method is powerful enough to generate a hierarchy of factorized S-matrices. It can
be applied to other symmetries such as GL(N), SU(N) etc. of the S-matrix.

Secondly, it is also effective to construct the Bethe states in the generalized
QISM. Although the procedure of calculations by the Bethe ansatz is cumbersome
(remember how simple the inversion method was!), there have been many innovations
and it is still a firm foundation of the theory. The quantum Gelfand-Levitan method
is one such thing for the correlation functions [11].

Thirdly, the Baxter hierarchy discussed in this note has all the properties of
the eight vertex model. It is expected to make as great a contribution to theoretical physics as the Baxter model does. In fact, as is mentioned in section 3, our
q-state vertex model is connected to the solid on solid model through the Wu-Kadanoff-
Wegner transformation which describes the roughening transition of the surface.
Furthermore it is, we expect, in the same universality class as the two-dimensional
XY model. To justify this the exact β-function is calculated and is in good
agreement with the renormalization group theory [6]. Our model will produce more
information about the Kosterlitz-Thouless transition.

Fourthly, we have computed the low-lying excitations of the Heisenberg XXZ model.
They contain the result of des Cloizeaux and Gaudin as a special case. Our result
is natural and sound for the symmetry consideration. It shows that nothing happens
as mysterious as Haldane's conjecture is saying.

Fifthly, it is shown that the same Bethe states are used to solve the
physics and particle physics. The central problem is concerned with the mass spectrum. Two types of dynamic mass generation are pointed out. The details of this
problem will be discussed elsewhere.

Last but not least, we may conclude that the QISM is the meeting place of
exact theories; we can solve them by the same procedure. This unity is the great
discovery of the QISM, and not only a reformulation of the Bethe ansatz.

References

1. M. Wadati, these proceedings.

2. H.J. De Vega, these proceedings.

3. K. Sogo, Y. Akutsu and T. Abe, Prog. Theor. Phys. $\underline{70}$, 730 (1983).

4. K. Sogo, Y. Akutsu and T. Abe, Prog. Theor. Phys. $\underline{70}$, 739 (1983).

5. K. Sogo, unpublished memorandum (1983).

6. K. Sogo, Phys. Lett. A (1984) in press.

7. F.D.M. Haldane, Phys. Lett. $\underline{93}$A, 464 (1983).

8. R. Botet and R. Jullien, Phys. Rev. B $\underline{27}$, 613 (1983); B $\underline{29}$, 5222 (1984).

9. J.C. Bonner and G. Müller, Phys. Rev. B $\underline{29}$, 5216 (1984).

10. B.M. McCoy and T.T. Wu, Phys. Lett. $\underline{87}$B, 50 (1979).

11. K. Sogo and M. Wadati, Prog. Theor. Phys. $\underline{69}$, 431 (1983).

Quantum Three Wave Interaction Models: Bethe Ansatz and Statistical Mechanics

Kenji Ohkuma

Institute of Physics, College of Arts and Sciences, University of Tokyo
Tokyo 153, Japan

1. INTRODUCTION

The Bethe Ansatz method was first used by Bethe to obtain the eigenstates of the Hamiltonian for the spin wave problem [1]. This method is also valid for many other models such as the quantum nonlinear Schrödinger model (delta-function interaction model) [2], the massive Thirring model [3], etc.. C.N.Yang and C.P.Yang exploit this method, and succeed in getting the equilibrium of the thermodynamics of the system with a finite temperature [4]. We find that the eigenstates of the quantum three wave interaction model can be derived by the Bethe Ansatz method [5]. We treat the thermodynamics in the case without the bound states in the eigenstates, as some difficulties occur with the existence of the bound states.

2. THE MODEL

The quantum three wave interaction (Q3WI, hereafter) model in 1-dimensional space is given by the Hamiltonian;

$$H = \int dx\{ \sum_{j=1}^{3} c_j Q_j^*(x)\frac{1}{i}\frac{\partial}{\partial x}Q_j(x)+g[Q_2^*(x)Q_3(x)Q_1(x)+Q_1^*(x)Q_3^*(x)Q_2(x)]\}, \qquad (2.1)$$

where c_j's are distinct constant velocities, "g" is the coupling constant, and Q_j^*'s and Q_j's are creation and annihilation operators, respectively. Eq.(2.1) means that the number of each particle is not conserved. Three choices of statistics can be considered;

1) Boson model   All fields are bosons, e.g. the equal time commutation relations:

$$[Q_j(x,t),Q_k^*(y,t)] = \delta_{jk}\delta(x-y), \text{ etc.} \qquad (2.2)$$

are satisfied.

2) Fermion model I   The fields Q_1 and Q_3 are fermions and the field Q_2 is a boson, e.g. the commutation relations ;

$$\{Q_j(x,t),Q_k^*(y,t)\} = \delta_{jk}\delta(x-y), \qquad j,k = 1,3 \quad \text{,etc.} \qquad (2.3)$$

are satisfied.

3) Fermion model II   The fields Q_1 and Q_2 are fermions and the field Q_3 is a boson, e.g. the commutation relations;

$$\{Q_j(x,t),Q_k^*(y,t)\} = \delta_{jk}\delta(x-y), \qquad j,k = 1,2 \quad \text{,etc.} \qquad (2.4)$$

are satisfied.

The equation of motion;

$$\frac{\partial}{\partial t}Q_j(x,t) = i [H,Q_j(x,t)] \qquad (2.5)$$

with the quantization rules yields

$$Q_{1_t} + c_1 Q_{1_x} = -i\sigma_1 g Q_3^* Q_2,$$

$$Q_{2_t} + c_2 Q_{2_x} = -i\sigma_2 g Q_3 Q_1,$$

$$Q_{3_t} + c_3 Q_{3_x} = -i\sigma_3 g Q_1^* Q_2, \tag{2.6}$$

where

1) $\sigma_1 = \sigma_2 = \sigma_3 = +1$ for Boson model and Fermion model II, $\tag{2.7a}$

2) $-\sigma_1 = \sigma_2 = \sigma_3 = +1$ for Fermion model I. $\tag{2.7b}$

This is the Q3WI equation, which describes the behavior of the three fields in a nonlinear medium and plays an important role in nonlinear optics, plasma physics, etc. [6].

3. THE BETHE ANSATZ

For the description of eigenstate we prepare the following notations. First we define the vacuum state $|0\rangle$ as

$$Q_j(x,t)|0\rangle = 0, \qquad j=1,2,3. \tag{3.1}$$

Ket states created by only one kind of field operators are expressed as

$$|\lambda_1,\cdots,\lambda_N\rangle = \int\cdots\int dx_1\cdots dx_N \theta(x_1>\cdots>x_N)\exp[i(p_1 x_1 + \cdots + p_N x_N)]$$
$$\times\, Q_1^*(x_1)\cdots Q_1^*(x_N)|0\rangle, \tag{3.2a}$$

$$|\mu_1,\cdots,\mu_N\rangle = \int\cdots\int dx_1\cdots dx_N \theta(x_1>\cdots>x_N)\exp[i(q_1 x_1 + \cdots + q_N x_N)]$$
$$\times\, Q_3^*(x_1)\cdots Q_3^*(x_N)|0\rangle, \tag{3.2b}$$

$$|\lambda_1+\mu_1,\cdots,\lambda_N+\mu_N\rangle = \int\cdots\int dx_1\cdots dx_N \theta(x_1>\cdots>x_N)$$
$$\times\, \exp\{i[(p_1+q_1)x_1 + \cdots + (p_N+q_N)x_N]\}Q_2^*(x_1)\cdots Q_2^*(x_N)|0\rangle, \tag{3.2c}$$

with

$$\theta(x_1>\cdots>x_N) = \theta(x_1-x_2)\theta(x_2-x_3)\cdots\theta(x_{N-1}-x_N),$$

$$\theta(x) = \begin{cases} 1 & \text{for } x > 0 \\ 1/2 & \text{for } x = 0 \\ 0 & \text{for } x < 0, \end{cases} \tag{3.3}$$

$$p_j = (c_2-c_3)\lambda_j, \qquad q_j = (c_1-c_2)\mu_j. \tag{3.4}$$

Argument λ_j (μ_j) means a Q_1- (Q_3-) particle with a wave number p_j (q_j). Similarly $\lambda_j+\mu_j$ means a Q_2-particle with a wave number p_j+q_k. Ket states with more than one kind of particle are introduced in a similar way, e.g.

$$|\lambda_1,\lambda_2+\mu_1,\mu_2\rangle = \int\int\int dx_1 dx_2 dx_3\, \theta(x_1>x_2>x_3)\exp\{i[p_1 x_1 + (p_2+q_1)x_2 + q_2 x_3]\}$$
$$\times\, Q_1^*(x_1)Q_2^*(x_2)Q_3^*(x_3)|0\rangle. \tag{3.5}$$

The Hamiltonian commutes with the number operators;

$$\hat{M} = \int dx(Q_1^* Q_1 + Q_2^* Q_2), \qquad \hat{N} = \int dx(Q_2^* Q_2 + Q_3^* Q_3). \tag{3.6}$$

Eigenvalues of these operators are non-negative integers. Therefore, we use them to classify the sector of the eigenstates. The eigenstate with eigenvalues M and N respectively is expressed $||M,N\rangle\rangle$, i.e.

$$\hat{M}||M,N\rangle\rangle = M||M,N\rangle\rangle, \qquad \hat{N}||M,N\rangle\rangle = N||M,N\rangle\rangle. \tag{3.7}$$

The terms in the state $||M,N\rangle\rangle$ are classified into $[\min(M,N)+1]$ kinds according to the number of Q_2^*-operators. The terms with ℓQ_2^*'s have $(M-\ell)Q_1^*$'s, and $(N-\ell)Q_3^*$'s, and ℓ satisfies the condition;

$$0 \le \ell \le \min(M,N). \tag{3.8}$$

The eigenstate is determined up to a constant factor by giving the quantum number M, N and the sets of the wave number $\{p_1,\cdots,p_M,q_1,\cdots,q_N\}$ with eq.(3.4). Hereafter we assume that $MN \neq 0$. We can write the state $||M,N\rangle\rangle$ in the case $M \geq N$ as follows

$$
\begin{aligned}
||M,N\rangle\rangle = \ &[\lambda_1,\cdots,\lambda_M,\mu_1,\cdots,\mu_N]|\lambda_1,\cdots,\lambda_M,\mu_1,\cdots,\mu_N\rangle \\
&+ [\lambda_1,\cdots,\mu_N,\mu_{N-1}]|\lambda_1,\cdots,\mu_N,\mu_{N-1}\rangle + \cdots \\
&+ [\mu_N,\cdots,\mu_1,\lambda_M,\cdots,\lambda_1]|\mu_N,\cdots,\mu_1,\lambda_M,\cdots,\lambda_1\rangle \\
&+ [\lambda_1,\cdots,\lambda_M+\mu_1,\cdots,\mu_N]|\lambda_1,\cdots,\lambda_M+\mu_1,\cdots,\mu_N\rangle + \cdots \\
&+ [\lambda_1+\mu_1,\cdots,\lambda_N+\mu_N,\lambda_{N+1},\cdots,\lambda_M]|\lambda_1+\mu_1,\cdots,\lambda_N+\mu_N,\lambda_{N+1},\cdots,\lambda_M\rangle \\
&+ \cdots.
\end{aligned}
\tag{3.9}
$$

The Bethe Ansatz implies that the above coefficients (square brackets) are related by the following rules;

$$
\begin{aligned}
&[\cdots,\mu_k,\lambda_j,\cdots] = S_1(\lambda_j-\mu_k)[\cdots,\lambda_j,\mu_k,\cdots] \\
&[\cdots,\lambda_k,\lambda_j,\cdots] = S_2(\lambda_j-\lambda_k)[\cdots,\lambda_j,\lambda_k,\cdots] \\
&[\cdots,\mu_k,\mu_j,\cdots] = S_3(\mu_j-\mu_k)[\cdots,\mu_j,\mu_k,\cdots] \\
&[\cdots,\lambda_j+\mu_k,\cdots] = S_+(\lambda_j-\mu_k)[\cdots,\lambda_j,\mu_k,\cdots],
\end{aligned}
\tag{3.10}
$$

where S_1,S_2,S_3,S_+ are expressed for each choice of statistics as follows

1) Boson model

$$S_1(\nu) = (\nu-i\kappa)/(\nu+i\kappa), \qquad S_2(\nu) = S_3(\nu) = (\nu+2i\kappa)/(\nu-2i\kappa),$$

$$S_+(\nu) = -2\kappa(c_1-c_3)/g(\nu+i\kappa), \tag{3.11}$$

$$\kappa = g^2/2(c_1-c_2)(c_2-c_3)(c_3-c_1). \tag{3.12}$$

2) Fermion model I

$$S_1(\nu) = -(\nu-i\kappa)/(\nu+i\kappa), \qquad S_2(\nu) = S_3(\nu) = -1,$$

$$S_+(\nu) = -2\kappa(c_1-c_3)/g(\nu+i\kappa), \tag{3.13}$$

3) Fermion model II

$$S_1(\nu) = (\nu-i\kappa)/(\nu+i\kappa), \qquad S_2(\nu) = -1,$$

$$S_3(\nu) = (\nu+2i\kappa)/(\nu-2i\kappa), \qquad S_+(\nu) = -2\kappa(c_1-c_3)/g(\nu+i\kappa), \tag{3.14}$$

For all models, the energy eigenvalue is

$$E = c_1(p_1+\cdots+p_M) + c_3(q_1+\cdots+q_N). \tag{3.15}$$

The number of terms with ℓQ_2^*'s is $(M+N-\ell)!\,_MP_\ell\cdot{}_NC_\ell$ for $0\leq\ell\leq\min(M,N)$.

Next we show the condition where bound states in the eigenstates occur [5].
1) Boson model
(1) Bound states of Q_1-particles occur when

$$(c_1-c_2)(c_3-c_1) < 0. \tag{3.16}$$

(2) Bound states of Q_3-particles occur when

$$(c_2-c_3)(c_3-c_1) < 0. \tag{3.17}$$

For distinct c_j's, at least one of eqs.(3.16) and (3.17) is always satisfied, which means, bound states can always exist in the Boson model.
2) Fermion model I
In this case, no bound states occur.
3) Fermion model II
(1) The bound state of Q_1-particles does not occur.
(2) The bound state of Q_3-particles occurs when

$$(c_2-c_3)(c_3-c_1) < 0. \tag{3.18}$$

4. THERMODYNAMICS

We treat the Fermion model I, as no bound states occur in the model. First we consider terms without Q_2-particles in the eigenstate $||M,N\gg$. From eqs.(3.10) and (3.13), they are expressed as

$$\alpha\{|\lambda_1,\cdots,\lambda_M,\mu_1,\cdots,\mu_N> - |\lambda_1,\cdots,\lambda_M,\lambda_{M-1},\cdots,\mu_N> + \cdots$$

$$+ \prod_{j,k}S_1(\lambda_j-\mu_k)|\mu_N,\cdots,\mu_1,\lambda_M,\cdots,\lambda_1>\}$$

$$= \alpha\int\cdots\int dx_1\cdots dx_M dy_1\cdots dy_N$$

$$\times \exp[i(p_1x_1+\cdots p_Mx_M+q_1y_1+\cdots+q_Ny_N)]$$

$$\times \{\theta(x_1>\cdots>x_M>y_1>\cdots>y_N) + \theta(x_1>\cdots>x_M>x_{M-1}>\cdots>y_N)$$

$$- S_1(\lambda_M-\mu_1)\,\theta(x_1>\cdots>y_1>x_M>\cdots>y_N) + \cdots$$

$$+ \prod_{j,k}[-S_1(\lambda_j-\mu_k)]\theta(y_N>\cdots>y_1>x_M>\cdots>x_1)\}$$

$$\times Q_1^*(x_1)\cdots Q_1^*(x_M)Q_3^*(y_1)\cdots Q_3^*(y_N)|0>. \tag{4.1}$$

We define χ as

$$\chi(x_1,\cdots,y_N) = \exp[i(p_1x_1+\cdots p_Mx_M+q_1y_1+\cdots+q_Ny_N)]$$

$$\times \{\theta(x_1>\cdots>x_M>y_1>\cdots>y_N) + \cdots$$

$$+ \prod_{j,k}[-S_1(\lambda_j-\mu_k)]\theta(y_N>\cdots>y_1>x_M>\cdots>x_1)\}, \tag{4.2}$$

and assume the boundary conditons;

$$\chi(x_j=0) = \chi(x_j=L), \quad \chi(y_j=0) = \chi(y_j=L). \tag{4.3}$$

Taking the logarithm of eq.(4.1), we obtain

$$p_j L = 2\pi I_j + \sum_{\ell=1}^{N} \phi(\lambda_j - \mu_\ell), \qquad q_j L = 2\pi J_j + \sum_{\ell=1}^{M} \phi(\mu_j - \lambda_\ell), \tag{4.4}$$

$$I_j \text{ and } J_j \text{ are integers}, \tag{4.5}$$

where

$$\phi(\nu) = -i\ell n[-S_1(\nu)] = -i\ell n[(\nu - i\kappa)/(\nu + i\kappa)], \qquad -\pi < \phi < \pi. \tag{4.6}$$

For the terms with Q_2-particles, the boundary conditions are automatically satisfied if we assume eq.(4.4), which can be understood from the relation (3.10). Given any set of I_j's and J_j's satisfying eq.(4.5), there is a unique set of p_j's and q_j's satisfying eq.(4.4) with every p_j and q_j being distinct, respectively.ly.

We now consider the limit: $M \to \infty$, $N \to \infty$ and $L \to \infty$ at a fixed density $D_1 = M/L$ and $D_3 = N/L$. For the ground state, I_j/L and J_j/L form a uniform distribution from $-\infty$ to a certain value corresponding to Fermi level. If an integer I is in the set of I_j's, the corresponding level is considered to be filled with a Q_1-particle. If not, the level is thought to be settled by a hole. In the large volume limit, we express the density of Q_1-particles in the width of the wave number dp as $L\rho_1(p)$. The density of Q_1-holes is $L\rho_1^h(p)$. Similarly for Q_3-particles, $L\rho_3(q)$ is a particle density and $L\rho_3^h(q)$ is a hole density. We define the sum of particle and hole densities as $f_1(p)$ and $f_3(q)$, i.e.

$$f_1(p) = \rho_1(p) + \rho_1^h(p), \qquad f_3(q) = \rho_3(q) + \rho_3^h(q). \tag{4.7}$$

Differentiating eq.(4.4) with p_j and q_j respectively and taking the large volume limit $L \to \infty$, we get

$$2\pi f_1(p) = 2\pi[\rho_1(p) + \rho_1^h(p)] = 1 - [2\kappa/(c_2 - c_3)]\int dq \rho_3(q)/g(p,q)$$

$$2\pi f_3(q) = 2\pi[\rho_3(q) + \rho_3^h(q)] = 1 - [2\kappa/(c_1 - c_2)]\int dp \rho_1(p)/g(p,q), \quad \text{where} \tag{4.8}$$

$$g(p,q) = 1/\{[p/(c_2 - c_3) - q/(c_1 - c_2)]^2 + \kappa^2\}. \tag{4.9}$$

The energies of Q_1-particles and Q_3-particles are

$$E_1 = L\int dp \rho_1(p) c_1 p, \qquad E_3 = L\int dq \rho_3(q) c_3 q, \tag{4.10}$$

and the particle densities are

$$D_1 = M/L = \int dp \rho_1(p), \qquad D_3 = N/L = \int dq \rho_3(q). \tag{4.11}$$

The entropies of the Q_1- and Q_3-particles are

$$S_1 = L\int dp[(\rho_1 + \rho_1^h)\ell n(\rho_1 + \rho_1^h) - \rho_1 \ell n \rho_1 - \rho_1^h \ell n \rho_1^h],$$

$$S_3 = L\int dq[(\rho_3 + \rho_3^h)\ell n(\rho_3 + \rho_3^h) - \rho_3 \ell n \rho_3 - \rho_3^h \ell n \rho_3^h]. \tag{4.12}$$

At a temperature T, the equilibrium is obtained by maximizing the partition function; $\exp[S_1 + S_3 - (E_1 + E_3)/k_B T]$ under the restriction (4.11). The result is

$$-A_1 + c_1 p + k_B T \ell n(\rho_1/\rho_1^h) + [\kappa/\pi(c_1 - c_2)]\int \frac{dq}{g(p,q)} \ell n[1 + \rho_3(q)/\rho_3^h(q)] = 0$$

$$-A_3 + c_3 q + k_B T \ell n(\rho_3/\rho_3^h) + [\kappa/\pi(c_2 - c_3)]\int \frac{dp}{g(p,q)} \ell n[1 + \rho_1(p)/\rho_1^h(p)] = 0, \tag{4.13}$$

where A_1, A_3 are Lagrange multipliers for the condition (4.11). Writing

$$\rho_1(p)/\rho_1^h(p) = \exp[-\varepsilon_1(p)/k_BT], \quad \rho_3(q)/\rho_3^h(q) = \exp[-\varepsilon_3(q)/k_BT], \tag{4.14}$$

we have

$$\varepsilon_1(p) = -A_1+c_1p+[k_BT\kappa/\pi(c_1-c_2)]\int\frac{dq}{g(p,q)}\ln\{1+\exp[-\varepsilon_3(q)/k_BT]\}$$

$$\varepsilon_3(q) = -A_3+c_3q+[k_BT\kappa/\pi(c_2-c_3)]\int\frac{dp}{g(p,q)}\ln\{1+\exp[-\varepsilon_1(p)/k_BT]\}, \tag{4.15}$$

and using eq.(4.8)

$$2\pi f_1(p) = 2\pi\rho_1(p)\{1+\exp[\varepsilon_1(p)/k_BT]\} = [2\kappa/(c_1-c_2)]\int\frac{dq}{g(p,q)}\rho_3(q)$$

$$2\pi f_3(q) = 2\pi\rho_3(q)\{1+\exp[\varepsilon_3(q)/k_BT]\} = [2\kappa/(c_2-c_3)]\int\frac{dp}{g(p,q)}\rho_1(p). \tag{4.16}$$

Eqs.(4.16) can be solved by iteration.

5.   CONCLUDING REMARKS

The excitation from the ground state can be studied along-side the scenario by Yang & Yang, and momentum difference and energy difference between the two states can be calculated.

Recently M.Wadati and M.Sakagami showed that the classical soliton of the non-linear Schrödinger model is a manifestation of the bound state of the quantum version [7]. Similarly, the soliton of the classical 3WI model should be derived from the bound state of the Q3WI model.

The eigenstates of the Q3WI model may be derived through the quantum inverse method.

6.   REFERENCES

1. H. A. Bethe: Z. Physik 71, 205 (1931).
2. C. N. Yang: Phys. Rev. Lett. 19, 1312 (1967).
3. F. A. Berezin and V. N. Sushko: JETP 21, 865 (1965),
   H. B. Thacker: Rev. Mod. Phys. 53, 253 (1981),
4. C. N. Yang and C. P. Yang: J. Math. Phys. 10, 1115 (1969).
5. M. Wadati and K. Ohkuma: J. Phys. Soc. Jpn. 53, 1229 (1984),
   K. Ohkuma and M. Wadati: J. Phys. Soc. Jpn. 53, 2899 (1984).
6. D. J. Kaup, A. Reiman and A. Bers: Rev. Mod. Phys. 51, 275 (1979).
7. M. Wadati and M. Sakagami: J. Phys. Soc. Jpn. 53, 1933 (1984),
   M. Wadati: in this volume.

Quantum and Classical Statistical Mechanics of the Non-Linear Schrödinger, Sinh-Gordon and Sine-Gordon Equations

R.K. Bullough and D.J. Pilling

Department of Mathematics, UMIST, P.O. Box 88, Manchester M60 1QD, Great Britain

J.T. Timonen

Department of Physics, University of Jyväskylä, SF-40100 Jyväskylä, Finland

1. INTRODUCTION

We are going to describe our work on the quantum and classical statistical mechanics of some exactly integrable non-linear one dimensional systems. The simplest is the non-linear Schrödinger equation (NLS)

$$i\psi_t = -\psi_{xx} + 2c\psi^\dagger\psi\psi \tag{1.1}$$

where c, the coupling constant, is positive. The others are the sine- and sinh-Gordon equations (sG and shG)

$$\phi_{xx} - \phi_{tt} = m^2 \sin \phi \tag{1.2}$$

$$\phi_{xx} - \phi_{tt} = m^2 \sinh \phi \quad . \tag{1.3}$$

These two are related by the transformation $\phi \to i\phi$ and $\gamma_0 \to -\gamma_0$. Here, m is a mass and γ_0 is a dimensionless coupling constant: $\hbar = c = 1$ (c = velocity). These equations can be solved completely by the classical inverse scattering method (CISM) [1,2,3,4]. An important result of this technique is as follows: (1.2) can be derived from the Hamiltonian

$$H[\phi] = \gamma_0^{-1} \int [\tfrac{1}{2}\gamma_0^2\Pi^2 + \tfrac{1}{2}\phi_x^2 + m^2(1 - \cos \phi)] \, dx \quad . \tag{1.4}$$

The CISM shows [3,4] that there is a canonical transformation to action angle variables in which (1.4) takes the form

$$H[p] = \sum_{i=1}^{N_k}(M^2+p_i^2)^{\frac{1}{2}} + \sum_{j=1}^{N_{\bar{k}}}(M^2+\bar{p}_j^2)^{\frac{1}{2}} + \sum_{\ell=1}^{N_b}(4M^2\sin^2\theta_\ell+\hat{p}_\ell^2)^{\frac{1}{2}} + \int_{-\infty}^{\infty}\omega(k)P(k)dk \; ; \tag{1.5a}$$

$$\omega(k) = (m^2 + k^2)^{\frac{1}{2}} \tag{1.5b}$$

and $M = 8m\gamma_0^{-1}$. The first three terms here correspond to the kink, anti-kink and breather solutions of (1.2). The last term is precisely the Hamiltonian of the free Klein-Gordon equation in action angle variables — a sum of an infinite number of 'phonon contributions' with dispersion relation $\omega(k)$. Similar results can be found for the shG and NLS. Neither of these has soliton solutions and so their Hamiltonians must just contain the last term of (1.5a); for the NLS the dispersion relation of (1.5b) is changed to $\omega(k) = k^2$. An important point is that the CISM is on an infinite support L with $L \to \infty$ and the solutions are (roughly speaking) assumed to decay exponentially at the boundaries.

The quantum equivalents of the models have been studied using two methods: the Bethe ansatz (BA) and the quantum inverse scattering method (QISM) [1,2,5]. In these quantum methods it is necessary to introduce periodic boundary conditions and to work on a lattice — despite the fact that at the end, the size of the box goes to infinity and the lattice spacing goes to zero. As we will see both of these features are crucial to the correct calculation of the statistical mechanics of these systems.

As to the thermodynamics of these systems — the situation is that both the QISM and the BA lead to equations from which the free energy can be found using a technique originally described by Yang and Yang for the NLS [6]. This has been adapted to give a formal solution for the statistical mechanics of the sG [7,8,9]. It seems very difficult to extract physical consequences from such solutions. So far there has been no analogue of this method for classical systems. There is however, a simple method for calculating their free energies — the transfer integral method (TIM) [10,11]. Again this does not make the underlying physics of the solution clear.

In this paper we will describe a technique which goes beyond those currently available. Essentially our method is based on using a transformation to action angle variables in a phase space path integral. Using this we can calculate both the quantum and (in the appropriate limit) classical partition functions.

The advantage of this method is that it allows us to keep track of the classical solutions of the equations and we can therefore avoid ending up with the integral equations of the BA with their non-physical excitations. In this direction, we should mention one result of THACKER [12]; he showed that it was possible to write the partition function of the NLS in terms of a system with decaying boundary conditions on infinite support and, by using a quantum form of the Gel'fand Levitan Marchenko inverse transform, re-derive the BA integral equation of YANG and YANG [6]. This suggests that the infinite support case already contains all necessary information for a proper thermodynamic limit. Certainly it suggests that action angle variables, which are easy to find for infinite support, can be used in the path integral.

2. STATISTICAL MECHANICS of the NLS

The quantum form of (1.1) corresponds to a first quantisation of the Schrödinger equation for a one-dimensional gas of N bosons with a repulsive δ-function interaction of strength c. This system is considered in a box of length L with periodic boundary conditions. The boundary condition becomes re-expressed in a condition for the wave functions expressed as super-positions of periodic modes labelled by allowed wave vectors k_j. The condition is an integral equation for the allowed k_j. It takes the form

$$Lk_j + \sum_{\ell \neq j} \theta_{j\ell} = 2\pi n_j \quad ; \quad n_j = \text{integer}$$

$$\theta_{ij} = -i \ln \left\{ \frac{(k_i - k_j) - ic}{(k_i - k_j) + ic} \right\} \quad . \tag{2.1}$$

In this θ_{ij} is the phase shift $\Delta(k_i - k_j)$ of the two particle S-matrix

$$\Delta(k) = -2 \tan^{-1}(ck^{-1}) \tag{2.2}$$

Following THACKER and others [5,6] Δ is chosen so that it covers the single branch $-2\pi \leq \Delta < 0$ for $-\infty < k < \infty$. As we shall see this imposes Fermi statistics on the particles in the problem. The zero temperature excitation spectrum of the BA equation was investigated by LIEB and LINIGER [13]. Later YANG and YANG [6] found the corresponding result for finite temperatures $T = \beta^{-1}$, namely

$$\varepsilon(k) = k^2 - \mu - \beta^{-1} \int_{-\infty}^{\infty} \frac{dk'}{2\pi} \frac{d\Delta(k-k')}{dk} \ln \left(1 + e^{-\beta \varepsilon(k')}\right) \tag{2.3}$$

in which μ is a chemical potential (which we set to zero henceforth). The derivative of the phase shift is

$$\Delta' = \frac{d\Delta(k-k')}{dk} = \frac{2c}{(k-k')^2 + c^2} \quad (> 0) \tag{2.4}$$

and exists for all k' in $-\infty < k' < \infty$. The solution of (2.3) determines the free energy per unit length by

$$F = \mu - (2\pi\beta)^{-1} \int_{-\infty}^{\infty} dk \, \ln \left(1+e^{-\beta\epsilon(k)}\right) . \qquad (2.5)$$

Note that when $c = 0$ then Δ' goes to a δ-function. Thus

$$\epsilon(k) = k^2 - \beta^{-1} \ln(1 + \exp - \beta\epsilon(k)) \quad , \qquad (2.6a)$$

and

$$F = (2\pi\beta)^{-1} \int_{-\infty}^{\infty} dk \, \ln \left(1 - e^{-\beta k^2}\right) \qquad (2.6b)$$

where the free energy is that of a non-interacting bose gas. Motivated by these results [6], we can bosonize (2.3) by setting

$$(1 + e^{-\beta\epsilon(k)}) = (1 - e^{-\beta\bar{\omega}(k)})^{-1} \quad ; \quad \omega(k) = k^2 \quad . \qquad (2.7)$$

From this one finds that

$$\bar{\omega}(k) = \omega(k) - \beta^{-1} \ln \left(1-e^{-\beta\bar{\omega}(k)}\right) - (2\pi\beta)^{-1} \int_{-\infty}^{\infty} dk' \, \frac{d\Delta(k-k')}{dk'} \, \ln \left(1-e^{-\beta\bar{\omega}(k')}\right)$$

$$= \omega(k) - (2\pi\beta)^{-1} \int_{-\infty}^{\infty} dk' \left[\frac{d\Delta}{dk'}(k-k') + 2\pi\delta(k-k') \right] \ln \left(1-e^{-\beta\bar{\omega}(k')}\right) \quad ; \qquad (2.8)$$

the square bracket is

$$\left[\frac{d}{dk'} \, \Delta(k-k') + 2\pi\delta(k-k') \right] = \frac{d}{dk'} \left[\Delta(k-k') - 2\pi\theta(k-k') \right] \qquad (2.9)$$

with θ the step function. For bosonization we adopt the convention that $\Delta \to 0$ for $k \to \pm\infty$. Then Δ is to be understood as the two branches of the $\tan^{-1}$ function with this property, and there is a jump of 2π on Δ at $k = k'$. Adopting the new definition one finds

$$\bar{\omega}(k) = \omega(k) + (2\pi\beta)^{-1} \int_{-\infty}^{\infty} dk' \Delta(k-k') \, \frac{d}{dk'} \, \ln \left(1-e^{-\beta\bar{\omega}(k')}\right) \quad . \qquad (2.10)$$

An integration by parts has been carried out. The $c = 0$ limit of the new phase shift is zero; thus we get the same result as before. By repeating Yang's and Yang's calculation with the new definition of the phase shift and treating the particles as bosons one again reaches (2.10).

The classical limit of (2.10) is

$$\bar{\omega}(k) = \omega(k) - (\pi\beta)^{-1} \int_{-\infty}^{\infty} dk' \, \frac{c}{(k-k')} \, \frac{d}{dk'} \, (\ln \beta\bar{\omega}(k')) \qquad (2.11a)$$

with

$$F = (2\pi\beta)^{-1} \int_{-\pi/a}^{\pi/a} dk \, \ln \beta\bar{\omega}(k) \qquad (2.11b)$$

To get this we have put the theory on a lattice of spacing a; this introduces the momentum cut off. It may not be obvious that this is the correct classical limit: this form was suggested to us by the expression for the classical phonon-phonon phase shifts in [14].

It is possible to apply the TIM to the NLS to find its classical free energy. Most of the necessary work for a complex field is described in [11]. Following this we have found

$$F = (\beta a^{-1}) \, \ln \beta a^{-2} + (1.49) \, \beta^{-4/3} \, c^{1/3} \, 2^{-2/3} \quad . \tag{2.12}$$

The numerical coefficient comes from solving a 2-dimensional radial Schrödinger equation for its lowest eigenvalue with no free parameters. As can be seen all the β and c dependences separate out. Note that the first term is the free energy of a classical phonon gas with the dispersion relation

$$\tilde{\omega}(k) = 4a^{-2} \sin^2 \frac{ka}{2} \quad . \tag{2.13}$$

This is the lattice form of the usual dispersion relation $\omega(k) = k^2$ for the NLS.

Now (2.12) should correspond to the solution of (2.11). However, all we have managed to do so far is to show that the solution of (2.11) can be written in the form

$$F = (\beta a^{-1}) \, \ln \beta a^{-2} + \alpha \beta^{-4/3} \, c^{1/3} \tag{2.14}$$

and we have not yet managed to find the coefficient α.

We have been able to do rather more with the classical limit of the shG.

3. CLASSICAL LIMIT of the SINH-GORDON BETHE ANSATZ

Using the TIM with the classical shG equation's Hamiltonian, we find it is possible to calculate the free energy as [15]

$$F = m\beta^{-1} \left[\tfrac{1}{4}(M\beta)^{-1} - \tfrac{1}{8}(M\beta)^{-2} + \tfrac{3}{16}(M\beta)^{-3} - \tfrac{53}{128}(M\beta)^{-4} \right]$$

$$+ \beta^{-1} a^{-1} (\ln(\beta a^{-1}) + \tfrac{1}{2} ma) \quad . \tag{3.1}$$

The last term in this corresponds [15] to a classical gas of non-interacting phonons with the dispersion relation

$$\tilde{\omega}(k) = \left[\left\{ \frac{\sin \tfrac{1}{2} ka}{\tfrac{1}{2} a} \right\}^2 + m^2 \right]^{\frac{1}{2}} \quad . \tag{3.2}$$

Again this is the lattice version of the usual form (quoted in eqn. (1.5b)).

The NLS can be thought of as the non-relativistic limit of the shG and it is no surprise that their quantum statistical mechanics are similar. For the shG the free energy is given by (2.5) with the replacement of k^2 by the form $(m^2+k^2)^{\frac{1}{2}}$ and with a change in the quantum phonon phase shift in (2.3). For the sG this has been calculated to be [16]

$$\Delta(k,k') = 2 \tan^{-1} \left\{ \frac{k\omega(k') - k'\omega(k)}{m^2 \sin(\gamma'/8)} \right\}^{-1}$$

$$\gamma' = 8\pi\gamma_0 / (8\pi - \gamma_0) \quad . \tag{3.3}$$

We get the shG form from this by changing the sign of γ_0. We find this then agrees with semi-classical calculations [17].

Carrying out precisely the same manipulations as for the NLS we get the classical limit

$$\bar{\omega}(k) = \omega(k) - (\pi\beta)^{-1} \int_{-\infty}^{\infty} dk' G(k,k') \frac{d}{dk'} (\ln \beta\bar{\omega}(k')) \quad ; \tag{3.4a}$$

$$G(k,k') = \frac{\gamma_0}{8} \frac{m^2}{k\omega(k') - k'\omega(k)} \quad ; \quad \omega(k) = (m^2+k^2)^{\frac{1}{2}} \quad ; \tag{3.4b}$$

$$F = (2\pi\beta)^{-1} \int_{-\pi/a}^{\pi/a} \ln \beta\bar{\omega}(k)\, dk \qquad . \tag{3.4c}$$

We can calculate the free energy by iterating the integral equation. In doing this iteration we can use the continuum dispersion relation. We get the free energy

$$F = m\beta^{-1}\left[\tfrac{1}{2}(M\beta)^{-1} - \tfrac{1}{8}(M\beta)^{-2} + \tfrac{3}{16}(M\beta)^{-3} - \tfrac{53}{128}(M\beta)^{-4}\right]$$

$$+ \beta^{-1}a^{-1}(\ln(\beta a^{-1}) + \tfrac{1}{2}ma) \quad . \tag{3.5}$$

Clearly there is good agreement here except for the first term which is a $\tfrac{1}{2}$ instead of the $\tfrac{1}{4}$ found by the TIM. To date we have not resolved this discrepancy. Apparently the classical limit of the free energy is not F given by (3.4c) but is $F - \tfrac{1}{4}mM^{-1}\beta^{-2}$.

4. SINH-GORDON STATISTICAL MECHANICS USING ACTION ANGLE VARIABLES

The partition function Z can be written as the trace of the Wick rotated propagator G as follows:

$$Z = \int G(\phi_0,\phi_0;T)\,\mathcal{D}\phi_0 \tag{4.1}$$

in which $\mathcal{D}$ means functional integration. Corresponding to this there will be a measure $\mathcal{D}\mu$ such that

$$Z = \int \mathcal{D}\mu \exp \hbar^{-1} S[p] \qquad . \tag{4.2}$$

Now

$$G(\phi,\phi_0;T) = \int_{\phi_0}^{\phi} \mathcal{D}\Pi\,\mathcal{D}\phi \exp \hbar^{-1} S[\phi] \tag{4.3}$$

where S is the Wick rotated classical action in Hamiltonian form

$$S[\phi] = \int_0^T dt \left\{ i \int_{-\infty}^{\infty} \Pi\phi_t\, dx - H[\phi] \right\} \tag{4.4}$$

with $T = \beta\hbar$; $H[\phi]$ is the classical Hamiltonian. For the shG we can make a transformation to action angle variables; in terms of these

$$S[p] = \int_0^T dt \left\{ i \int_{-\infty}^{\infty} P(k)Q_t(k)\, dk - H[p] \right\} \quad . \tag{4.5}$$

This appears to be separable and if this were the case the functional integral would be easy to perform. In terms of the new variables Z is apparently given by

$$Z = \int \mathcal{D}P\,\mathcal{D}Q \exp \hbar^{-1} S[p] \qquad . \tag{4.6}$$

This would be because the transformation is canonical so that its Jacobian is unity. Thus the obvious choice $\mathcal{D}\mu = \mathcal{D}P\,\mathcal{D}Q$ has been made for the measure. It can be shown, however [15], that this partition function is exactly that of a *free* quantum Klein-Gordon system.

This result shows that the choice for the measure cannot be correct for, as the BA analysis has demonstrated, the shG partition function is not that of a gas of free particles. It is therefore necessary to go beyond these action angle variables. An important point is that, although the canonical transformation formally has a Jacobian of unity, it is well known that such a change of variables under a functional integral cannot always be performed in this naive way [18,19].

In the classical limit $\hbar \to 0$, $\hbar^{-1}S[\phi] \to -\beta H[\phi]$ and $\hbar^{-1}S[p] \to -\beta H[p]$ [15]. The classical partition function is then given by the functional integral

$$Z = \int \mathcal{D}\pi \, \mathcal{D}\phi \, \exp - \beta H[\phi] \tag{4.7}$$

We now evaluate this functional integral in terms of action angle variables.

The TIM works directly from (4.7). Results are obtained in an explicit thermodynamic limit in which boundary conditions of period L are imposed and then $L \to \infty$. It therefore becomes necessary to include this feature in the action angle variables used to evaluate the functional integral. We have used Floquet theory[15,20] to show that there are analytic functions of ζ, a_L, b_L which, for $L \to \infty$, go over to the usual transmission coefficient a and reflexion coefficient b/a of infinite support classical inverse scattering theory: ζ is the spectral parameter and $a(\zeta)$ is analytic in the upper half plane; $b(\zeta)$ has no simple analyticity but plays its usual role on the real axis k of ζ. The condition for period L solutions is $\Delta(\zeta) = 2$; Δ is the discriminant. This leads to the result that on the real axis periodic solutions labelled by $k = \tilde{k}$ must satisfy [15,20]

$$1 = \cos(\tilde{k}L - \Xi(\tilde{k})) \tag{4.8}$$

in which $\Xi = \arg a_L(\zeta)$. Consequently

$$\tilde{k}_n = k_n + L^{-1}\Xi(\tilde{k}_n) \tag{4.9}$$

where $k_n = 2\pi n L^{-1}$. (In [15] we adopt a different sign in defining Ξ).

We are concerned with $n, L \to \infty$ in such a way that the k_n become dense on the real axis $-\infty < k < \infty$. Since $L^{-1}\Xi(\tilde{k}_n)$ is $O(L^{-1})$ if $\Xi \overset{?}{=} O(1)$, it seems sufficient to use for $\Xi(\tilde{k}_n)$ a form very little different from $\arg a(k_n)$. Consider the sG equation for completeness. In light cone coordinates one shows that on the real axis ξ of the corresponding ζ-plane.

$$a(\xi) = |a(\xi)| \prod_{j=1}^{N} \frac{(\xi - \zeta_j)}{(\xi - \zeta_j^*)} \exp \left[-\frac{P}{\pi i} \left\{ \xi \int_{-\infty}^{\infty} \frac{d\xi' \ln|a(\xi')|}{\xi^2 - \xi'^2} \right\} \right] \tag{4.10}$$

in which ζ_j are the zeros of $a(\zeta)$ in the upper half plane and P the principle value integral. From this ones shows [15] that in laboratory co-ordinates

$$\arg a(k) = \sum_{i=1}^{N_k} \Delta_k(k,p_i) + \sum_{j=1}^{N_{\bar{k}}} \Delta_{\bar{k}}(k,\bar{p}_j) + \sum_{\ell=1}^{N_b} \Delta_b(k,\hat{p}_\ell,\theta_\ell) + \frac{1}{8}\gamma_0 m^2 \int_{-\infty}^{\infty} \frac{dk' \, P(k')}{k\omega(k') - k'\omega(k)} \tag{4.11}$$

The momenta $\{p_i, \bar{p}_j, \hat{p}_\ell, 4\gamma_0^{-1}\theta_\ell, P(k)\}$ are the action variables of (1.5) The quantities $\Delta_k(k,p_i)$ and $\Delta_{\bar{k}}(k,\bar{p}_j)$ are given by

$$-\Delta_k(k,p_i) = 2 \tan^{-1} \{m^2[k\omega(p) - p\omega(k)]^{-1}\} \tag{4.12}$$

where $p = mV_i\gamma_i$, $\omega(p) = (m^2+p^2)^{\frac{1}{2}}$, $\gamma_i = (1-V_i^2)^{-\frac{1}{2}}$ and $MV_i\gamma_i = p_i$: $\Delta_{\bar{k}}$ is the same expression with $j,\bar{p}_j$ replacing i,p_i. The quantity Δ_b is

$$-\Delta_b(k,\hat{p}_\ell,\theta_\ell) = 4 \tan^{-1} \{m^2 \sin \theta [k\omega(p) - p\omega(k)]^{-1}\} \quad ; \tag{4.13}$$

p now refers to the translational velocity V of the breather. These are precisely the phase shifts computed by CURRIE [21], induced in vanishingly weak harmonic solutions of the sG in laboratory co-ordinates by kinks, antikinks and breathers. Curiously [15] we still have some problems over factors of 2 in arg a(k).

For the shG, there will be no soliton phase shifts and we have the result that

$$\arg a(k) = \frac{-\gamma_0 m^2}{8} \, P \int_{-\infty}^{\infty} \frac{1}{k\omega(k')-k'\omega(k)} \, P(k')dk' \quad .$$
(4.14)

And, since for finite period L we expect the action variable $P_n \longleftrightarrow 2\pi L^{-1}P(k_n)$, the condition (4.9) becomes [15]

$$\tilde{k}_n = k_n - \frac{1}{L} \frac{\gamma_0 m^2}{8} \sum_{m\neq n} \frac{P_m}{\tilde{k}_n \omega(\tilde{k}_m) - \omega(\tilde{k}_n)\tilde{k}_m}$$
(4.15)

This is a peculiar form of the BA integral equation [2,7,15] for allowed k_n vectors under periodic boundary conditions. Reference to (3.4b) shows that it corresponds to the classical limit for the shG BA equations. This equation thus demonstrates that we have used the correct classical limit for G, the phonon-phonon phase shifts, in the preceding section.

We shall now use (4.15) to develop the functional integral method for the shG. For finite support the shG Hamiltonian can be interpreted as

$$H[p,N] = \sum_{-\frac{1}{2}N}^{\frac{1}{2}N} \tilde{\omega}(\tilde{k}_n)P_n$$
(4.16)

where $\tilde{\omega}$ is given by (3.2) and L = Na (a is a lattice spacing). Since from (4.15) the Hamiltonian depends on the P_n through $\tilde{\omega}(\tilde{k}_n)$ it is not separable. To $O(L^{-1})$ we have that

$$H[p,N] = \sum_{-\frac{1}{2}N}^{+\frac{1}{2}N} \tilde{\omega}_n P_n - \frac{1}{L} \sum_{-\frac{1}{2}N}^{+\frac{1}{2}N} \sum_{m\neq n} \frac{\partial \tilde{\omega}_n}{\partial k_n} G(k_n,k_m)P_n P_m \quad ;$$
(4.17)

the notation is $\tilde{\omega}_n \equiv \tilde{\omega}(k_n)$, and $G(k_n,k_m) \equiv G(n,m)$ is given by (3.4)

The corresponding *discretized functional integral* will be

$$Z = \left\{ \int_0^{2\pi} \int_0^{\infty} \right\} \prod_n (2\pi)^{-1} \, dQ_n \, dP_n \, \exp - \beta H[p,N]$$
(4.18)

and $N,L \to \infty$ with $a \to 0$ (if possible). Thus to $O(L^{-1})$ we get

$$Z = \left\{ \int_0^{\infty} \right\} \prod_n dP_n \left[1 + \beta L^{-1} \sum_{m\neq n} \frac{\partial \tilde{\omega}_n}{\partial k_n} G(n,m) P_n P_m \right] \exp \left[- \beta \sum_n \tilde{\omega}_n P_n \right] \quad ;$$
(4.19)

and from this we find $Z = \exp - \beta LF$ where

$$\beta LF - \sum_{n=-\frac{1}{2}N}^{+\frac{1}{2}N} \ln \beta\tilde{\omega}_n - (\beta l)^{-1} \sum_{m\neq n} \omega_n^{-1} G(n,m) \frac{\partial}{\partial k_m} \ln \omega_m \quad .$$
(4.20)

Then for $N,L \to \infty$ at fixed a (a determines only the cut off and $\tilde{\omega}$)

$$F = (2\pi\beta)^{-1} \int_{-\pi/a}^{+\pi/a} dk \, \ln (\beta\tilde{\omega}(k)) - (2\pi\beta)^{-2} \int dk(\omega(k))^{-1} \int dq \, G(k,q) \frac{\partial}{\partial q} \ln \beta\omega(q)$$
(4.21)

$$= \beta^{-1}a^{-1} (\ln (\beta a^{-1}) + \tfrac{1}{2}ma + \beta^{-1}m [\tfrac{1}{4}(M\beta)^{-1} + \cdots])$$
(4.22)

111

Again the continuum ω is used in all but the leading term.

The result (4.22) agrees identically with that from the TIM at that order. The
result agrees also with the classical BA result except that it gets the $\frac{1}{4}$ factor
which that method failed to produce. At the time of writing, we have not calculated
the next term at $O((M\beta)^{-2})$.

5. SINE-GORDON STATISTICAL MECHANICS

Calculating the statistical mechanics of the sG system is a much more difficult task
than that for the NLS and shG systems described in the previous sections. The reason
for this is that the sG has soliton solutions, which the other two equations do not.
There is not enough space here to cover all the intricacies surrounding these calculations for the sG: we will just report the progress we have made.

By applying the TIM to the sG system it is possible [22] to show that the free
energy can be written in the form

$$F = -\beta^{-1} m 8 (m\beta/\pi\gamma_0)^{\frac{1}{2}} e^{-\beta M} \left[1 - \frac{7}{8}(M\beta)^{-1} + \cdots \right]$$

$$-\beta^{-1} m \left[\frac{1}{4}(M\beta)^{-1} + \frac{1}{8}(M\beta)^{-2} + \frac{3}{16}(M\beta)^{-3} + \frac{53}{128}(M\beta)^{-4} \right]$$

$$+\beta^{-1} a^{-1} (\ln \beta a^{-1} + \tfrac{1}{2} ma) \quad . \tag{5.1}$$

The exponentially damped term is assumed to come from the kinks and antikinks which
have the finite mass M. The last term is as usual given by the free Klein-Gordon
modes. It is plausible that the power series in the middle is generated by the
breather solutions which, since their masses range down to zero, are not expected
to be exponentially damped. However, it is worth noting that this power series simply
corresponds to the shG TIM result with the appropriate change in the sign of the
coupling constant.

In calculating the sG partition function the change to action angle variables
described in the previous section can be repeated. However it is necessary to take
into account the soliton degrees of freedom. First we construct a measure for the
functional integral. The key remark is that the functional integral is defined on
finite support $Na = L$ with N, L $\to \infty$ (and $a \to 0$). There are 2N degrees of freedom for each finite N. Consequently, there is a constraint:

$$N_k + N_{\bar{k}} + 2N_b + N_{ph} = N \tag{5.2}$$

where the N's on the left represent the numbers of kinks, antikinks, breathers and
phonons. The proper measure seems to be

$$\mathcal{D}\mu = (N_k! N_{\bar{k}}! N_b!)^{-1} \prod_{i=1}^{N_k} (2\pi)^{-1} dp_i \, dq_i \prod_{j=1}^{N_{\bar{k}}} (2\pi)^{-1} d\bar{p}_j \, d\bar{q}_j$$

$$\prod_{\ell=1}^{N_b} (2\pi)^{-2} d\hat{p}_\ell \, d\hat{q}_\ell \, 4\gamma_0^{-1} d\theta_\ell \, d\Phi_\ell \, \mathcal{D}P \, \mathcal{D}Q \tag{5.3}$$

in the $N \to \infty$, $a \to 0$ limit. We will refer to the discretized phonon Hamiltonian
(4.16) as H_{KG}, since this corresponds to the Klein Gordon solutions. The $\tilde{k}_n$ are
now determined by the integral equation (4.15) generalised to include the soliton
phase shifts. From (4.11) this is

$$\tilde{k}_n = k_n + \frac{1}{L} \left\{ \sum_{i=1}^{N_k} \Delta(k, p_i) + \sum_{j=1}^{N_{\bar{k}}} \Delta_{\bar{k}}(k, \bar{p}_j) \right.$$

$$+ \sum_{\ell=1}^{N_b} \Delta (k, \hat{p}_\ell, \theta_\ell) + \frac{\gamma_0 m^2}{8} \sum_{m \neq n} \frac{P_m}{\tilde{k}_n \omega(\tilde{k}_m) - \omega(\tilde{k}_n) \tilde{k}_m} \Bigg\} \tag{5.4}$$

We define the Hamiltonian $H[N_k, N_{\bar{k}}, N_b]$ by

$$H[N_k, N_{\bar{k}}, N_b] = H[p,N] - H_{KG}[p, N_{ph}] \quad . \tag{5.5}$$

Here $H[p,N]$ is $H[p]$, (1.5) with the Klein Gordon contribution replaced by (4.16) and the constraint (5.2) imposed. Thus $H[N_k, N_{\bar{k}}, N_b]$ is $H[p]$ once the KG contribution is removed. Reasons why this is so are given in [17]. We normalise the problem about the free KG so that

$$Z/Z_{KG} = \lim_{N \to \infty} \sum_{N_k, N_{\bar{k}}, N_b} \int \mathcal{D}\mu \, \exp \{ - \beta H [N_k, N_{\bar{k}}, N_b] + \sigma [N_k, N_{\bar{k}}, N_b] \} \tag{5.6}$$

Without the quantity σ, the partition function would now be that of a gas of free non-interacting kinks and breathers. As calculations have shown [15,22], this does not reproduce the TIM result. Thus σ should provide the necessary corrections.

It is clear that σ arises through a comparison between KG partition functions one computed with $H_{KG}[p, N_{ph}]$ with shifted modes $\tilde{k}_n$ as N_{ph} and $N \to \infty$; the other computed with $H_{KG}[p,N]$ and unshifted modes k_n. The shifted modes $\tilde{k}_n$ are related to the unshifted modes k_n by (5.4). Because (for $L \to \infty$) the phase shifts in (5.4) are additive; one finds for example [17] that instead of a free particle contribution, the kinks provide the following in the free energy

$$F_{kink} = -\beta^{-1} L^{-1} Z_k$$

$$= -(2\pi\beta)^{-1} \int_{-\infty}^{\infty} \exp [- \beta(M^2 + p^2)^{\frac{1}{2}} + \sigma_k(p) + \ln \beta m] \, dp \quad . \tag{5.7}$$

The $\ln \beta m$ comes from the difference $N - N_{ph} = N_k + N_{\bar{k}} + 2N_b$ in the number of modes contributing to the free KG(N) and the coupled one (N_{ph}). Without the phonon term in (5.4)

$$\sigma_k(p) = \sigma_{\bar{k}} = \frac{1}{2\pi} \int_{-\pi/a}^{\pi/a} \left[\frac{\ln \omega(k) d\Delta_k(k,p)}{dk} \right] dk$$

$$= \ln [\gamma^{-1}(1+\gamma)] \quad ; \quad \gamma \equiv (1 - V^2)^{-\frac{1}{2}} = p(VM)^{-1} \tag{5.8}$$

Thus

$$F_{kink} = -\beta^{-1} m4 \left\{ \beta^{\frac{1}{2}} m^{\frac{1}{2}} \pi^{-\frac{1}{2}} \gamma_0^{-\frac{1}{2}} \right\} e^{-\beta M} \left\{ 1 + \frac{1}{8} (M\beta)^{-1} + \cdots \right\} \tag{5.9}$$

When this is doubled to include antikink contributions, the result differs from the first term in (5.1) only in the coefficient $\frac{1}{8}$ in the term $\frac{1}{8}(M\beta)^{-1}$.

These results are very satisfactory and arise entirely through the periodic boundary condition imposed via (5.4). We have also shown (up to some factors of 2 still to be resolved) that, by including the phonon term in (5.4), the $\frac{1}{8}(M\beta)^{-1}$ is changed into $-\frac{7}{8}(M\beta)^{-1}$ in complete agreement with (5.1) at that order. The uncertainty in factors of 2 seems to lead to an inconsistency with the work on the shG. What is certain, however, is that phonon-phonon phase shifts lead to terms of $O((M\beta)^{-1})$ in F_{kink} and so must be included in calculations.

The situation surrounding the breathers is very complicated. As we pointed out above, the part of the TIM result one would expect the breathers to generate is in fact given by the phonon-phonon phase shifts. However, there is an explicit breather series and this is again dressed by the phonon modes. This has both the odd and even powers of the TIM result [15]. It is obvious that the contributions of small mass breathers and phonons are overlapping in a way still to be understood: the constraint (5.2) should have eliminated this problem.

Some of the problems raised in this paper are described at greater length in the lectures [15]. We hope to solve them and report all our work in complete detail elsewhere in due course.

References

1. L.D. Faddeev: "Integrable Models in 1+1 Dimensional Quantum Field Theory", in Proc. of the Ecole d'Eté de Physique Theorique, Les Houches 1982, eds. R. Stora and J.B. Zuber (North-Holland, Amsterdam 1983)
2. P.P. Kulish, E.K. Sklyanin: "Quantum Spectral Transform Method. Recent Developments", in Proc. of the Tvärminne Symposium , Finland 1981., eds. J. Hietarinta and C. Montonen (Springer Verlag, Heidelberg 1982)
3. L.A. Takhtadzhyan, L.D. Faddeev: Teor. Mat. Fiz. : 21, 160 (1974)
4. R.K. Dodd, R.K. Bullough: in Synergetics A Workshop, ed. H. Haken (Springer-Verlag, Heidelberg 1977)
5. H.B. Thacker: Rev. Mod. Phys. 53, 253 (1981)
6. C.N. Yang, C.P. Yang: J. Math. Phys. 10, 1115 (1969)
7. M. Fowler, X. Zotos: Phys. Rev. 24, 2634 (1981)
8. M. Imada, K. Hida, M. Ishikawa: J. Phys. C16, 35 (1983)
9. M. Takahashi, M. Suzuki: Prog. Theor. Phys. 48, 2187 (1972)
10. N. Gupta, B. Sutherland: Phys. Rev. A14, 1790 (1976)
11. D.J. Scalapino, M. Sears, R.A. Ferrel: Phys. Rev. B6, 3409 (1972)
12. H.B. Thacker: Phys. Rev. D16, 2515 (1977)
13. E.H. Lieb, W. Liniger: Phys. Rev. 130, 1605 (1963)
14. P.P. Kulish, S.V. Manakov, L.D. Faddeev: Teor. Mat. Fiz. 28, 38 (1976)
15. R.K. Bullough: Lectures at the Latin American School of Physics ELAF '84, Santiago, Chile, 16th July - 3rd August 1984 to be published. (Springer-Verlag, Heidelberg 1985) ed. F. Claro
16. V.E. Korepin: Teor. Mat. Fiz. 41, 169 (1979)
17. S.N. Vergeles, V.M. Gryanik: Sov. J. Nucl. Phys. 23, 704 (1976)
18. S.F. Edwards, Y.V. Gulyaev: Proc. Roy. Soc. London A279, 229 (1964)
19. A.M. Arthurs: Proc. Roy. Soc. London A313, 445 (1969)
20. H. Flaschka, D.W. McLaughlin: Prog. Theor. Phys. 55, 438 (1976)
21. J.F. Currie: Phys. Rev. A16, 1692 (1977)
22. R.K. Bullough, D.J. Pilling, J. Timonen: in Nonlinear Waves, ed. L. Debnath (CUP, New York 1983), pp. 326 - 355

Classical Statistical Mechanics of Integrable Systems

Nikos Theodorakopoulos

Max-Planck-Institut für Festkörperforschung, Heisenbergstraße 1
D-7000 Stuttgart 80, Fed. Rep. of Germany

I. Introduction

The aim of this contribution is to present a brief account of recent developments
in the area of soliton thermodynamics, which demonstrate the validity of the
soliton paradigm [1] in a statistical-mechanical context and establish a link
between three seemingly disparate lines of nonlinear development: Inverse scattering theory (IST), transfer integral (TI) method and the Bethe Ansatz (BA).

According to the IST [2] the options of a classical nonlinear integrable system,
regarding its dynamical evolution, are exhausted by solitons and (nonlinear)
phonons. It is therefore not unreasonable to expect that a complete description
of thermodynamic properties (known by TI methods) and -perhaps- of the theoretically less trivial, experimentally accessible [3] dynamical correlation functions, could be achieved in terms of indestructible solitons, phonons and the
various phase shifts which characterize interactions among them (soliton-phonon
phenomenology). A convincing demonstration of this expectation [4,5] would indeed
extend the domain of the soliton paradigm from dynamical evolution to statistical
mechanics and serve to legitimize the soliton as a <u>normal mode</u> of a classical
nonlinear many-body system. It will be shown below that an exact reconstruction
of the sine-Gordon (SG) free energy is possible if all interactions reducing
available phase space [5] are taken into account [5-7]. These involve not only
clasically nonvanishing phase shifts, originating in classically nonvanishing
thermal population of solitons, but also phase shifts of order $\hbar$, originating in
classically divergent, i.e. $O(T/\hbar)$, phonon populations. Soliton-phonon phenomenology can thus be viewed as a vehicle relating physical quantities pertinent to
the dynamical evolution and thermodynamics - or IST and TI (cf. Fig. 1).

A different and, at first glance, independent line of approach towards the excitation spectrum of an integrable system is provided by the Bethe Ansatz [8]. It has
been applied to the thermodynamics of the massive Thirring/sine-Gordon model [9].
However, the minimal requirement of exact agreement between TI and classical-
limit-BA thermodynamics has not been fully satisfied [9-11]. It thus appears
preferable to establish [12] the missing explicit link (cf. Fig. 1) in the
context of a discrete integrable system, the Toda lattice [13]. This circumvents
difficulties related to the absence of a natural cutoff in SG theory and renders
the BA "commensurable" with the intrinsically discrete TI method. Moreoever the
simplicity of the model allows us to gain some further insight into the interre-
lationship between the soliton-phonon picture and the classical limit of the BA.
It will namely be seen that the former can be thought of as the low-temperature
asymptotic expansion of the latter [12].

The paper is organized as follows: In Section II the general relationship between
dynamics (phase shifts) and statistics (counting of states) is reviewed. Section
III describes the formalism of soliton/phonon phenomenology in a schematic fashion and states its results for the SG model. Finally Section IV presents a brief
account of BA thermodynamics of the Toda lattice and a discussion of its relationship to the phenomenological approach.

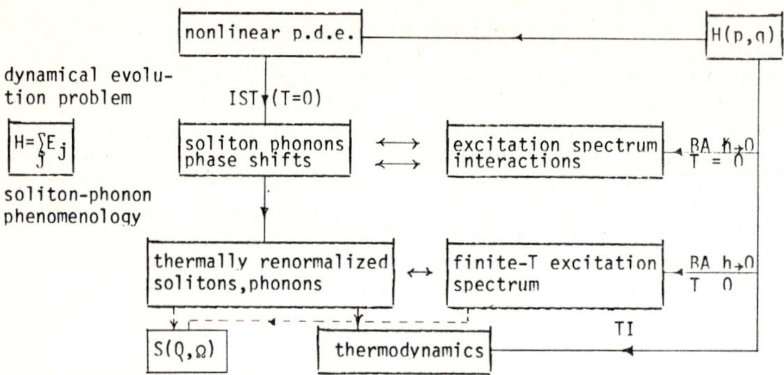

Fig. 1: Flow chart illustrating the interrelationships between inverse scattering theory (IST), Bethe Ansatz (BA) and the transfer integral (TI) methods. The connecting element is soliton-phonon phenomenology. Thermodynamics can be shown to be identical within all three approaches. The extension to dynamical correlations (broken line) is tentative.

II. Phase Shifts and Counting of States

Scattering between solitons is strictly elastic in integrable systems. The only (asymptotically) relevant effect of the interaction is a phase shift, which may be directly obtained from the two-soliton solution. Relations between phase shifts reflect the underlying symmetries of the equation of motion, e.g.

$$\sum_i E_i \Delta_i = 0 \quad [\text{Lorentz invariance (SG)}], \tag{1a}$$

where E_i and Δ_i are the energy and spatial shift of the ith soliton (or nonlinear phonon), respectively. Alternatively,

$$\sum_i M_i \Delta_i = 0 \quad [\text{Galilean invariance (Toda)}] \quad , \tag{1b}$$

where M_i is the mass of the ith soliton (or nonlinear phonon).

The relationship between phase shifts and changes in the density of states for any given excitation can be formulated within the framework of the semiclassical Bohr/Sommerfeld quantization scheme, under periodic boundary conditions in a one-dimensional ring of length L. The action

$$W(E) = \int_0^L dx \cdot P = nh \tag{2}$$

is an integral multiple of h. Here x represents the classical position coordinate, E the energy and P the canonical momentum, i.e. $\partial E/\partial P = \dot{x} \equiv V$. It follows directly that the derivative $\partial W/\partial E = \int_0^L dx/\dot{x}$ is equal to the travel time $\tau(E)$ of our excitation. On the other hand as the quantum numbers n increase (semiclassical limit), we can identify $(E_{n+1}-E_n)^{-1} = (\partial W/\partial E)_{E=E_n}/h$ with $\tilde{\rho}(E_n)$, the density of states in energy space. We are thus led to conclude that

$$\tilde{\rho}(E) = \frac{1}{2\pi\hbar} \tau(E) \quad . \tag{3}$$

If no other objects are present, the free travel time $\tau_0 = L/V$ leads to the free density of states $\tilde{\rho}_0$. If a time delay $\Delta\tau$ occurs (equivalent to a spatial shift $\Delta x/V$) due to interaction with another excitation, there is a change $\Delta\tilde{\rho}(E) = \Delta\tau/2\pi\hbar$ corresponding to a change in the d.o.s. in <u>momentum</u> space

$$\Delta\rho(P) = -\frac{1}{2\pi\hbar} \Delta x \quad , \tag{4}$$

where the fact that $\rho(P) = \tilde{\rho}(E) \cdot \partial E/\partial P$ has been used. The relationship (4) is central to the arguments that follow, which is the reason I chose to elaborate on it. A series of further comments should be made at this point. Firstly, the arguments given above are independent of the dispersion relation $E(P)$ obeyed by the elementary excitation [e.g. relativistic energy-momentum relation such as in the SG or ϕ^4 soliton and phonon case, acoustic phonon (Toda) or implicit monotonic function of P (Toda soliton)]. Secondly, the free d.o.s in momentum space is equal to $\rho_0(P) = L/2\pi h$. Thirdly, the arguments used assumed a classical particle-like structure with a position coordinate. It is immediately evident that the spatial shift of a phonon mode of wave vector q, meaningful only in the limiting form of a wave packet, is given by the derivative of the phase shift, i.e. $\Delta_q = -\delta'(q)$. The resulting change in the phonon d.o.s. satisfies Levinsons theorem:

$$\int dq\; \Delta\rho(q) = -n_B \tag{5}$$

where n_B is the number of bound states corresponding to the phase-shifting excitation (Goldstone mode(s) plus any internal oscillation modes).

It is now expedient to anticipate some of the results to be presented below. By virtue of Eq. (1a) a soliton recoils after collision with a phonon wave packet. Furthermore, the amount of recoil is proportional to the phonon energy and vanishes in the classical limit. This argument, while perfectly acceptable in the case of a single phonon-like excitation, fails to take proper account of populations typical of a thermal phonon bath, i.e. $\bar{n}_q = k_B T/h\omega_q$ in the classical limit. A SG kink for example loses a finite fraction of states that would otherwise be available to it, when placed in a thermal phonon bath:

$$\Delta R_k(P) = -\frac{1}{2\pi h} \sum_q \Delta_k(q)\bar{n}_q$$
$$\cong -\frac{L}{2\pi h}\frac{k_B T}{E_k(P)} \quad , \tag{6}$$

where $E_k(P)$ is the total energy of a moving kink. The total density of states available to the kink is now given by

$$R(P) = R_0 \left[1 - \frac{k_B T}{E_k(P)} + \ldots\right] \quad , \tag{7}$$

which we interpret as an asymptotic low-temperature expansion (cf. Section III). This classically significant effect which originates in miroscopic space shifts of order h attests to the importance of the reciprocity of phase-shifting interactions [6].

Having disposed of the essentials of phonon-soliton interactions we might now ask ourselves how - if at all - phonons feel each other's presence in a thermal ensemble. After all, plane waves do not exactly satisfy the underlying nonlinear field equation but only its linearized version (small oscillations). Again, it turns out [7] that the phase shift $\delta(q,q')$ of the phonon (q) due to the presence of (q') is of order $\hbar$. For the same reason given above, effects of the phonon-phonon interaction are expected to be nonvanishing in a thermodynamic context. However, it should be noted that the resulting change in the phonon d.o.s.

$$\Delta\rho(q) = \frac{1}{2\pi}\frac{\partial}{\partial q}\delta(q;q') \tag{8}$$

involves only a rearrangement and no net gain or loss of phonon states. In other words, Levinson's rule is satisfied in the form $\int dq \Delta\rho(q) = 0$; there are no phonon bound states in the classical limit.

III. Soliton-Phonon Phenomenology

For any given configuration involving n_q phonons in the state q and N_s soliton in the state s, the free energy is assumed to be

$$F(\{n_q\},\{N_s\}) = \sum_q F_q + \sum_s F_s \equiv F_{ph} + F_{sol} \quad , \tag{9}$$

where

$$F_q = - T \left[(1+n_q)\ln (1+n_q) - n_q \ln n_q - \beta\hbar\omega_q n_q \right] \tag{10a}$$

$$F_s = - T \left[N_s - N_s \ln N_s - \beta E_s N_s \right] \tag{10b}$$

and the equilibrium value of F is to be determined by simultaneous minimization
with respect to soliton and phonon populations, i.e.,

$$\frac{\delta F}{\delta n_q} = \frac{\delta F}{\delta N_s} = 0 \quad . \tag{11}$$

The symbol s represents collectively all soliton variables necessary to
characterize a given state (e.g. momentum plus any internal oscillation
frequency). Furthermore, we have suppressed the appearance of more than one type
of solitons in Eqs. (9-12) for the sake of notational simplicity.

If solitons and phonons were noninteracting entities, the implementation of (9-
11) would be trivial since the substitution $\sum_q \to \int dq \rho(q)$ could be justified with a
constant density of states ρ_0. The most probable value $\bar{n}_k = 1/\beta\hbar\omega_q$ would reflect
the Bose-Einstein statistics of (10a) in the classical limit. More generally, in
the case of interacting soliton and phonons we expect phase-space sharing to take
place via the constitutive equations

$$\rho(q) = \rho_0 + \frac{1}{2} \sum_{q'} \Delta\rho(q,q')n_{q'} + \sum_{s'} \Delta\rho(q,s')N_{s'} \tag{12a}$$

$$R(s) = R_0 + \sum_{q'} \Delta R(s,q')n_{q'} + \frac{1}{2} \sum_{s'} \Delta R(s,s')N_{s'} \quad , \tag{12b}$$

where the kernels $\Delta\rho(q,q')$, $\Delta\rho(q,s)$,... are given by Eq. (4) i.e. the corresponding phonon-phonon, phonon-soliton,... spatial shift. Equations (9-12) formally
define a complete configurational phenomenology. The leading effects of the
interactions may be incorporated by thermally renormalizing soliton [5] and pho-
non [7] energies - in the sense that the functional forms $\bar{n}_q = 1/\beta\hbar\tilde{\omega}_q$, $N_s = \exp-\beta E_s$ remain invariant whereas the effective excitation energies become
renormalized due to thermal fluctuations (cf. below), while the effective phonon
density of states remains equal to ρ_0 - and restoring consistency [6] in the
counting of nonlinear modes by applying (6).

We may achieve this asymptotic low-temperature result by the following heuristic
approach, which has the advantage of representing a physical "realization" of
thermodynamics, while still keeping track of counting requirements (Levinson's
rules). As a first step we fill up our systems with phonons, reserving solitonic
effects for later. This means that the "total" free energy is given by (9), (10)
with all the N_s set equal to zero. The extremum condition $\delta F/\delta n_q = 0$ then leads to $\bar{n}_q = 1/\beta\hbar\tilde{\omega}_q$, $\tilde{\omega}_q = \omega_q + \Delta\omega_q$, where the thermal renormalization of phonon energies is given
by

$$\hbar\Delta\omega_q = - \frac{1}{2} \int dq' \Delta\rho(q,q') F(\bar{n}_q) \quad . \tag{13}$$

The explicit form is given in Table I for the SG model [7], along with the
corresponding free energy of the phonon sector.

The effect of solitons on phonons may be taken into account as follows: For any
given phonon configuration we minimize the total free energy with respect to all
$\{N_s\}$. This leads to most probable values $N_s = \exp\left[-\beta(E_s + \Delta F_s)\right]$ and defines soliton
energies E_s thermally renormalized [5] by an amount

$$\Delta E_s = \int dq \, \Delta\rho(q,s) F(n_q) \quad , \tag{14}$$

where the most probable values $\bar{n}_q$ defined by (13) can be inserted as the phonon configuration in the r.h.s. of (14). Thus, the disappearance of phonon states which must occur as we proceed to fill up our system with solitons is effectively described - in the sense that its effect on the free energy is automatically considered - by the thermal renormalization of soliton energies [5].

The total free energy is now given by

$$F = \int dq \; \rho_0 \; F_q(\bar{n}_q) + \sum_s F_s \qquad (15)$$

with $F_q = T \ln \beta \hbar \tilde{\omega}_q$ and $F_s = \exp(-\beta E_s)$. Noting that, formally, all phonon states appear in the free energy, we proceed to restore counting consistency as we substitute the $\sum_s$ by an integral $\int ds R(s)$, i.e. use a $R(s)$ which does not include soliton states that must be removed in the presence of the full phonon gas (cf. (6), except that phonon energies should be renormalized). The expressions for the effective density of states of SG kinks and breathers are also given in Table I, along with the respective contributions to the free energy. The sum of phonon, breather and kink terms is in full agreement with the leading orders of the asymptotic expansion of the Mathieu equation [14]. This detailed agreement, although falling short of an exact solution of (9)-(12), can be regarded as a formal demonstration of the soliton paradigm within classical statistical mechanics. It vindicates the original conjecture of KRUMHANSL and SCHRIEFFER [4], the implementation of which had been prevented by difficulties related to the role of nontopological solitons (breathers). In the next section it will be seen that the soliton-phonon picture is intimately related to the Bethe Ansatz.

Table I. Thermally renormalized energies of phonons (13) and solitons (14), along with consistent densities of states in momentum space (6) and contributions of the various sectors to the free energy (15) of the SG model. The calculation has been carried to order T^2 [7] in the reduced temperature (measured in units of the kink energy E_0). f_0 represents the free energy of a gas of harmonic phonons with dispersion relation $\omega_q^2 = \omega^2 + c_0^2 q^2$, measured in units of E_0/d, where $d = c_0/\omega_0$ is a measure of the kink's spatial extent. A soliton's state is characterized by the Lorentz factor $\gamma = (1 - v^2/c_0^2)^{-1/2}$ and (breather) internal oscillation frequency $\omega_0 \cos\alpha$.

	$\beta\Delta E$	$\Delta R/R_0$	f
phonon	$-\hbar\omega_0^2/2\omega_q$	0	$f_{ph} = f_0 - T^2/4$
breather	$-2\ln[\beta\hbar\omega_0(\gamma+\sin\alpha)]$ $+2T(1+\gamma\sin\alpha)/(\gamma+\sin\alpha)^2$	$-T\gamma\sin\alpha[1-$ $T(1+\gamma\sin\alpha)^2]$	$f_b = 0$
kink	$-\ln[\beta\hbar\omega_0(1+\gamma)]$ $+T/(1+\gamma)$	$-\dfrac{T}{\gamma}[1-T/(1+\gamma)]$	$f_k = -(2T/\pi)^{1/2}$ $e^{-1/T}[1-\tfrac{7}{8}T]$

IV. Finite-T Excitation Spectrum of Toda Lattice; Relationship of Soliton-Phonon Phenomenology to the Bethe-Ansatz

The application of BA methods to the Toda lattice was initiated by SUTHERLAND [15], who calculated the zero temperature properties in the classical limit. In close analogy with earlier results on the δ-function Bose gas [16] he established that particles occupy a "Fermi sea" $|k|<k_F$ in the space of single particle momenta. Furthermore particle-like and hole-like excitations correspond to solitons and phonons, respectively, in the sense that they exhibit the correct [13] dispersion relations. It can further be shown [17] that the equivalence of the classical (IST) description to the BA is in fact complete at T=0 (cf. Fig. 1) by comparing the change in the density of states that results from interaction between elementary excitations, as derived from the BA, with the classical phase shift of IST.

For nonzero temperatures it is possible to apply the procedure of YANG and YANG [18] and obtain the densities of occupied and omitted states $\rho(k)$ and $\rho(k')$, respectively [12]. The ratio $\varepsilon(k) = -_k T\ln(\rho/\rho_h)$ can then be interpreted as a quasiparticle energy, whereas $h(k)=\int_0^k dk' \, \rho_h(k')$ denotes the quasiparticle momentum.

In the remainder of this section we summarize the contents and implications of [12]. The following statements hold for any finite value of T:
a) The classical excitation spectrum consists of particles over all k-space, in the sense that $\rho/\rho_h=0(\hbar)$ for all k.
b) It is possible to "absorb" the singular T-dependence of quasiparticle energies by making the Ansatz $\varepsilon(k)=T\ln\beta\hbar\omega_0 + \tilde{\varepsilon}(k)$; $\tilde{\varepsilon}$ is then a smooth function of k.
c) The free energy derived via the BA agrees within numerical accuracy with the TI result (cf. Introduction and Fig. 1)
d) The statistics can be reconstructed exactly in terms of the numerical solution for ε and ρ, in the following sense: contributions to the particle and entropy density from any interval dk are identical with those of a (ficticious) classical gas of noninteracting particles with energies $\varepsilon(h)$ in an interval $dh/2\pi$.
e) The phenomenology of Section III leads to renormalized soliton (phonon) energies which are in good agreement with $\varepsilon(k)$ for $|k| \gg (\ll)k_F$. It develops (unphysical) singularities in the neighborhood of k_F.

In view of (a)-(e) above it is possible to use the BA as an exact standard in order to assess the validity of the soliton-gas picture. We note that the failure of the latter to describe the physics in the vicinity of k_F (where phonons and

nontopological solitons become practically indistinguishable long-wavelength excitations) is related to its very nature as an asymptotic expansion in powers of $1/\beta$. For much the same reasons (divergences due to large occupation numbers of gapless phonons) the finite-T BA equations for the Toda lattice, although solvable numerically, are not amenable to asymptotic treatment. We therefore suggest that:
(i) Low-temperature soliton-phonon phenomenology should be regarded as an asymptotic approach to the classical limit of the BA.
(ii) Conversely, the BA formalism might be interpreted in terms of the soliton-phonon picture. The temperature dependence of quasiparticle energies and momenta may be regarded as being due to thermal fluctuations, as the - presumably exact-scheme (9-12) suggests.
The detailed comparison of soliton-phonon phenomenology with the Bethe-Ansatz has revealed an intimate link between these two approaches at a classical level. An immediate challenge is to explore the consequences for the calculation of dynamical correlation functions.

References

1.   J.A. Krumhansl: in "Solitons and Condensed Matter Physics", A.R. Bishop and T. Schneider (Eds.), Springer, Berlin 1978
2.   e.g. as reviewed by R.K. Bullough and R.K. Dood in Ref. 1
3.   a recent review is given by M. Steiner: J. Magn. Magn. Mater $\underline{31-34}$, 1277 (1983)
4.   J.A. Krumhansl and J.R. Schrieffer: Phys. Rev. B $\underline{11}$, 3535 (1975)

5.   J.F. Currie, J.A. Krumhansl, A.R. Bishop and S.E. Trullinger: Phys. Rev. B 22, 477 (1980)
6.   N. Theodorakopoulos: Z. Phys. B 46, 367 (1982)
7.   N. Theodorakopoulos: Phys. Rev. B 30, (1984)
8.   H.B. Thacker, Rev. Mod. Phys. 53, 253 (1981)
9.   M. Fowler and X. Zotos: Phys. Rev. B 25, 5806 (1982)
10. S.G. Chung: Phys. Lett. 89A, 363 (1982)
11. M. Imada, K. Hida and M. Ishikawa: J. Phys. C 16, 35 (1983)
12. N. Theodorakopoulos: Phys. Rev. Lett. 53, 871 (1984)
13. M. Toda: Theory of Nonlinear Lattices (Springer, Berlin 1981)
14. N. Gupta and B. Sutherland: Phys. Rev. A 14, 1790 (1976)
     J. Meixner and F.W. Schäfke: Mathieusche Funktionen und Sphäroidfunktionen (Springer, Berlin 1954)
15. B. Sutherland: Rocky Mountain J. Math. 8, 413 (1978)
16. E.H. Lieb and W. Liniger: Phys. Rev. 130, 1605 (1963),
     E.H. Lieb, ibid., 1615 (1963)
17. N. Theodorakopoulos, to be published
18. C.N. Yang and C.P. Yang, J. Math. Phys. 10, 1115 (1969)

Correlation Functions of the Non-Ideal Gas of Sine-Gordon Solitons

Kazuo Sasaki

Department of Physics, Faculty of Science, Tohoku University, Sendai 980, Japan

1. Introduction

Extensive studies on statistical mechanics of one-dimensional soliton-bearing systems have revealed that solitons, as well as linear excitations such as phonons, can be considered as elementary excitations [1]. Soliton contributions to various thermodynamic quantities and correlation functions have been calculated within the ideal soliton gas approximation [2,3,4]. In this theory the soliton-phonon interaction is taken into account, and the interaction between solitons is neglected. This approximation is valid at low temperatures. Recently the importance of the soliton-soliton interaction has been implied in connection with the anomaly of specific heat [5,6]. In this paper we study the effect of the interaction between solitons on various correlation functions of the sine-Gordon model. These quantities are of particular interest, since they are accessible by neutron scattering and NMR experiments on one-dimensional magnets such as $CsNiF_3$ and $(CH_3)_4NMnCl_3$ (TMMC).

2. Free Energy and Soliton Density

In this section we review a virial expansion theory for the free energy of a soliton gas proposed by the present author [7]. It is applied to the calculation of correlation functions in the next section.

We consider the sine-Gordon model described by the following Hamiltonian in the dimensionless form:

$$H = \frac{1}{8} \int dx \left[\frac{1}{2} \left(\frac{\partial\phi}{\partial t}\right)^2 + \frac{1}{2} \left(\frac{\partial\phi}{\partial x}\right)^2 + (1 - \cos\phi) \right], \tag{1}$$

where the factor 1/8 is introduced for convenience so that the rest energy of soliton is unity.

Let Z_{Tot} be the partition function of this model. It can be factorized as $Z_{Tot} = Z_0 \cdot Z$, where Z_0 is the partition function of the phonon gas in the absence of solitons and Z is that of the soliton gas. Let q and p be position and momentum of soliton. Then Z is given by

$$Z = e^{-\beta F} = \sum_{N=1}^{\infty} \frac{2^N}{N!} \int \prod_{i=1}^{N} d\Gamma_i \; \exp[-\beta E^*(p_i)] \; R(1,\cdots,N), \tag{2}$$

where $\beta = 1/T$, $d\Gamma_i = dq_i dp_i/2\pi\hbar$, and $\hbar$ is the Planck constant in our dimensionless units. The factor 2^N means that solitons and antisolitons contribute equally to Z. The thermally renormalized soliton energy is given by

$$E^*(p) = 1 + \frac{1}{2} p^2 - T \ln\left(\frac{2\hbar}{T}\right) + O(p^4, p^2 T, T^2), \tag{3}$$

where T-dependent terms represent the effect of the soliton-phonon interaction [2]. The effect of the soliton-soliton interaction is included in $R(1,\cdots,N)$ which represents the restriction on the phase space available for solitons in the presence

of N solitons (R = 1 for an ideal gas). By expanding the logarithm of (2) and noticing that the soliton density $n_S^{(0)}$ of the ideal gas is given by

$$n_S^{(0)} = L^{-1} \int d\Gamma \, \exp[-\beta E^*(p)] = (2/\pi T)^{1/2} e^{-1/T} [1 + O(T)] \tag{4}$$

(L is the length of the system), the free energy F of the soliton gas can be obtained
in a power series of $n_S^{(0)}$ (the virial expansion).

To obtain the correction of $O[(n_S^{(0)})^2]$ (the second virial correction) we must
know R(1,2). In Fig. 1 trajectories of solitons in a two-soliton collision are
sketched. They are analogous to those in a collision of hard rods of length $\Delta(p_1,p_2)$.
This analogy leads us to assume that

$$R(1,2) = \theta(|q_1 - q_2| - \Delta(p_1,p_2)), \tag{5}$$

where $\theta(q)$ is the Heaviside step function. The "excluded length" $\Delta(p_1,p_2)$ can be
deduced from the two-soliton solution as $\Delta(p_1,p_2) = 2\ln(2/|p_1 - p_2|)$ for $p_i^2 \ll 1$
(i = 1,2). Equation (5) is the fundamental assumption in our theory.

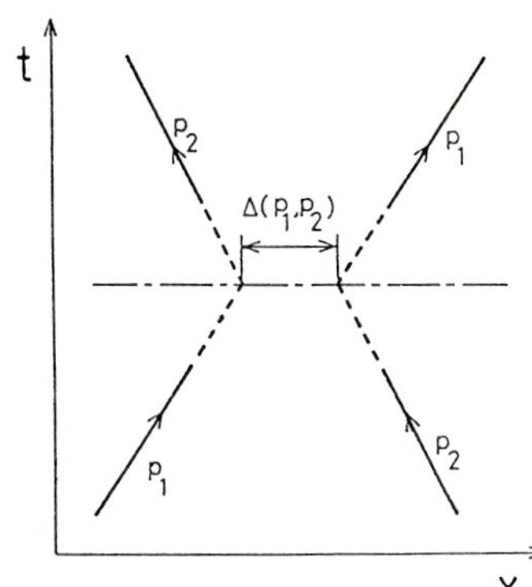

Fig.1 Trajectories of solitons in a two-
soliton collision for $p_i^2 \ll 1$ (i = 1,2)

Within the second virial approximation we get the following results for the free
energy F and the soliton density n_S of the soliton gas [7]:

$$F = -LT \cdot 2n_S^{(0)}(1 - 2\bar{\Delta}n_S^{(0)}), \tag{6}$$

$$n_S = n_S^{(0)}(1 - 4\bar{\Delta}n_S^{(0)}), \tag{7}$$

where $\bar{\Delta}$ is the average of $\Delta(p_1,p_2)$ in the following sense:

$$\bar{\Delta} = \int dp_1 dp_2 \, P^{(0)}(p_1)P^{(0)}(p_2) \, \Delta(p_1,p_2) = \ln(4\gamma/T), \quad \gamma = 1.781\cdots. \tag{8}$$

Here $P^{(0)}(p)$ is the momentum distribution function of solitons in the ideal gas:
$P^{(0)}(p) = (2\pi T)^{-1/2}\exp(-p^2/2T)$. The interaction between solitons reduces the soliton density. The reduction is 25% at T = 0.22.

3. Correlation Functions

We consider soliton contributions to the following correlation functions:

$$\chi_{CC}(x,t) = \langle\cos\phi(x,t)\cos\phi(0,0)\rangle - \langle\cos\phi\rangle^2, \tag{9}$$

$$\chi_{ss}(x,t) = <\sin\phi(x,t)\sin\phi(0,0)>, \tag{10}$$

$$\chi_{c/2,c/2}(x,t) = <\cos[\phi(x,t)/2]\cos[\phi(0,0)/2]>. \tag{11}$$

Thermal average of a quantity Q is obtained by

$$<Q> = e^{\beta F} \sum_{N=0}^{\infty} \frac{2^N}{N!} \int \prod_{i=1}^{N} d\Gamma_i \, \exp[-\beta E^*(p_i)] \, R(1,\cdots,N) \, Q(1,\cdots,N), \tag{12}$$

where $Q(1,\cdots,N)$ denotes the space-time dependence of Q in the presence of N solitons, and q_i in $d\Gamma_i$ means the position of soliton at $t = 0$. For simplicity we have neglected the contribution from phonons to Q. It is not difficult to generalize (12) to include this contribution.

Two kinds of virial expansion for $<Q>$ can be made:

$$<Q> = Q_0 + \sum_{n=1}^{\infty} \frac{2^n}{n!} J_n \, (n_S^{(0)})^n, \tag{13}$$

$$\ell n<Q> = \ell n \, Q_0 + \sum_{n=1}^{\infty} \frac{2^n}{n!} K_n \, (n_S^{(0)})^n, \tag{14}$$

where Q_0 is the value of Q in the absence of solitons and the expansion coefficients are given by

$$J_1 = <Q(1) - Q_0>_1, \qquad K_1 = <Q(1)/Q_0 - 1>_1,$$

$$J_2 = <R(1,2) \, [Q(1,2) - Q_0] - 2[Q(1) - Q_0]>_2,$$

$$K_2 = <R(1,2) \, [Q(1,2)/Q_0 - 1] - [Q(1)Q(2)/Q_0^2 - 1]>_2, \tag{15}$$

etc. We have introduced the notation $<A>_N$ defined by

$$<A>_N = \int \prod_{i=1}^{N} dq_i dp_i \, P^{(0)}(p_i) \, A(1,\cdots,N). \tag{16}$$

The first formula (13) is useful to calculate χ_{cc} and χ_{ss}, whereas the second one (14) is appropriate for $\chi_{c/2,c/2}$.

In calculating the correlation functions (9)-(11), we adopt the following approximation for $\cos\phi$, $\sin\phi$ and $\cos(\phi/2)$ in the presence of n solitons and m antisolitons:

$$\cos\phi = 1 - \sum_{i=1}^{N} 2\mathrm{sech}^2 X_i, \qquad \sin\phi = (- \sum_{i=1}^{n} + \sum_{i=n+1}^{N}) \, 2\mathrm{th}X_i \, \mathrm{sech}X_i, \tag{17}$$

$$\cos(\phi/2) = \prod_{i=1}^{N} \mathrm{th}X_i, \tag{18}$$

where $N = n + m$, $X_i = x - q_i(t)$, and $q_i(t)$ is the position of the i-th soliton (antisoliton) at time t. It should be noted that $q_i(t) \neq q_i(0) + v_i t$ (v_i is the velocity of soliton) except for the case of $N = 1$, because solitons suffer phase shifts when they collide with other solitons. In (17) and (18) we have assumed $v^2 \ll 1$ and neglected the "Lorentz contraction" of solitons.

Now we present the results for the above correlation functions in the second virial approximation. Details of calculation will be published elsewhere.

124

(A) $\chi_{cc}(x,t)$ and $\chi_{ss}(x,t)$

We have not obtained the second virial correction for $t \neq 0$ in simple form because of the phase-shift effect on soliton motion. The results for $t = 0$ are summarised as follows:

$$S_{\alpha\alpha}(q) \equiv \int dx \; e^{iqx} \; \chi_{\alpha\alpha}(x,0) = F_\alpha(q) \; [2n_s - C(\alpha)\cdot 8(n_s^{(0)})^2 \cdot \frac{1}{q} \overline{\sin(q\Delta)} \;], \qquad (19)$$

where $\alpha = c$ or s, $C(c) = 1$, $C(s) = 0$, and the bar denotes the average in the sense of (8). The soliton density n_s is given by (7) and the soliton form factors are given by $F_c(q) = (4Q/shQ)^2$ and $F_s(q) = (4Q/chQ)^2$ with $Q = q\pi/2$. The second term of (19) represents the effect of correlation between positions of two solitons. This term does not appear in $S_{ss}(q)$, since contributions from soliton-soliton and soliton-antisoliton correlations cancel out. For $S_{cc}(q)$ both the correlations give the same contributions, so that we can find this effect in (19). The interaction between solitons reduces $S_{cc}(q)$ and $S_{ss}(q)$. The reduction is 50% for $S_{cc}(0)$ and 25% for $S_{ss}(q)$ at $T = 0.22$.

(B) $\chi_{c/2,c/2}(x,t)$

Substituting $Q = \cos[\phi(x,t)/2]\cos[\phi(0,0)/2]$ with (18) into (15), we get

$$K_1(x,t) = -2 \int dp \; P^{(0)}(p) \; |x - vt|, \qquad (20)$$

$$K_2(x,t) = 8|t| \int_{x/t}^{\infty} dp_2 \int_{-\infty}^{x/t} dp_1 \; P^{(0)}(p_1) \; P^{(0)}(p_2) \; \Delta(p_1,p_2) \; |v_1 - v_2|$$

$$+ 4(\int\int_{-\infty}^{x/t} + \int\int_{x/t}^{\infty})dp_1 dp_2 \; P^{(0)}(p_1) \; P^{(0)}(p_2) \; \Delta^2(p_1,p_2) \qquad (21)$$

for $x^2 + v_0^2 t^2 \gg 1$, where $v_0 = (2T/\pi)^{1/2}$ is the thermal velocity of soliton. Within the ideal gas approximation our result agrees with the earlier one [3,4]. The second virial correction (21) becomes

$$K_2(x,t) = 4\overline{\Delta^2} = 4[(\overline{\Delta})^2 + \pi^2/4] \qquad (22)$$

for $|x| \gg |v_0 t|$ and

$$K_2(x,t) = 4v_0|t|[\; \overline{\Delta} + \ln 2 - 2\sqrt{2} \; \ln(1 + \sqrt{2})] + 4f(T) \qquad (23)$$

for $x = 0$, where the bars again denote the averages in the sense of (8) and $f(T)$ is defined by

$$f(1) = \frac{4}{\pi} \int_0^{\infty} du \int_0^u dv \; (uv)^{-1/2} \; e^{-u-v} \; [\ln(vT)]^2. \qquad (24)$$

From (20) and (23) we get the inverse correlation time Γ:

$$\Gamma = 4v_0 n_s^{(0)}[1 - 2n_s^{(0)}(\overline{\Delta} + \ln 2 - 2\sqrt{2}\ln(1 + \sqrt{2}))]. \qquad (25)$$

The interaction between solitons does not affect the correlation length (see (22)) but reduces the inverse correlation time Γ. The reduction is 6% at $T = 0.22$.

4. Concluding Remarks

We have estimated the second virial corrections to correlation functions of the soliton gas of the sine-Gordon model. The result (19) may be used to analyze the integrated intensity of the central peak observed by the neutron scattering experiment on the ferromagnetic chain $CsNiF_3$ [8]. Our theory shows that the interaction between solitons considerably reduces the intensity. It should be noted that the interference effect between solitons and phonons, which is neglected in the present analysis, also reduces the intensity [9]. The inverse correlation time (25) can be observed by the NMR experiment on the antiferromagnetic chain TMMC [10]. The effect of the soliton-soliton interaction on this quantity is rather small.

Acknowledgement

This work is partially supported by the Grant-in-Aid for Fundamental Scientific Research from the Ministry of Education, Science and Culture and by Japan Society for the Promotion of Science.

Appendix

Here we show that the present result (19) for $S_{cc}(q)$ is consistent with the exact result for the free energy obtained by the transfer integral method. Let $F_{Tot}(h)$ be the free energy of the sine-Gordon model (1) with $(1 - \cos\phi)$ replaced by $h(1 - \cos\phi)$. Since $S_{cc}(0)$ is related with response to the "external field" h, it can be calculated from $F_{Tot}(h)$ as

$$S_{cc}(0) = - \frac{T}{L} \cdot 64 \frac{\partial^2}{\partial h^2} F_{Tot}(h) \Big|_{h=1} . \tag{A1}$$

The transfer integral result for $F_{Tot}(h)$ is as follows [7]:

$$F_{Tot}(h)/L = h \{ \frac{1}{2} \tau - \frac{1}{4} \tau^2 + O(\tau^3) - 2\sqrt{\frac{2\tau}{\pi}} e^{-1/\tau} [1 + O(\tau)]$$

$$+ \frac{8}{\pi} e^{-2/\tau} [\ln(\frac{4\gamma}{\tau}) + O(\tau)] + O(e^{-3/\tau})\}, \tag{A2}$$

where $\tau = T/\sqrt{h}$. We have neglected a term independent of h in (A2). Substitution of (A2) into (A1) yields

$$S_{cc}(0) = 8T^2 + O(T^4) + 16 \{2\sqrt{\frac{2}{\pi T}} e^{-1/T} [1 + O(T)] - \frac{32}{\pi T} e^{-2/T} [\ln(\frac{4\gamma}{T}) + O(T)]\}. \tag{A3}$$

In (A3) the terms expressed in powers of T only can be interpreted as the contribution from phonons, which we have neglected in (19). Substituting (7) and (8) into (19), we find that (19) for $S_{cc}(0)$ is consistent with the exact result (A3).

References

1. For a review see, for example, A.R. Bishop: in _Physics in One Dimension_, ed. J. Bernasconi and T. Schneider (Springer, Berlin, 1981)
2. J.F. Currie, J.A. Krumhansl, A.R. Bishop and S.E. Trullinger: Phys. Rev. B22, 477 (1980)
3. H.J. Mikeska: J. Phys. C11, L29 (1978); C13, 2913 (1980)

4. H. Takayama and K. Maki: Phys. Rev. $\underline{B21}$, 4558 (1980)
5. H. Takayama and G. Satō: J. Phys. Soc. Jpn. $\underline{51}$, 3120 (1982)
6. K. Sasaki and T. Tsuzuki: J. Magn. Magn. Mater. $\underline{31\text{-}34}$, 1283 (1983)
7. K. Sasaki: Prog. Theor. Phys. $\underline{70}$, 593 (1983); $\underline{71}$, 1169 (1984)
8. M. Steiner, K. Kakurai and J.K. Kjems: Z. Phys. $\underline{B53}$, 117 (1983)
9. E. Allroth and H.J. Mikeska: Z. Phys. $\underline{B43}$, 209 (1981)
10. J.P. Boucher and J.P. Renard: Phys. Rev. Lett. $\underline{45}$, 486 (1980)

Classical Solutions for the Grassmannian Sigma Models and Their Supersymmetric Versions

Ryu Sasaki

Research Institute for Theoretical Physics, Hiroshima University, Takehara
Hiroshima 725, Japan

1. Introduction

In this short note we will present a short introduction to our recent work [1], [2], [3] on the problem of solving the Grassmannian sigma models and their supersymmetric versions in euclidean two dimensional space. Exactly integrable field theoretical models in two dimensions such as the sine-Gordon equation, the massive Thirring models and the O(n) sigma model have provided an interesting theater for discussing the complicated nonlinear interactions in particle physics. Among them the Grassmannian sigma models [4] are particularly interesting because of their many common features with the four dimensional gauge theories and of the simpler structure.

The Grassmannian sigma models and the CP^{N-1} models [5] are a massless free scalar field theory with their field variables constrained on the complex Grassmann manifold or the complex projective space. To put it mathematically, solving the Grassmannian sigma models corresponds to finding explicit *harmonic maps* from the two dimensional euclidean space to the Grassmann manifold. As is well known the harmonic map from the two dimensional space to an arbitrary linear space is given by the real and imaginary parts of a holomorphic function. Here we show that a fairly wide class of *classical solutions (harmonic maps) of the Grassmannian sigma models can be constructed explicitly and elementarily* starting from arbitrary holomorphic functions. The same method also solves the supersymmetric Grassmannian sigma models, provided that the fermion fields are treated as commuting numbers. We believe that these models give the first nontrivial example of a relativistic (susy) field theory whose extensive set of classical solutions can be constructed explicitly and elementarily.

It should be noted that the present solution method apparently has no direct connection with the conventional soliton techniques such as the linear scattering problem or the Riemann-Hilbert transform. However, since we have explicit solutions and the solutions of the linear scattering problem for them can also be obtained explicitly, we can construct and discuss the explicit forms of the Bäcklund transformations for them. Thereby we might hope to get some insight into the action of the Bäcklund transformations as well as the infinite dimensional group of symmetry transformations inherent to the integrable models.

2. Complex Grassmannian Sigma Models

Let us denote the complex Grassmann manifold by $G(N,m)$, which can be written as a quotient space

$$G(N,m) = U(m+n)/U(m) \times U(n), \quad N = m+n. \tag{2.1}$$

Here $U(m)$ is the group of $m \times m$ unitary matrices. Let $g = g(x)$, $x = (x_1, x_2) \epsilon R^2$ be an element of $U(N)$ and we decompose it into two parts

$$g = (X,Y), \quad X = (z_1,\ldots,z_m), \quad Y = (z_{m+1},\ldots z_N), \tag{2.2}$$

in which each z_i ($i=1,\ldots,N$) is an N-component column vector. We consider the N×m matrix $X=X(x)$ as a dynamical variable with the constraint († means hermitian conjugation)

$$X^\dagger X = 1_m, \text{ m×m unit matrix}, \tag{2.3}$$

due to the unitarity of g. In terms of the covariant derivative for X,

$$D_\mu X = \partial_\mu X - XX^\dagger \partial_\mu X, \quad \mu = 1.2, \tag{2.4}$$

the Lagrangian density and the action are defined as

$$L = \text{Tr}\{(D_\mu X)^\dagger (D_\mu X)\} \; , \quad S = \int d^2 x L \; , \tag{2.5}$$

which show clearly that the model is a *massless free field theory with nonlinear constraints*. The above Lagrangian is invariant under the local U(m) as well as the global U(N) transformation

$$X \to X'=g_o X h, \quad h = h(x)\epsilon U(m), \quad g_o \epsilon U(N) \text{ constant.} \tag{2.6}$$

The equation of motion is obtained by minimizing the action

$$D_\mu D_\mu X + X(D_\mu X)^\dagger (D_\mu X) = 0. \tag{2.7}$$

This can also be written in a gauge invariant form as

$$[\partial_\mu \partial_\mu P, P] = 0 \; , \tag{2.8}$$

in which P is an N×N projection matrix

$$P = XX^\dagger = \sum_{i=1}^{m} z_i z_i^\dagger, \quad P^\dagger = P = P^2, \quad \text{Tr}P = m. \tag{2.9}$$

For later calculation we introduce the complex coordinates of the two dimensional euclidean space

$$x_+ = x_1 + ix_2, \quad x_- = x_1 - ix_2, \quad \partial_+ = \partial/\partial x_+, \quad \partial_- = \partial/\partial x_- \; , \tag{2.10}$$

and (2.7) is rewritten in two equivalent forms

$$D_+ D_- X + X(D_- X)^\dagger (D_- X) = 0, \quad \text{or} \quad D_- D_+ X + X(D_+ X)^\dagger (D_+ X) = 0. \tag{2.11}$$

Likewise we have the gauge invariant equation

$$[\partial_+ \partial_- P, P] = 0. \tag{2.12}$$

As is clear from (2.11), two special classes of solutions are obtained from those of the following first order equations

$$0 = D_- X \equiv \partial_- X - XX^\dagger \partial_- X, \quad 0 = D_+ X = \partial_+ X - XX^\dagger \partial_+ X. \tag{2.13}$$

These solutions are called *instantons* or *anti-instantons*, respectively.

For the simplest case of $m=1$, the Grassmann manifold G(N,m) reduces to the complex projective space, $G(N,1)=CP^{N-1}$, and the corresponding sigma model is the well known CP^{N-1} model. In this case the dynamical variable X is a simple N-component unit column vector z, $|z|^2 = z^\dagger \cdot z=1$. Therefore the instanton equation looks particularly simple,

$$0 = D_- z = \partial_- z - z(z^\dagger \partial_- z).$$

That is, the x_- derivative of the N-component column vector z is proportional to itself. A solution is provided by an *arbitrary holomorphic N-component vector* f through normalization [6]

$$z = f/|f|, \quad (\partial/\partial x_-)f = 0. \tag{2.14}$$

Any instanton solution can also be written in the above form up to a phase factor, which is irrelevant because of the U(1) gauge symmetry. Likewise an anti-instanton solution is obtained from an anti-holomorphic vector.

In analogy with (2.14) we expect that an instanton solution of the general G(N,m) model is obtained from a set of m linearly independent holomorphic vectors $f_1, f_2, \ldots, f_m$, through some kind of orthonormalization in order to satisfy the constraint (2.3). Let us denote by F an N×m matrix consisting of the holomorphic vectors and an m×m hermitian matrix α,

$$F = (f_1, f_2, \ldots, f_m), \quad \partial_- F = 0, \quad \alpha = F^\dagger F . \tag{2.15}$$

Because of the linear independence of the vectors, α is positive definite and invertible. The hermitian square root matrices $\alpha^{1/2}$ and $\alpha^{-1/2}$ exist uniquely. Now it is easy to see that

$$X = F\alpha^{-1/2} , \tag{2.16}$$

which is a simple generalization of (2.14), satisfies the instanton equation as well as the constraint [7].

Next we show that a fairly wide class of solutions of the G(N,m) model (generic solutions) can be obtained by generalizing the above instanton construction method [1]. Let us introduce as in (2.15) *m linearly and functionally independent holomorphic N-component vectors*

$$f_1, f_2, \ldots, f_m, \quad \partial_- f_i = 0 , \tag{2.17}$$

and define another set of N-component vectors $f_{m+1}, \ldots, f_N$ by differentiation with respect to x_+,

$$f_{m+1} = \partial_+ f_1, \quad f_{m+2} = \partial_+ f_2, \ldots, f_{2m} = \partial_+ f_m ,$$

$$f_{2m+1} = \partial_+^2 f_1, \ldots, f_{3m} = \partial_+^2 f_m, \ldots, f_{m+i} = \partial_+ f_i, \ldots, f_N. \tag{2.18}$$

Assuming their linear independence, we orthonormalize these vectors $f_1, \ldots, f_N$ by the Gram-Schmidt procedure in this order and obtain an orthonormal basis of C^N,

$$e_1, e_2, \ldots, e_N, \quad e_i^\dagger \cdot e_k = \delta_{ik} . \tag{2.19}$$

By picking up *m consecutive orthonormal vectors*, we define the following N×m matrices,

$$X_{(j)} = (e_j, e_{j+1}, \ldots, e_{j+m-1}) , \quad 1 \leq j \leq N-m+1. \tag{2.20}$$

All of them *satisfy the equation of motion* (2.11) as well as the constraint (2.3). In particular, the first one $X_{(1)}$ is an instanton and the last $X_{(N-m+1)}$ is an anti-instanton. The series of solutions $X_{(j)}$ (j=1,\ldots,N-m+1) are all characterized by the input holomorphic functions $f_1, \ldots, f_m$ (2.17), which are completely arbitrary except for the linear independence. Since these solutions are closely related to each other, we expect that there should exist some discrete transformations which map one member of the series $X_{(j)}$ into another $X_{(j')}$. In fact such transformations do exist and moreover they can be defined as *discrete symmetry transformations (Σ transformations)*

mapping an arbitrary solution X into another. Needless to say,
these symmetry transformations also characterize the system dynami-
cally. The proof that (2.20) really satisfies the equation of
motion is quite elementary [1]. The same method with smaller number
of input holomorphic functions also gives a solution (degenerate
solution). For more details and other types of solutions we refer
to [1].

3. Supersymmetric Grassmannian Sigma Models

The supersymmetric (susy) version of the $G(N,m)$ model is expected to
provide an interesting example of integrable boson-fermion interactions with an internal non-abelian gauge symmetry. We might hope
that the models give some qualitative understanding of the strong
interactions of the quark-gluon system. In this section it is shown
that the classical solutions of the susy Grassmannian sigma models
can also be constructed in almost the same way as the bosonic case,
provided that the classical fermion solutions are treated as ordinary c-number fields [2]. The susy $G(N,m)$ model can be constructed
in a standard manner [8]. Let us denote by ϕ an N×m matrix of
superfields

$$\phi(x,\theta) = X(x) + i\theta_\alpha \chi_\alpha(x) + \frac{i}{2}\theta\gamma_5\theta F(x) \ , \tag{3.1}$$

in which θ_α ($\alpha=1,2$) is a two component Majorana spinor. The two
component spinor field χ_α is the susy partner of the scalar field X
and F is an auxiliary field. The action is defined as

$$S = \int d^2x d\theta\gamma_5 d\theta \operatorname{Tr}\{(\nabla\phi)^\dagger \gamma_5 (\nabla\phi)\}, \tag{3.2}$$

in which

$$\nabla\phi = \mathcal{D}\phi - \phi\phi^\dagger \mathcal{D}\phi, \quad \mathcal{D} = \frac{\partial}{\partial\theta} + i\gamma_\mu \partial_\mu \theta \ . \tag{3.3}$$

By integrating out the auxiliary field, we obtain

$$\begin{aligned}
S = \int d^2x \operatorname{Tr}\{ &2(D_+X)^\dagger(D_+X) + 2(D_-X)^\dagger(D_-X) \\
&- 4i(\Psi^{(-)\dagger}D_+\Psi^{(+)} + \Psi^{(+)\dagger}D_-\Psi^{(-)}) \\
&+ 4(\Psi^{(+)\dagger}\Psi^{(+)}\Psi^{(-)\dagger}\Psi^{(-)} - \Psi^{(+)\dagger}\Psi^{(-)}\Psi^{(-)\dagger}\Psi^{(+)})\} \ .
\end{aligned} \tag{3.4}$$

Here the physical fermion field Ψ and its chiral components $\Psi^{(+)}$,
$\Psi^{(-)}$ are defined

$$\Psi = (1-XX^\dagger)\chi, \quad \Psi = \binom{1}{-i}\Psi^{(+)} + \binom{1}{i}\Psi^{(-)} \ . \tag{3.5}$$

It obeys the constraint due to (3.5)

$$X^\dagger\Psi = 0 \ , \tag{3.6}$$

and its covariant derivative is defined by

$$D_\mu\Psi = \partial_\mu\Psi - \Psi X^\dagger \partial_\mu X \ . \tag{3.7}$$

The equations of motion read

$$\begin{aligned}
&D_+D_-X + X(D_-X)^\dagger(D_-X) + i(\Psi^{(+)}\Psi^{(-)\dagger}\partial_+X + \Psi^{(-)}\Psi^{(+)\dagger}\partial_-X) \\
&\qquad - i((D_+X)\Psi^{(-)\dagger}\Psi^{(+)} + (D_-X)\Psi^{(+)\dagger}\Psi^{(-)}) = 0 \ , \\
&(1-XX^\dagger)D_+\Psi^{(+)} + i(\Psi^{(-)}\Psi^{(+)\dagger}\Psi^{(+)} - \Psi^{(+)}\Psi^{(+)\dagger}\Psi^{(-)}) = 0 \ , \\
&(1-XX^\dagger)D_-\Psi^{(-)} + i(\Psi^{(+)}\Psi^{(-)\dagger}\Psi^{(-)} - \Psi^{(-)}\Psi^{(-)\dagger}\Psi^{(+)}) = 0 \ .
\end{aligned} \tag{3.8}$$

A simple class of solutions are constructed as follows. We pre-
pare *2m linearly and functionally independent holomorphic N-component*

vectors (depending on x_+ only)

$$f_1, f_2, \ldots, f_{2m} \, , \quad \partial_- f_i = 0 \, , \tag{3.9}$$

and define another set of vectors $f_{2m+1}, \ldots, f_N$ by differentiation with respect to x_+,

$$f_{2m+1} = \partial_+ f_1, \ldots, f_{3m} = \partial_+ f_m, \ldots, f_{4m} = \partial_+ f_{2m} \, ,$$
$$\ldots, f_{2m+i} = \partial_+ f_i, \ldots, f_N. \tag{3.10}$$

The input holomorphic functions (3.10) are completely arbitrary. For each set of functions we obtain a series of solutions as shown below. Supposing that the vectors $f_1, \ldots, f_N$ are linearly independent, we orthonormalize them in this order by the Gram-Schmidt procedure and obtain a basis of C^N. From these orthonormal vectors, we pick up 3m consecutive vectors, characterized by an integer j, and group them into the following three N×m matrices

$$X_{(-1)} = (e_{j-m}, \ldots, e_{j-1}), \; X_{(o)} = (e_j, \ldots, e_{j+m-1}),$$
$$X_{(1)} = (e_{j+m}, \ldots e_{j+2m-1}). \tag{3.11}$$

A solution of the coupled equation (3.8) characterized by a set of functions (3.9) and by an integer j is given by

$$X = X_{(o)},$$
$$\psi^{(+)} = c_+ X_{(-1)} (a^\dagger_{(-1)})^{-1} a^\dagger_{(o)}, \quad \psi^{(-)} = c_- X_{(1)} a_{(1)} a_{(o)}^{-1}, \tag{3.12}$$

in which $c_\pm$ are arbitrary constants satisfying $c_+ \overline{c}_- = -i$. The nonsingular m×m matrices $a_{(-1)}$, $a_{(o)}$ and $a_{(1)}$ will be defined shortly. Let us first define the following N×N projection matrices

$$P_{(K)} = X_{(K)} X^\dagger_{(K)}, \; K = -1, 0, 1, \text{ and } Q = \sum_{k=1}^{j-m-1} e_k e_k^\dagger, \tag{3.13}$$

which are orthogonal to each other by construction. Next we introduce the following three N×m matrices

$$F_{(-1)} = (f_{j-m}, \ldots, f_{j-1}), \; F_{(o)} = (f_j, \ldots, f_{j+m-1}),$$
$$F_{(1)} = (f_{j+m}, \ldots f_{j+2m-1}). \tag{3.14}$$

Here $F_{(-1)}$ and $F_{(1)}$ are functionally related by definitions of the vectors $f_1, \ldots, f_N$; $F_{(1)} = \partial_+ F_{(-1)}$. Each holomorphic vector f_k in (3.9) and (3.10) is spanned by $e_1, e_2, \ldots, e_k$, because of the Gram-Schmidt procedure, thus $F_{(K)}$ and $X_{(K)}$ are related as

$$F_{(-1)} = X_{(-1)} a_{(-1)} + Q b_{(-1)}, \; F_{(o)} = X_{(o)} a_{(o)} + (Q + P_{(-1)}) b_{(o)},$$
$$F_{(1)} = X_{(1)} a_{(1)} + (Q + P_{(-1)} + P_{(o)}) b_{(1)} \, , \tag{3.15}$$

in which $a_{(K)}$ and $b_{(K)}$ are m×m and N×m matrices, respectively. In other words, the m×m matrix $a_{(K)}$ is the transformation matrix between the holomorphic vectors and the orthonormal basis, therefore it is nonsingular. For more details and other types of solutions we refer to [2].

A related problem of interest, i.e., the linearized Dirac equation with a background bosonic solution $X = X_{(j)}$ (2.20)

$$(1 - XX^\dagger) D_+ \psi^{(+)} = 0, \quad (1 - XX^\dagger) D_- \psi^{(-)} = 0 \, , \tag{3.16}$$

can be solved in a similar way [2]. This is related to the Atiyah-
Singer index theorem [9].

4. Bäcklund Transformations

It is well known that the equation of the G(N,m) model (2.8) gives
the integrability condition for the following set of linear equations

$$\partial_+ \psi(x;\lambda) = 2t[\partial_+ P,P]t\psi(x;\lambda)/1+\lambda \ ,$$

$$\partial_- \psi(x;\lambda) = 2t[\partial_- P,P]t\psi(x;\lambda)/1-\lambda, \tag{4.1}$$

in which ψ is an $N \times N$ matrix function of x and the spectral para-
meter λ and $t = \begin{pmatrix} I_n & 0 \\ 0 & -I_m \end{pmatrix}$. It is interesting to note that for the
solution P given in section 2 the above linear scattering problem
can also be solved explicitly [3], [10]. To be more specific, if
we take as a solution

$$P = P_{(j)} = X_{(j)}X_{(j)}^\dagger, \tag{4.2}$$

then

$$\psi = tU, \quad U(x;\lambda) = 1 + \frac{2}{\lambda-1}P + \frac{4\lambda}{(\lambda-1)^2}Q \ ,$$

$$Q = \sum_{k=1}^{j-1} e_k e_k^\dagger \ , \tag{4.3}$$

solves (4.1). A general solution of (4.1) is obtained from (4.3) by
multiplying an arbitrary x-independent (λ-dependent) matrix from the
right. Since we have a general enough class of starting solutions
and the solutions for the corresponding linear scattering problems,
an infinite set of new solutions can be generated by means of the
multi-Bäcklund transformations of HARNAD et al. [11]. The transformation involves only the linear algebraic procedure, namely, the
linearization of the G(N,m) model is achieved. However, in the process of the Bäcklund transformation (B. T. for short) one has to
invert a big matrix function, at least $K \times K$ for a K-ple B. T., which
is highly nontrivial. In [3] **we obtained the explicit forms of**
the Bäcklund transformations for the G(N,m) model by restricting
ourselves to the simplest case of the B. T.. A particularly simple
interpretation of the B. T. for the CP^{N-1} model is obtained.

References

1. R. Sasaki, Phys. Lett. 130B (1983) 69, Zeit. f. Phys. C, 24
    (1984) in press.
2. K. Fujii, T. Koikawa and R. Sasaki, Prog. Theor. Phys. 71 (1984)
    338, K. Fujii and R. Sasaki, Prog. Theor. Phys. 71 (1984) 831.
3. R. Sasaki, RITP preprint RRK 84-14, July 1984.
4. H. Eichenherr and M. Forger, Nucl. Phys. B155 (1979) 381.
5. H. Eichenherr, Nucl. Phys. B146 (1978) 215.
6. A. D'Adda, P. di Vecchia and M. Lüscher, Nucl. Phys. B146 (1978)
    63. A. M. Din and W. J. Zakrzewski, Nucl. Phys. B174 (1980)
    397.
7. A. J. MacFarlane, Phys. Lett. 82B (1979) 239.
8. A. D'Adda, P. di Vecchia and M. Lüscher, Nucl. Phys. B152 (1979)
    125.
9. A. M. Din and W. J. Zakrzewski, Phys. Lett. 101B (1981) 166.
10. A.M. Din and W. J. Zakrzewski, Nucl. Phys. B237 (1984) 461.
11. J. Harnad, Y. Saint-Aubin and S. Snider, Comm. Math. Phys. 92
    (1984) 329.

Einstein Equations, Non-Linear Sigma Models and Self-Dual Yang-Mills Theory

Norma Sanchez

ER 176 CNRS - Département d'Astrophysique Fondamentale, Observatoire de Meudon
F-92195 Meudon Principal Cedex, France

We analyze the connection between non-linear sigma models, self-dual Yang-Mills
theory and General Relativity (Self-dual and non self-dual, with and without
Killing vectors), both at the level of the equations and at the level of the different type of solutions (solitons, instantons and calorons) of these theories. Until
now no systematic procedure has been found to solve the highly non-linear four
dimensional Einstein equations even in the presence of one Killing vector field
("three-dimensional reduced" Gravity). Only the two Killing vector case (two
dimensional reduced Gravity) is known to admit a systematic resolution, (although
the general solution is not yet known). The complete integrability of the Einstein
equations without any Killing vector is still an open problem. In gravity, examples
of **solitons** are provided by **black-holes**, which are the analogues of electric types
monopoles (which may rotate) and **Taub-NUT** metrics which may be thought of as magnetic type gravitational monopoles. In the Euclidean (imaginary time $t = i\tau$) regime
these solutions exist as gravitational instantons (complete non singular solutions
of the Einstein equations with (++++) signature). Black-holes have finite tempe-
rature ($T = \beta^{-1}$, $0 \leq \tau \leq \beta$), which can be expressed [1] as the differential winding
number of the mapping $\gamma_k + i\,\tau_k = f(\gamma^* + i\tau)$ defining the maximal analytic
extension of the manifold, namely

$$T = \frac{1}{2\pi} \frac{\partial}{\partial \tau} \text{Im Log } f(\gamma^* + i\tau).$$

Here (γ_k, τ_k) are Kruskal type coordinates and (γ^*, τ) are of Schwarzschild
type. For Black-holes, $f(\gamma^* \pm i\tau) = \exp 2\pi\beta^{-1}(\gamma^* \pm i\tau)$ and so $T = \beta^{-1}$ where

$$\beta(4\pi)^{-1} = \begin{cases} 2M & \text{for Schwarzschild} \\ M + M^2 (M^2 + 1^2 + Q^2)^{-1/2} & \text{for Kerr-Newmann} \end{cases}$$

The euclidean action $I = \beta^2(4\pi)^{-1}$ is finite, interpreted as the intrinsic entropy
of the solutions and related to its Euler number. The Riemann curvature is non-
self-dual. A different class of finite temperature metrics for which the (periodic)
imaginary time coordinate is not cyclic is given in [2]. Such metrics are the
gravitational analogues of the "caloron" solutions of the Yang-Mills field. The
caloron generalization of the multi Taub-Nut metric is also given.

I - FIELD EQUATIONS

O(2,1) non-linear sigma model :

The Lagrangian is defined by

$$L_\sigma = \frac{1}{2} (\partial_\mu \sigma^a)(\partial^\mu \sigma^a) - \frac{1}{2} \Lambda [(\sigma^a)^2 - 1],$$

Λ being a Lagrangian multiplier. The σ field $\sigma = (\sigma^1, \sigma^2, \sigma^3)$ is constrained
to satisfy $(\sigma^1)^2 + (\sigma^2)^2 - (\sigma^3)^2 = 1$. The equation of motion is given by

$$\Box^2 \sigma^a = (\sigma^b \Box^2 \sigma^b) \sigma^a \tag{1}$$

By defining

$$V = \frac{1}{\sigma^1 + \sigma^3} \, , \qquad\qquad \phi = \frac{i\sigma^2}{\sigma^1 + \sigma^3} \, , \qquad\qquad (2)$$

The Lagrangian reads

$$L_{\sigma 0(2,1)} = \frac{1}{V^2} \, [(\nabla V)^2 - (\nabla \phi)^2]$$

and the corresponding equation of motion can be written as

$$\nabla^2 \varepsilon - \frac{1}{(Re\,\varepsilon)} \, (\nabla \varepsilon)^2 = 0 \qquad\qquad (3)$$

where $\varepsilon = V + i\,\phi$

Non Self-dual Einstein equations without Killing vectors :

In order to write the "R gauge" parametrisation and to recognize some kind of duality, namely the "magnetic" and "electric" aspects of the gravitational field, it is convenient to perform a 3+1 decomposition of the Einstein eqs.

$$R_{\mu\nu} = 0 \qquad\qquad (4)$$

which is invariant under the transformation

$$\begin{cases} \tilde{X}^o = \tilde{X}^o \, (X^o, \, X^i) \\ \tilde{X}^i = \tilde{X}^i \, (X^i) \end{cases} \qquad (i = 1, \, 2, \, 3) \qquad\qquad (5)$$

i.e. it is chronometric invariant. We introduce the notation

$$V = g_{oo}; \qquad \omega_j = g_{oj}\big/g_{oo}; \qquad h_{ij} = g_{ij} - g_{oi}\,g_{oj}\big/g_{oo}$$

Then,

$$dS^2 = V(dx^o + \omega_j dx^j)^2 - h_{ij}\, dx^i\, dx^j$$

and eqs (4) can be written as [3]

$$A_j A^j + {}^*\nabla_j E_j - E_j E^j + 2{}^*\nabla_j \left(\frac{A^j}{2V}\right) + {}^*\partial_o D + D_{j1} D = 0 \qquad\qquad (6)$$

$${}^{(3)}R_{ik} \quad 3A_{ij}A^j{}_k + E_i E_k + \frac{1}{2}({}^*\nabla_i E_k + {}^*\nabla_k E_i)$$

$$- \, (D_{ij} + A_{ij})(D_k{}^j + A_k{}^j) + DD_{ik} - D_{ij}\, D_k{}^j + {}^*\partial_o D_{ik} = 0 \qquad\qquad (7)$$

where

$$A_{ik} = \frac{1}{2}\,\sqrt{V} \quad ({}^*\partial_k \omega_i - {}^*\partial_i \omega_k)$$

$$E_i = -{}^*\partial_i \ln \sqrt{V} + (1/\sqrt{V})\, {}^*\partial_o (V\omega_i)$$

$$\begin{aligned} {}^*\partial_o &= (1/\sqrt{V})\,\partial_o & [{}^*\partial_i, \, {}^*\partial_o] &= E_i\, {}^*\partial_o \\ & & \text{such that} & \\ {}^*\partial_i &= \partial_i - \omega_i\, \partial_o & [{}^*\partial_i, \, {}^*\partial_k] &- 2\Lambda_{ik}\, {}^*\partial_o{}^o \end{aligned}$$

A_{ik} and E_k reflects the magnetic and electric parts of the gravitational field respectively, ${}^{(3)}R_{ik}$ is the three-dimensional Ricci tensor with respect to hij and

$$D_{ik} = \frac{1}{2}\,{}^*\partial_o\, hik, \qquad\qquad D = {}^*\partial_o \ln \sqrt{\|h\|}$$

It is possible to define the potentials ϕ and $({}^*\partial_o \rho_k)$ such that

$$A_j = \frac{1}{2V} \; (*\nabla_j \phi + *\partial_o \rho_j)$$

where

$$*\nabla_j \; (h^{ij}D - D^{ij}) = e^{ijk} \, *\nabla_j \; (*\partial_o \rho_k)$$

Then, in terms of the metric $\gamma_{ij} = V \, h_{ij}$, eq (6) can be written as

$$*\nabla^2 \; (V + i\phi) - \frac{1}{V} \, [*\nabla \; (V + i\phi)]^2 + *\nabla_k \; (*\partial_o \rho_k)$$

$$+ \frac{1}{2V^2} \, [(*\partial_o \rho_k)^2 + (*\partial_o \rho_k)(*\partial_k \phi) + \frac{1}{2} D_k \, D^k + \frac{1}{4} D_{ik} \, D^{ik} = 0 \qquad (8)$$

This generalizes the Ernst equation to the case where there are no symmetries in the space-time. Besides the potentials (V, ϕ) this equation involves other fields $(*\partial_o \rho_k)$ and D_{ik} which vanishes when stationarity is required.

With one Killing vector :

In this case, the transformation (5) is given by

$$\begin{cases} \tilde{X}^o = X^o + \alpha(X^j) \\[1mm] \tilde{X}^j = X^j \end{cases}$$

and eqs (6) and (7) take the form

$$2 \, A_j \, A^j + \nabla_j \, E_j - E_j \, E^j + 2 \, \nabla_j \, \frac{A_j}{2V} = 0$$

$$^{(3)}R_{ik} = 2 \, A_{ij} \, A_k^{\;j} + E_i \, E_k + \frac{1}{2} \, (\nabla_i \, E_k + \nabla_k \, E_i) = 0$$

where

$$A_{ik} = \frac{1}{2} \, \sqrt{V} \, (\partial_k \, \omega_i - \partial_i \, \omega_k) = \frac{1}{2V} \, e_{ikj} \, \nabla^j \phi$$

$$E_i = - \, \partial_i \, \ln \sqrt{V}$$

eq (8) reduces to

$$\nabla^2 \; (V + i\,\phi) - \frac{1}{V} \, [\nabla \; (V + i\phi)]^2 = 0$$

$$\nabla^2 \varepsilon - \frac{1}{(\mathrm{Re}\,\varepsilon)} \, (\nabla\varepsilon)^2 = 0 \qquad (9)$$

In this case,

$$^{(3)}R_{ik} = \frac{1}{2V^2} \, [\nabla_i \, V \, \nabla_k \, V + \nabla_i \phi \; \nabla_k \phi]$$

$$(\nabla \times \omega)_k = \frac{\nabla_k \phi}{V^2} \qquad (10)$$

With two Killing vectors :

The metric $g_{\mu\nu}$ can be written in the canonical form as

$$ds^2 = \lambda \, V \, (dX^1 + \omega_2 \, dX^2)^2 + \frac{\gamma_{ij}}{V} \, dX^i \, dX^j$$

where $\gamma_{ij} dX^i dX^j = e^{2\gamma} [(dX^3)^2 - \lambda(dX^4)^2 + S^2(dX^2)^2]$

$\lambda = \pm 1$. V, ω and S depend only on X^3 and X^4.

In this case eqs (9) and (10) read

$$[\lambda\partial_3^2 - \partial_4^2 + \lambda(\frac{\partial_3 S}{S})\partial_3 - (\frac{\partial_4 S}{S})\partial_4]\epsilon - \frac{1}{(Re\epsilon)}[\lambda(\partial_3\epsilon)^2 - (\partial_4\epsilon)^2] = 0$$

$$^{(3)}R_{33} = \lambda\partial_4^2\gamma - \partial_3^2\gamma + \frac{\partial_3 S}{S}\partial_3\gamma + \lambda\frac{\partial_4 S}{S}\partial_4\gamma - \frac{\partial_3^2 S}{S}$$

$$^{(3)}R_{34} = \frac{\partial_3 S}{S}\partial_4\gamma + \frac{\partial_4 S}{S}\partial_3\gamma - \frac{\partial_{3,4}^2 S}{S}$$

$$^{(3)}R_{44} = \lambda\partial_3^2\gamma - \partial_4^2\gamma + \frac{\partial_4 S}{S}\partial_4\gamma + \lambda\frac{\partial_3 S}{S}\partial_3\gamma - \frac{\partial_4^2 S}{S}$$

$$^{(3)}R_{2j} = 0 \qquad (j = 2, 3, 4) \quad\Longrightarrow\quad \partial_3^2 S - \lambda\partial_4^2 S = 0$$

These eqs must be compatible with

$$^{(3)}R_{33} = \frac{1}{2V^2}[(\nabla_3 V)^2 + (\nabla_3\phi)^2]$$

$$^{(3)}R_{34} = \frac{1}{2V^2}[\nabla_3 V \nabla_4 V + \nabla_3\phi \nabla_4\phi]$$

$$^{(3)}R_{44} = \frac{1}{2V^2}[(\nabla_4 V)^2 + (\nabla_4\phi)^2]$$

Self-dual Yang-Mills and Einstein equations without Killing vectors :

In complex space $(y, \bar{y}, z, \bar{z})$ the (anti) self-dual Einstein eqs.

$$R_{\mu\nu\alpha\beta} = -\frac{1}{2}\sqrt{g}\,\epsilon_{\mu\nu\alpha\beta}R^{\gamma\sigma}{}_{\alpha\beta} \tag{11}$$

can be written as [4]

$$g^{y\bar{y}}\partial_{\bar{y}}(g^{-1}\partial_y g) + g^{z\bar{z}}\partial_{\bar{z}}(g^{-1}\partial_z g) = 0 \qquad \text{with} \tag{12}$$

$$\partial_\mu g_{\alpha\bar{\beta}} = \partial_\alpha g_{\mu\bar{\beta}} , \qquad c.c. \tag{13}$$

$$g_{\mu\nu} = \begin{pmatrix} 0 & g_{\bar{\alpha}\beta} \\ \hline g_{\bar{\beta}\alpha} & 0 \end{pmatrix} \qquad \text{is the (complex) metric tensor}$$

Here 0 is the 2x2 null matrix ; $g = \begin{pmatrix} g_{\bar{y}y} & g_{\bar{y}z} \\ g_{\bar{z}y} & g_{\bar{z}z} \end{pmatrix}$

and det g = 1.

For Hermitian g, eq.(12) is that of a (modified) self-dual Yang-Mills theory with the additional condition (13). This condition defines a Kähler type metric. In 2^n dimensions, O is the n x n null and g is n x n. Even if eq. (11) is meaningless for $n \neq 2$, g provides a Kähler type metric with vanishing Ricci tensor in 2^n dimensions and the corresponding eqs are those of a (modified) SU(n) theory with the additional condition (13). In order to also allow complex solutions we consider in general $g^t \neq \bar{g}$ and parametrise g in terms of three complex functions (V, π, C) namely

$$g = \begin{pmatrix} V^{-1} & C V^{-1} \\ V^{-1} \pi & V + \pi C V^{-1} \end{pmatrix}$$

This corresponds to the triangular gauge $g = r\, e$

$$r = (\sqrt{V})^{-1} \begin{pmatrix} 1 & 0 \\ \pi & V \end{pmatrix} , \quad e = (\sqrt{V})^{-1} \begin{pmatrix} 1 & C \\ 0 & V \end{pmatrix}$$

Hence the length element $dS^2 = g_{\alpha\bar{\beta}}\, dX^\alpha\, dX^{\bar{\beta}}$ takes the form

$$dS^2 = V^{-1}\, dy\, d\bar{y} + (V + \pi C V^{-1})\, dz\, d\bar{z} + C V^{-1}\, dy\, d\bar{z} + \pi V^{-1}\, d\bar{y}\, dz$$

Equations (12), (13) can be expressed in terms of the complex functions

$$\varepsilon_1 = V + i\pi , \qquad \varepsilon_2 = V - i\pi , \qquad \varepsilon_3 = V + i C , \qquad \varepsilon_4 = V - i C$$

If $C = \bar{\pi}$ and V is real (i.e. $\varepsilon_1 = \bar{\varepsilon}_4$, $\varepsilon_2 = \bar{\varepsilon}_3$) dS^2 is real with (++++) signature and eqs (12) are generalized Yang's type equations. In this case

$$e = (\sqrt{V})^{-1} \begin{pmatrix} 1 & \bar{\pi} \\ 0 & V \end{pmatrix} = r^+ .$$ This can be considered as the "R gauge parametrisation" of the self-dual gravitational field.

If $C = \pi$ and V are real (i.e. $\varepsilon_1 = \varepsilon_3 = \bar{\varepsilon}_2 = \bar{\varepsilon}_4 = \varepsilon$), eqs (12), (13) reduce to

$$(\varepsilon\bar{\varepsilon})\, [(\mathrm{Re}\,\varepsilon)\, \partial_y \partial_{\bar{y}}\varepsilon - \partial_y\varepsilon\, \partial_{\bar{y}}\varepsilon] + [(\mathrm{Re}\,\varepsilon)\, \partial_z \partial_{\bar{z}}\varepsilon - \partial_z\varepsilon\, \partial_{\bar{z}}\varepsilon] = 0$$

$$(\varepsilon\bar{\varepsilon})\, [\partial_y\varepsilon\, \partial_{\bar{y}}\bar{\varepsilon} - \partial_{\bar{y}}\varepsilon\, \partial_y\bar{\varepsilon}] + [\partial_z\varepsilon\, \partial_{\bar{z}}\bar{\varepsilon} - \partial_{\bar{z}}\varepsilon\, \partial_z\bar{\varepsilon}] = 0 \quad \text{and} \tag{14}$$

$$\varepsilon\, \partial_y \bar{\varepsilon} = -i\, \partial_z \bar{\varepsilon}$$

$$\bar{\varepsilon}\, \partial_y \varepsilon = i\, \partial_z \varepsilon \qquad\qquad , \quad \text{c.c.} \tag{15}$$

Eq. (14) is a (modified) Ernst equation. Note that here this is a four dimensional equation. We do not assume the existence of any Killing vector field.

II - SOLUTIONS

Sigma instantons :

Holomorphic (anti-holomorphic) mappings provide self-dual solutions for both O(3) and O(2,1) sigma models in the sense that for these solutions the sigma field $(\sigma^1, \sigma^2, \sigma^3)$ satisfies

$$\partial_\mu \sigma^\alpha - \varepsilon_{\mu\nu}\, \varepsilon_{\alpha\beta\gamma}\, \sigma^\beta\, \partial_\nu \sigma^\gamma = 0 \tag{16}$$

This condition defines the σ configurations with lowest euclidean action in each homotopic class. By the change of parametrisation given by eq. (2), condition (16) takes the form of the Cauchy-Riemann equations

$$\partial_{X_1} V = \partial_{X_2} \Phi \quad , \quad \partial_{X_2} V = - \partial_{X_1} \Phi ,$$

whose general solution is given by

$$V + i \, \phi = \epsilon \, (X_1 + i \, X_2) \equiv \epsilon \, (y)$$

The choice

$$\epsilon = (y - a_0) \ldots (y - a_q)/(y - b_0) \ldots (y - b_q) \qquad (17)$$

gives (multi)-instanton solutions ; $a_0, \ldots, a_q$ being the positions where the σ instantons are centered. The σ instantons are defined in two-dimensional euclidean space .

Solutions of the Einstein eqs without Killing vectors :

We find solutions of the form

$$dS^2 = V dz^2 + 2 \, dy \, d\bar{y} + 2 \, dz d\bar{z} + G dy^2 \qquad (18)$$

where

$$\sqrt{2} \, y = X_1 + i \, X_2 , \qquad \sqrt{2} \, z = X_3 + i \, X_4$$
$$\sqrt{2} \, \bar{y} = X_1 - i \, X_2 , \qquad \sqrt{2} \, \bar{z} = X_3 - i \, X_4 \qquad \text{and}$$

$$\left\{ \begin{array}{l} V(y,\bar{y},z,\bar{z}) = A_1(y,z)\bar{z} + A_2(y,\bar{y},z) \\ G(y,\bar{y},z,\bar{z}) = B_1(y,z)\bar{y} + B_2(y,z,\bar{z}) \end{array} \right.$$

satisfy

$$\partial_y A_1 + \partial_z B_1 - (\partial_{\bar{y}} B_2) = 0$$
$$- \partial_z \partial_{\bar{z}} B_2 + \frac{1}{2} A_1 \partial_{\bar{z}} B_2 + \frac{1}{2} (\bar{z} A_1 + A_2) \partial_{\bar{z}}^2 B_2 = 0$$
$$- \partial_y \partial_{\bar{y}} A_2 + \frac{1}{2} B_1 \partial_{\bar{y}} A_2 + \frac{1}{2} (\bar{y} B_1 + B_2) \partial_{\bar{y}}^2 A_2 = 0$$

The ansatz ($\partial_{\bar{z}} B_2 = 0$) gives

$$A_1(y,z) = \int^y d\eta \, \partial_z B_1 \, (\eta, z) + h(z)$$
$$A_2(y,\bar{y},z) = K_1 \exp \{ \mathbf{C} \, \bar{y} \, b \, (y,z) + \int^y d\eta \, B_2 \, (\eta,z) \, b(\eta,z) + K_2 \}$$

Here h is an arbitrary function of z, $\mathbf{C}$, K_1, K_2 are arbitrary constants and

$$b(y,z) = \exp \int^y \, d\xi \, B_1 \, (\xi,z)$$

By giving B_1, B_2 (arbitrary functions on y, z) one obtains in this way A_1 and A_2. The same type of solution is obtained if we start with the ansatz ($\partial_{\bar{y}} A_2 = 0$), because of the symmetry of the metric under the transformation

$$\{ y \leftrightarrow z , \bar{y} \leftrightarrow \bar{z} , V \leftrightarrow G \}.$$

On the other hand, the ansatz

$$\left\{ \begin{array}{l} (\partial_{\bar{y}} A_2) \equiv \text{independent of } \bar{y} \\ (\partial_{\bar{z}} B_2) \equiv \text{independent of } \bar{z} \end{array} \right.$$

gives

$$\left\{ \begin{array}{l} V = A_1(y,z) \, \bar{z} + a_1(y,z)y + \tilde{a}_1(y,z) \\ G = B_1(y,z) \, \bar{y} + b_1(y,z)z + \tilde{b}_1(y,z) \end{array} \right.$$

$$A_1(y,z) = 2 \, \partial_z \, \ell n \, F(y,z) - \int^y d\eta \, \partial_z B_1 \, (\eta ,z) + f(z)$$
$$b_1(y,z) = F(y,z) f(z) \exp [- \int^y d\eta \, B_1 \, (\eta ,z)]$$
$$a_1(y,z) = f(z)^{-1} \exp [\int^y d\eta \, B_1 (\eta ,z)]$$

where F(y,z) is the solution of the (complex) Liouville equation

$$\partial_y \, \partial_z \, \ell n \, F(y,z) = F(y,z),$$

i.e. $\quad F(y,z) = \dfrac{2C'(y)\, D'(z)}{[C(y) + D(z)]}$

This determines A_1, a_1, b_1, by giving an arbitrary function $B_1(y,z)$; $f(z)$, $\tilde{a}_1$ and $\tilde{b}_1$ are arbitrary functions. These solutions do not exhibit any space time symmetry. The Killing equations $\quad \xi_{i;j} + \xi_{j;i} = 0$ give $(\, \xi^y, \, \xi^{\bar{y}}, \, \xi^z, \, \xi^{\bar{z}}) = 0$.

These metrics have **non- self dual Riemann curvature** and **(++++) signature**. For a discussion of the properties see ref [5].
A different class of **non-self dual** metrics without Killing vectors is given by [4]

$$dS^2 = V^{-1} \, dy \, d\bar{y} - V \, \pi^{-1} \, dy \, d\bar{z} + \pi \, V^{-1} \, d\bar{y} \, dz \tag{19}$$

with
$$\begin{cases} V = P(y, \, \bar{y}, \, \bar{z}) \, \pi \\ \pi^{-1} = -\, z \, \partial_y \, [\ln \, P(y, \, \bar{y}, \, \bar{z}) + Q(y, \, \bar{y}, \, \bar{z})] \end{cases}$$

P and Q being arbitrary functions of the complex variables indicated.

With one Killing vector :

If $G = 0$ in eq (18), we obtain the solution
$$dS^2 = Vdz^2 + 2 \, dy \, d\bar{y} + 2 \, dz \, d\bar{z} \tag{20}$$

where $V = V(y, \, \bar{y}, \, z)$ satisfies
$$\partial_y \, \partial_{\bar{y}} \, V(y, \bar{y}, z) = 0, \text{ i.e.}$$
$$V = \varepsilon_1 \, (y, \, z) + \varepsilon_2 \, (\bar{y}, \, z)$$

This metric has a null Killing vector (∂_z).
The Wick rotation $\tau = $ it allows us to obtain a subclass of real solutions with (-+++) signature. In particular, the choice

$$\varepsilon_2 \, (\bar{y}, \, X_3 - t) = \bar{\varepsilon}_1 \, (\bar{y}, \, X_3 - t)$$

gives the so called pp wave solutions : plane fronted gravitational waves with parallel rays. (Petrov N type). These solutions can be generalized to be solutions of the Einstein-Maxwell equations by taking

$$V = \varepsilon_1 \, (y, \, X_3 - t) + \bar{\varepsilon}_1 (\bar{y}, \, X_3 - t) + \mathscr{H}_o \, F(y, \, X_3 - t) \, \bar{F}(\bar{y}, \, X_3 - t)$$

where $F \, (\bar{F})$ are holomorphic functions in $y(\bar{y})$ and depend arbitrarily on ($X_3 - t$). For instance $\varepsilon_1 = A(X_3 - t) \, y^2$, $F = \sqrt{B(X_3 - t)} \, y$ with $A(X_3 - t) = $ const. $A(X_3 - t)$ and B real gives linearly polarised pp waves.

On the other hand, the ansatz $- \, C = V = \pi$ satisfy both eqs (12) and (13) for one (null) Killing vector (∂_z). This gives the **self-dual metric**

$$dS^2 = V^{-1} \, dy \, d\bar{y} - d\bar{y} \, dz - dy \, d\bar{z},$$

$V^{-1} = \varepsilon_1 \, (Y, \, \bar{Z}) + \varepsilon_2 \, (\bar{Y}, \, \bar{Z})$. This solution can be obtained from eq (19) by assuming $P = $ constant and one (null) Killing vector (∂_z). These solutions can be generalized to include finite temperature metrics which satisfy

$V(\vec{x}, \, \tau) = V \, (\vec{x}, \, \tau + \beta)$ where $(\vec{x}, \, \tau) \equiv (x_1, \, x_2, \, x_3, \, \tau)$ and $\beta^{-1} = kT$ is the temperature of the solutions.

With two Killing vectors :

For V independent of ($X_3 - t$), the metric (20) gives the solution of the Einstein eqs associated to the holomorphic solutions of the σ model. For this metric

$$\gamma_{ij} \, dX^i \, dX^j = [\, \frac{\varepsilon_1(y) + \varepsilon_2(\bar{y})}{2} \,] \, dy \, d\bar{y} + (dX^3)^2$$

$$\omega = 1/V$$

If $\varepsilon_1 = \bar{\varepsilon}_2 = \varepsilon$ is given by eq (17) this is the metric associated to the multi-instantons of the σ model. This metric has non-self dual curvature and (-+++)

signature (it cannot be defined as a gravitational instanton). This metric describes non-axially symmetric stationary fields. (The Pohlmeyer's reduction of the
σ model to the Sine-Gordon or to the Liouville equation does not hold for these
solutions [3]. Solutions describing axially symmetric stationary fields, colliding
plane waves or cylindrical waves are associated to solutions of the σ model not
given by holomorphic functions and the Pohlmeyer's reduction to the Sine-Gordon
or to the Liouville equations holds for them. (i.e. they can be parametrised by
Painlevé transcendents or Liouville type solutions).

On the other hand, if $\varepsilon_1 = 0$ or $\varepsilon_2 = 0$, then $V = \pm i\Phi$, γ_{ij} is flat and
$\nabla \times \omega = (1/V)$ (which implies in this case $\nabla^2(1/V) = 0$). This gives a metric
of Taub-NUT type with **self dual curvature**.

References

1 - N. Sánchez, in Proc. of the 2nd Marcel Grossmann Meeting, North Holland 1982,
    pp 501-518.

2 - N. Sánchez, Phys. Lett $\underline{125B}$, 403, (1983).

3 - N. Sánchez, Phys. Rev. $\underline{26D}$, 2589 (1982).

4 - N. Sánchez, Phys. Lett. $\underline{B}$, (1984) to appear.
    See also Proc. of the "XII Internation. Conference on Differential Geometric
    Methods in Physics", Lect. Notes in Mathematics, Springer Verlag (1984, to
    appear).

5 - N. Sánchez, Phys. Lett. $\underline{94A}$, 125, (1983).

Part III

Solitons in Plasma Physics and Hydrodynamics

Soliton Resonance in Plasmas

Nobuo Yajima

Research Institute for Applied Mechanics, Kyushu University, Kasuga 816, Japan

Summary report is presented on soliton processes in real plasma physics. In particular, the resonance phenomena in two-dimensional interaction of ion-acoustic solitons and the nonlinearly coupled explosive mode between slow and ion-acoustic waves in an ion-beam plasma are discussed.

1. Introduction

For nearly two decades, various studies on "solitons" have been made in the vast field of natural sciences, from pure mathematics to biological physics. These studies have revealed the characteristics of solitons, that is the dependence of shape and velocity of solitons on their amplitude, preservation of identities of solitons through their collisions, and formation of solitons from initial disturbances. As a result, the solitons are now regarded as nonlinear normal modes which play an essential role in the evolution of nonlinear dispersive systems. However, it remains yet to be studied whether solitons play a similar role in real and non-integrable systems as they do in integrable but idealized systems. In this review, we discuss some of the recent progress in soliton phenomena which are expected to occur in multi-dimensional systems and in multi-mode systems.

Most solitons are one-dimensional entities which maintain their stability by balancing nonlinear steepening with wave dispersion. What happens when such one-dimensional solitons interact two-dimensionally with each other? In some cases, their interaction is so strong that the formation of a triple soliton structure is observed, where a new soliton branches out from the point at which the original solitons meet. This is called the resonance interaction of solitons. In §2 a brief review on resonance of ion-acoustic solitons is made.

Resonant interaction of solitons is also possible in a system with one spatial dimension. This example is given in §3, in which low frequency waves in a plasma with an ion beam are considered. Such a system possesses three kinds of solitons corresponding to the fast, slow and ion-acoustic mode. In this case, however, the resonant interaction between them is not stable enough to grow into the nonlinear explosion mode. The localization of disturbance is still kept, though the coupled system of solitons blows up.

144

2. Resonant Interaction of Solitons

The one-dimensional motion of ion-acoustic waves of small amplitude and long wavelength is governed by the Korteweg-de Vries (KdV) equation /1/, which is expressed in a nondimensional form as

$$\frac{\partial u}{\partial t} + \frac{\partial u}{\partial x} + u \frac{\partial u}{\partial x} + \frac{1}{2} \frac{\partial^3 u}{\partial x^2} = 0 \tag{1}$$

where u is the dimensionless electrostatic potential normalized by T/e, T is electron temperature, -e is the electron charge, x is normalized by the Debye length, and t is multiplied by the ion plasma frequency. Equation (1) has a soliton solution

$$u = 6K^2 \mathrm{sech}^2(Kx - \Omega t) \quad \text{with} \tag{2}$$

$$\Omega = K(1 + 2K^2). \tag{3}$$

In isotropic media, solitons can propagate in every direction , and they interact obliquely with each other. A typical example of oblique interaction of two solitons was photographed, off the Oregon coast, by T. Toedtemeier /2/. In the photograph, two wedges made by interacting soliton crests are connected with an intermediate wave crest. Such interactions can be studied by using the Kadomtsev-Petviashvili (KP) equation /3/,

$$\frac{\partial}{\partial x} \left[\frac{\partial v}{\partial t} + \frac{\partial v}{\partial x} + v \frac{\partial v}{\partial x} + \frac{1}{2} \frac{\partial^3 v}{\partial x^3} \right] + \frac{1}{2} \frac{\partial^2 v}{\partial y^2} = 0 \ . \tag{4}$$

This is an extension of the KdV equation to a weakly two-dimensional case. A one soliton solution of (4) takes the form

$$v = 6k^2 \mathrm{sech}^2(Kx + Ly - \Omega t) \ , \quad \text{where} \tag{5}$$

$$\Omega = K(1 + 2K^2 + L^2/2K^2) \ . \tag{6}$$

Multi-soliton solutions of the KP equation were obtained by using Hirota's solution method /4/. Miles applied such solutions to the study of oblique reflection of shallow water solitons on a rigid wall and obtained satisfactory results /5/. He has shown that three solitons strongly interact when the resonance conditions between $K_i = (K_i , L_i)$, i = 1, 2, 3,

$$\Omega_3 = \Omega_1 \pm \Omega_2 \quad \text{and} \quad K_3 = K_1 \pm K_2 \tag{7}$$

hold, where each wave satisfies the soliton dispersion relation (6). Under the condition (7), three solitons resonantly couple to each other, such that the three crests come from one point with a Y-shaped structure. The concept of soliton resonance was successfully used to explain the Mach reflection of solitons, where "the apex of the incident and reflected solitons moves away from the wall at a constant angle and is joined to the wall by a third soliton (the Mach stem)" /6/.

The KP equation is derived under the assumption of weak two-dimensionality, that is, $(L/K)^2 \sim K^2 \ll 1$. This assumption can be easily removed by using the two-dimensional Boussinesq equations /7/;

$$\frac{\partial^2 f}{\partial t^2} - \nabla^2 f + \frac{\partial}{\partial t} (\nabla f)^2 - \frac{\partial^2}{\partial t^2} \nabla^2 f = 0 \qquad (8)$$

which describes two-dimensional motions of ion-acoustic waves and f is the normalized velocity potential of the ion fluid. In (8) there is no restriction on the direction of the soliton wave-vector, whilst the wave-vector in (4) must be nearly parallel to the x-axis. The soliton solution for (8) is the same as (2), except for the point that the product Kx should be replaced by $\mathbf{K \cdot x}$ and Ω by

$$\Omega = \pm |K| / (1 - 4K^2)^{1/2} . \qquad (9)$$

The dispersion relation (9) agrees with (3) in the case of small dispersion, $|K| \ll 1$. If we put the x- and y-component of K to be K and L, respectively, and assume $|L| \ll |K|$, we can substitute $|K| \sim |K|(1+(L/K)^2/2)$ to obtain (6).

We can directly show that the KP equation (4) is derived from (8) under the assumption of nearly unidirectional wave motions: When the wave is propagated nearly to the positive x-direction, the hyperbolic operator in (8) can be approximated as

$$\frac{\partial^2}{\partial t^2} - \nabla^2 \sim -2 \frac{\partial}{\partial x} \left(\frac{\partial}{\partial t} + \frac{\partial}{\partial x} \right) - \frac{\partial^2}{\partial y^2}$$

and the y-dependence in the nonlinear term and the higher derivative terms can be neglected, that is,

$$\frac{\partial}{\partial t} (\nabla f)^2 - \frac{\partial^2}{\partial t^2} \nabla^2 f \sim - \frac{\partial}{\partial x} \left(\frac{\partial f}{\partial x} \right)^2 - \frac{\partial^4}{\partial x^4} f$$

where $\partial/\partial t \sim -\partial/\partial x$ is used. Substituting these into (8) and putting $\partial f/\partial x = v$ immediately yields (4).

Exact solutions, other than one-soliton solution, i.e., (2) with (9), have not been found for (8), unlike for (1) and (4). However, N-soliton solutions for (8) can be *approximately* obtained by assuming small nonlinearity and long wave-length in Hirota's algorithm /7,8/. This assumption is the same as that just made above in deriving (4) from (8). As an example, the approximated two-soliton solution is illustrated in Fig.1. When (7) is satisfied, the internal wave crest elongates unrestrictedly and triple wave structure appears (see Fig.2).

If the interaction angle of solitons is between two resonance angles, which are determined from (7) corresponding to its double sign, the two-soliton solution obtained by Hirota's method becomes singular in the sense that solitons with sech^2-profile are transformed after the interaction into anti-solitons with $(-\mathrm{cosech}^2)$-profile. This implies that such a phase-locked solution for the two-soliton problem has no physical meaning in this case, but a time dependent treatment is needed. Numerical calculations for the initial value

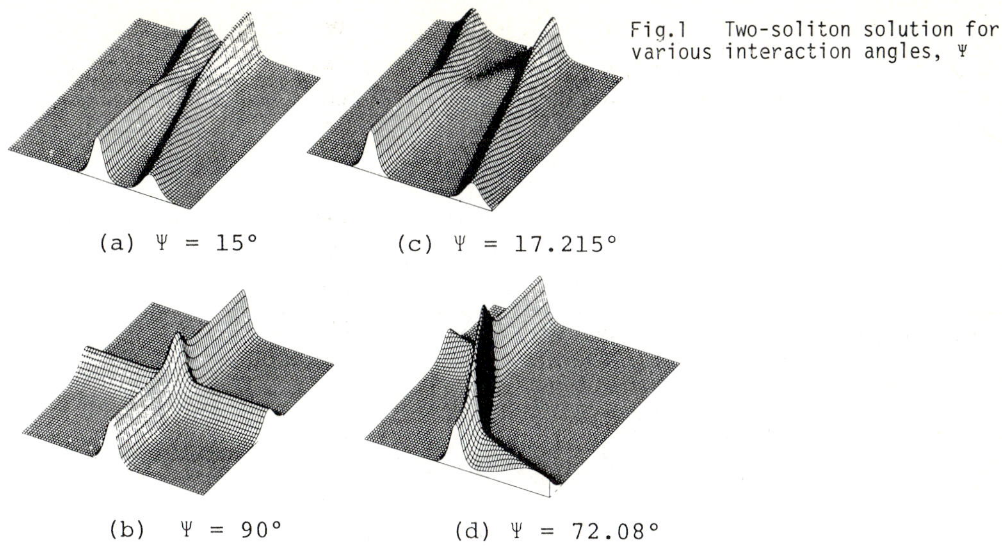

Fig.1   Two-soliton solution for various interaction angles, Ψ

problem of (8) show that the soliton interaction has no simple structure such as Fig.1, but a trapezoidal structure made by internal soliton crests appears to join the two colliding solitons and that non-genuineness of internal solitons brings a slight mismatch in the resonance condition (7) to be satisfied at each apex of the trapezoid (see Fig.3) /8/.

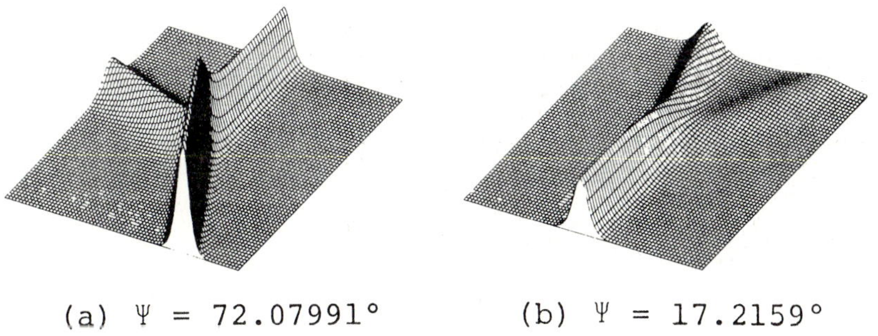

Fig.2 Resonant interaction of solitons

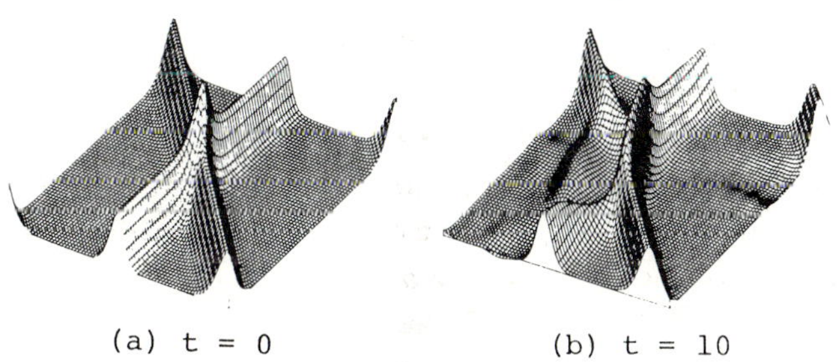

Fig.3 Soliton interaction with anangle between two resonance angles

Experimental studies on oblique interaction of two ion-acoustic solitons were made by Nishida and Nagasawa /9/, Tsukabayashi and Nakamura /10/, and Khazei, Bulson and Lonngren /11/. They have confirmed experimentally the above theoretical predictions. Typical wave patterns observed in their experiments are shown in Fig.4 /9/ and Fig.5 /12/.

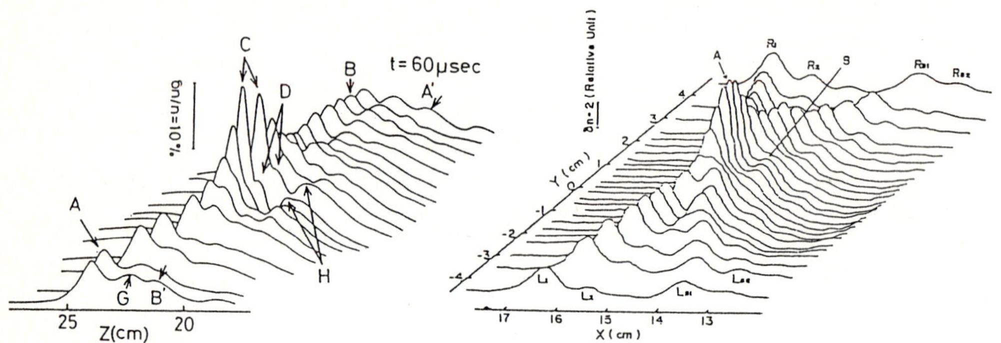

Fig.4 Experimental observation of oblique interaction of ion acoustic solitons, /9/

Fig.5 Experimental observation of oblique interaction of ion acoustic solitons, /12/

3. Solitons in a Plasma with an Ion-Beam

The system considered in §2 is linearly stable, and is composed of a single mode only, that is, an ion-acoustic mode. In multi-mode systems a wide variety of phenomena would be anticipated because of mode-mode coupling therein. As an example of such a system an ion-beam plasma system is considered in /13/. The KdV equation should be modified under the effect of an ion-beam /14/, i.e.,

$$\frac{\partial u}{\partial t} - \frac{\partial u}{\partial x} + u\frac{\partial u}{\partial x} + \frac{\partial^3 u}{\partial x^3} + \frac{a^3}{2}\frac{\partial v}{\partial x} = 0 \tag{10a}$$

$$\frac{\partial v}{\partial t} + \frac{\partial w}{\partial x} = 0 \tag{10b}$$

$$\frac{\partial w}{\partial t} + \frac{\partial u}{\partial x} = 0 \tag{10c}$$

where u is the nondimensional electrostatic potential, v the normalized beam density and w the normalized beam velocity. The parameter a stands for the strength of ion-beam, proportional to the cubic root of beam density. This was numerically solved by Ostrovsky, Petrukhina and Fainstein /14/.

The linear dispersion equation of (10) is

$$(\omega/k)^3 + (1 + k^2)(\omega/k)^2 - a^3/2 = 0 \tag{11}$$

There are three modes corresponding to the fast beam mode (F-mode), slow beam mode (S-mode) and ion-acoustic mode (A-mode). Stability criterion of such

148

waves is given as

$$k^2 > 3a/2 - 1 \tag{12}$$

Therefore, the case $a < 2/3$ is absolutely stable.

Equation (10) has a stationary soliton solution,

$$u = 12K^2 \mathrm{sech}^2(Kx - \Omega t)$$
$$w = (K/\Omega)v = (K/\Omega)^2 u \tag{13}$$

where Ω is determined by

$$(\Omega/K)^3 + (1 - 4K^2)(\Omega/K)^2 - a^3/2 = 0 \ . \tag{14}$$

For $a < 2/3$, three soliton modes are possible, that is, fast (F) soliton, slow (S) soliton and ion-acoustic (A) soliton. However, for $K > K_c$, where

$$K_c = \sqrt{(1-3a/2)} \ /2 \ , \tag{15}$$

A- and S- solitons can not exist. For $a > 2/3$ only F-solitons are allowed.

As long as the amplitude is small enough, the characteristics of solitons are very similar to those in the KdV system; any mode of solitons is stable even through the collision processes. In Fig.6, an example of collsion of A- and S-solitons is depicted. It follows from this that nonlinear processes for the case of small amplitudes can be described as a superposition of mutually independent KdV solitons.

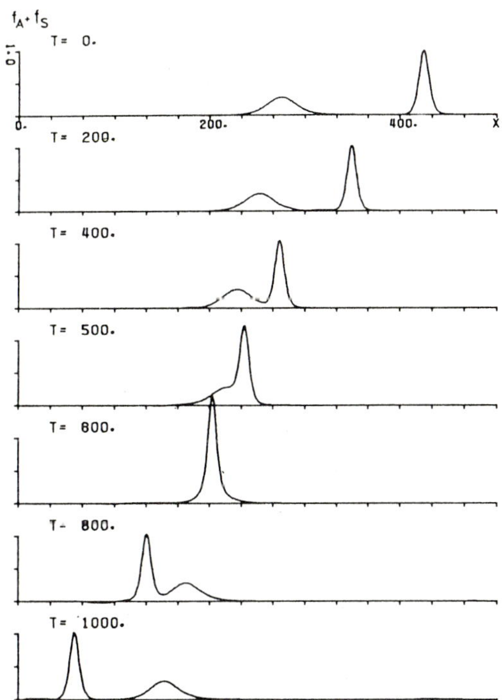

Fig.6 Pass-through collision between A- and S-solitons

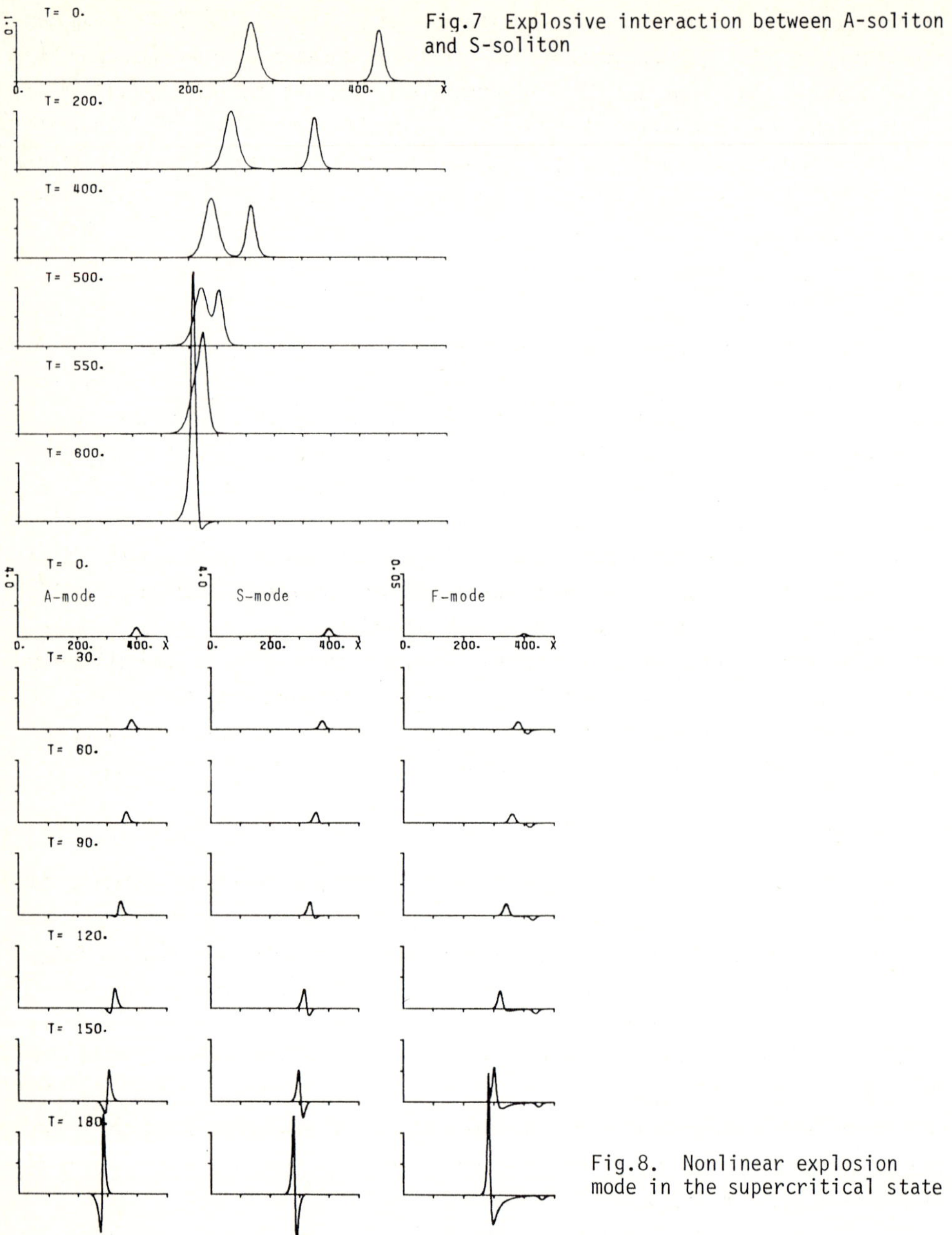

Fig.7 Explosive interaction between A-soliton and S-soliton

Fig.8. Nonlinear explosion mode in the supercritical state

When amplitudes of colliding solitons are increased, an explosion phenomenon is observed at the point where two solitons join together, even though they themselves are stable (see Fig.7). This is because the amplitude of the intermediate state of the collision becomes large enough to exceed the critical value beyond which the solution explodes. Figure 8 illustrates an example of such a coupled nonlinear explosion mode.

For small amplitude waves, the reductive perturbation method /15/ applies the system of (10), to yield a set of three KdV equations which describe the A-, S- and F-mode waves. This approximation explains well the qualitative results from the numerical computations. On the other hand, the nonlinear explosion mode can be analytically treated only in the vicinity of the critical point, a = 2/3.   Singular perturbation approach with multi-time scales leads to the Boussinesq like equation;

$$\frac{\partial^2 g}{\partial \tau^2} + \epsilon \frac{\partial^2 g}{\partial \eta^2} + 6 \frac{\partial^2 g^2}{\partial \eta^2} + \frac{\partial^4 g}{\partial \eta^4} = 0 \tag{16}$$

where the small parameter ϵ denotes the deviation of the parameter 'a' from 2/3, i.e., $\epsilon \propto (3a/2) - 1$.   The positive and negative of ϵ correspond to unstable and stable systems, respectively. The function g denotes the coupled modes of S- and A-modes near $K = K_c$. By using Hirota's method the N-soliton solution of (8) for $\epsilon < 0$ can be constructed. Particularly the two soliton solution is known to be singular under certain conditions /16/, which is related to the explosion phenomena of Figs.7 and 8. The solution which corresponds to the nonlinear explosion mode is also constructed in /13,17/. This has the sech^2-shape but blows up at a finite time.

Experimental studies on amplification of ion-acoustic solitons under linearly unstable condition were made by Okutsu, Nakamura, Nakamura and Itoh /18/. It is not clear whether their experiment provides an evidence of nonlinear explosion modes or not. More careful study is needed before drawing a hasty conclusion.

4. Summaries

In this article, recent developments of strong interactions of solitons in plasmas are briefly described:
(1) In a system with two spatial dimensions, the soliton resonance is possible; when the resonance condition holds among three solitons, a coupled system of two solitons moves with the same velocity   as the other soliton.   These three solitons then form a phase-locked system in which the Y-shaped structure is made by three wave crests.
(2) Owing to the occurrence of soliton resonances, the nonlinear evolution of an ion acoustic wave system with two-dimensional system is not so simple. In some cases the recurrence phenomenon is incomplete /7,8/.
(3) An explosive type instability with localized space structure is found in the case of an ion-beam plasma. The nonlinear explosion process is connected to the merging phenomena of S- and A-modes.   This process is expected to be saturated by the nonlinear or trapping effects of beam particles, though this problem is still open.

The possibility and importance of soliton resonance have now been proven by many experimental results.   However, the role of soliton resonance in the onset of plasma turbulence and the possibility of the nonlinear explosion mode are yet to be studied experimentally.

References

1. H.Washimi and T.Taniuti: Phys. Rev. Lett. 17, 996 (1966).

2. M.J.Ablowbitz and H.Segur: "Solitons and the Inverse Scattering Transform" (SIAM, Philadelphia 1981) p.291.

3. B.B.Kadomtsev and V.I.Petviashvili: Sov. Phys. Doklady 15, 539 (1970).

4. J.Satsuma: J. Phys. Soc. Jpn. 40, 286 (1976).

5. J.W.Miles: J. Fluid Mech. 79, 157 and 171 (1977).

6. J.W.Miles: J. Fluid Mech. 79, 171 (1977).

7. N.Yajima, M.Oikawa and J.Satsuma: J. Phys. Soc. Jpn. 44, 1711 (1978).

8. F.Kako and N.Yajima: J. Phys. Soc. Jpn. 49, 2063 (1980), and 51, 311 (1982).

9. Y.Nishida and T.Nagasawa: Phys. Rev. Lett. 45, 1626 (1980).
    T.Nagasawa and Y.Nishida: Phys. Rev. A28, 3043 (1983).

10. I.Tsukabayashi and Y.Nakamura: Phys. Lett. 85A, 152 (1981).

11. K.Khazei, J.Bulson and K.E.Lonngren: Phys. Fluids 25, 759 (1982).

12. I.Tsukabayashi, Y.Nakamura, F.Kako and K.E.Lonngren: Phys. Fluids 26, 790 (1983).

13. N.Yajima, M.Kono and S.Ueda: J. Phys. Soc. Jpn. 52, 3414 (1983).

14. L.A.Ostrovsky, V.I.Petrukhina and S.M.Fainstein: Sov. Phys. JETP 42, 1041 (1976).

15. M.Oikawa and N.Yajima: J. Phys. Soc. Jpn. 37, 1093 (1973).

16. M.Tajiri and T.Nishitani: J. Phys. Soc. Jpn. 51, 3720 (1982).

17. N.Yajima: Prog. Theor. Phys. 69, 678 (1983).

18. E.Okutsu, M.Nakamura, Y.Nakamura and T.Itoh: Plasma Phys. 20, 561 (1978).

Equilibrium Solutions in a Nonlinear Dispersive System with Instability and Damping

Takuji Kawahara and Sadayoshi Toh

Department of Physics, Faculty of Science, Kyoto University, Kyoto 606, Japan

1. Introduction

The problem of the evolution of instability waves in fluid systems has received considerable attention in recent years. One typical governing equation of such a problem is the nonlinear Schrödinger equation with complex coefficients(envelope equation) and its properties have been investigated from various aspects[1-6]. In this paper, we take up another fundamental equation of the form

$$u_t + uu_x + \nu u + \alpha u_{xx} + \beta u_{xxx} + \gamma u_{xxxx} = 0, \tag{1}$$

where ν, α, β, and γ are positive constants characterizing respectively damping, instability, dispersion, and dissipation. This equation not only models instability waves in nonlinear dispersive systems but also governs asymptotic behaviours of long waves in actual physical problems, for example, the unstable drift waves in a plasma [7] and other problems as special cases[8-11].

Substitution of $u \propto \exp(ikx + \sigma t)$ into eq.(1) yields the linear dispersion relation

$$\sigma = -\nu + \alpha k^2 - \gamma k^4 + i\beta k^3. \tag{2}$$

The damping term($\nu \neq 0$) uniformly reduces the linear growth rate and the system becomes linearly stable for $\nu \geq \nu_c \equiv \alpha^2/4\gamma$. The maximum growth rate is given at the wavenumber $k_{max} = (\alpha/2\gamma)^{1/2}$. If we take ν as the control parameter, we may regard eq.(1) as a model which governs nonlinear instability waves near to or far from the marginal stability.

We investigate the properties of eq.(1) both numerically and theoretically paying attention to the effects of dispersion. For a strongly dispersive case(large β), multiple equilibrium solutions exist under the periodic boundary conditions and the periodic solutions agree well with the cnoidal wave perturbation. For nondispersive case($\beta = 0$), chaotic behaviours or two-mode equilibria are observed numerically, and the latter agree with the analytical two-mode solutions.

2. Initial Value Problem

Imposing periodic boundary conditions on the interval $[0,L]$, we integrate eq.(1) using a finite difference method in space and the Runge-Kutta-Gill method in time. Spatial mesh-points were 200 in the periodicity length $L = 2$. Throughout this paper, fixed values $\alpha = 1.0 \times 10^{-2}$ and $\gamma = 5.066 \times 10^{-6}$ are used, so that k_{max} is taken as 10π. Numerical integrations were done for a variety of parameters β and ν and the initial conditions $b\cos N\pi x$ with various values of b and N.

Consequences of the initial value problem are summarized as follows. In contrast to the Fermi-Pasta-Ulam recurrence in the case of the Korteweg-de Vries equation[12], multiple equilibrium states are observed for strongly dispersive case depending on the initial conditions[13]. The equilibrium solutions become periodic as the number of pulses in L increases or as ν increases[14]. Each pulse becomes solitonlike as ν decreases and the equilibrium amplitudes become large as β increases. The waves diminish their amplitudes and become sinusoidal as ν increases. On the other hand, for small β or for nondispersive cases, ultimate states are either chaotic or two-mode equilibria depending on the initial conditions. Some examples of temporal evolutions of waveforms are shown in Figs.1 and 2.

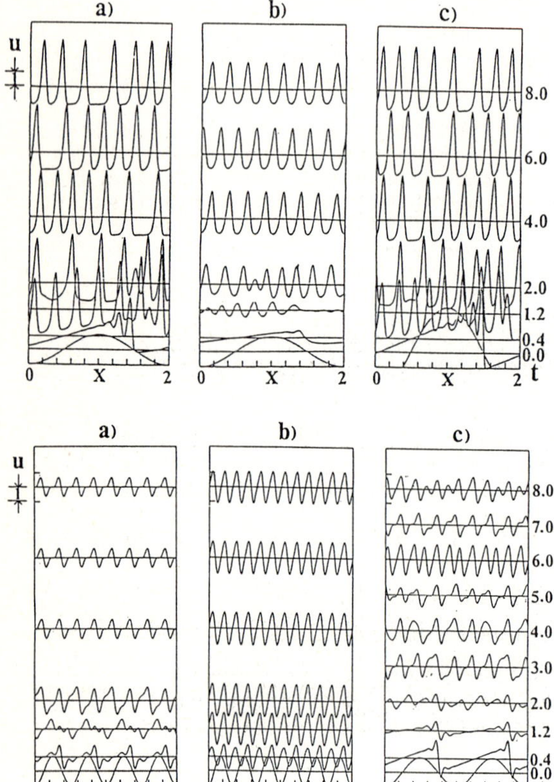

Fig.1 Temporal evolutions of waveforms for $\beta = 4.84 \times 10^{-4}$. Initial condition and damping rate: a) $\cos\pi x$, $\nu = 0$, b) $\cos\pi x$, $\nu = 2.5$, c) $3\cos\pi x$, $\nu = 0$.

Fig.2 Two-mode equilibria and chaotic evolution for $\beta = \nu = 0$. Initial condition: a) $\cos 4\pi x$, b) $\cos 6\pi x$, c) $\cos 2\pi x$.

3. Cnoidal Wave Perturbation

For strongly dispersive cases with $\nu = 0$, the temporal evolution is characterized by the formation of a row of solitonlike pulses. The width of each pulse is determined by the relative importance of instability and dissipation, i.e., $(\gamma/\alpha)^{1/2}$ and the amplitude is proportional to $\alpha\beta/\gamma$. The stability of equilibrium solitonlike pulses is investigated by the linear stability analysis[15]. It is found that there is a critical distance between adjacent pulses for stability. A configuration with distance longer than the critical is unstable and further pulse generation occurs, leading to an equilibrium with larger pulse number. Numerical simulations reveal that aperiodic pulse trains can stably exist if the pulse configuration is appropriate.

The equilibrium pulse trains acquire further periodicity as the damping effect is increased and recent numerical simulations indicate that the pulse trains attain further periodicity in a long time evolution through pulse interactions. This fact motivated the perturbation analysis of the cnoidal wave solution[14].

If $\nu = \alpha = \gamma = 0$, eq.(1) reduces to the Korteweg-de Vries equation and admits a cnoidal wave solution

$$u = u_0 + a \cdot cn^2(A\xi, m), \quad \xi \equiv x - [u_0 + (2 - m^{-2})a/3]t, \tag{3}$$

with $A = (a/12\beta m^2)^{1/2}$, $A\lambda = 2K(m)$, and

$$u_0 = - a[E(m) + (m^2 - 1)K(m)]/m^2 K(m). \tag{4}$$

Here m^2 is the modulus of the Jacobi elliptic function, $K(m)$ and $E(m)$ denote the complete elliptic integral of the first and second kind, λ is the wavelength, and u_0 is the uniform level of the equilibrium pulses being determined by the condition $\int_0^L udx = 0$.

When the perturbation term is of the order of ε, changes of the quantity A with respect to the slow time scale $T \equiv \varepsilon t$ can be given in the form $dA/dT = g(A)$ in terms of a standard perturbation procedure [14]. If N pulses are formed in L, then $N\lambda = L$ and we get the equilibrium value of A for N pulses in L, so that equilibrium value A and m^2 are determined for given ν and N. Once A and m^2 are determined, we can obtain the pulse amplitude $a = 12\beta m^2 A^2$ and u_0. The expected equilibrium amplitudes $(a + u_0, u_0)$ will be compared with the numerical results in Fig.3.

4. Two-Mode Equilibrium

Examinations of the Fourier amplitudes of the numerical solutions reveal the existence of two-mode equilibria for particular initial conditions. If the initial condition is such that one mode quickly grows and predominates over other modes due to the mode selection mechanism, there occurs a balance between the linearly unstable mode and its stable harmonics.

Now we derive analytically the two-mode solution. Substituting

$$u(x, t) = A_1 exp(ikx) + c.c. + A_2 exp(2ikx) + c.c. , \tag{5}$$

into eq.(1) and introducing $A_1 = a_1 exp(i\phi_1)$, $A_2 = a_2 exp(i\phi_2)$ and $\Phi \equiv \phi_2 - 2\phi_1$, we obtain

$$\dot{a}_1 - \sigma_r(k)a_1 - ka_1a_2\sin\Phi = 0, \quad \dot{a}_2 - \sigma_r(2k)a_2 + ka_1^2\sin\Phi = 0,$$

$$\dot{\Phi} + 2\sigma_i(k) - \sigma_i(2k) + (k/a_2)(a_1^2 - 2a_2^2)\cos\Phi = 0, \tag{6}$$

where the dot denotes the time derivative and σ_r and σ_i are the real and the imaginary part of σ.

The equilibrium solution is given by

$$a_1^2 = - \sigma_r(k)\sigma_r(2k)k^{-2}(1 + cot^2\Phi), \quad a_2^2 = \sigma_r(k)^2k^{-2}(1 + cot^2\Phi),$$

$$cot\Phi = [\sigma_i(2k) - 2\sigma_i(k)]/[\sigma_r(2k) + 2\sigma_r(k)], \tag{7}$$

where $\sigma_r(k) = - \nu + \alpha k^2 - \gamma k^4 > 0$, $\sigma_r(2k) = - \nu + 4\alpha k^2 - 16\gamma k^4 < 0$,

and $\sigma_i(2k) - 2\sigma_i(k) = 6\beta k^3$.

The two-mode equilibrium solution is given by

$$u = 2a_1\cos\theta + 2a_2\cos(2\theta + \Phi), \quad \theta \equiv kx + \phi_1. \tag{8}$$

For a nondispersive case, we have $\Phi = - \pi/2$, so that $cot\Phi = 0$. For a dispersive case, the equilibrium amplitude increases as β increases and the phase deviation Φ depends on the value of β.

5.   Envelope Equation

Asymptotic behaviours of amplitude evolutions near marginal stability
are governed by the envelope(Ginzburg-Landau or Stewartson-Stuart)
equation.   We assume that the control parameter ν is smaller than ν_c
by a small positive parameter ε^2, i.e., $\nu = \nu_c - \delta\varepsilon^2$ with $\nu_c \equiv \alpha^2/4\gamma$
and $k_c = (\alpha/2\gamma)^{1/2}$.   Then the governing equation for the complex am-
plitude of the fundamental component can be obtained by means of the
perturbation technique, giving rise in unstretched variables to

$$[i(\frac{\partial}{\partial t} - \frac{3\alpha\beta}{2\gamma}\frac{\partial}{\partial x}) - \frac{\alpha}{2}(3\sqrt{2}\eta + 4i)\frac{\partial^2}{\partial x^2}]\varepsilon A$$

$$+ \frac{2(-2\sqrt{2}\eta + 3i)}{3\alpha(9 + 8\eta^2)}|\varepsilon A|^2\varepsilon A = i(\nu_c - \nu)\varepsilon A \, , \tag{9}$$

where $\eta \equiv \beta/(\alpha\gamma)^{1/2}$.

The uniform solution to eq.(9)

$$\varepsilon A = [\alpha(9 + 8\eta^2)(\nu_c - \nu)/2]^{1/2}\exp[-i(2\sqrt{2}/3)\eta(\nu_c - \nu)t], \tag{10}$$

is modulationally unstable for the wavenumber $\varepsilon k < [8(\nu_c - \nu)(\eta^2 - 1$
$)/\alpha(9\eta^2 + 8)]^{1/2}$ if $\eta^2 > 1$, so that the modulational instability can
occur for a strongly dispersive case.
For $\beta = 0$, eq.(9) reduces to the Newell-Whitehead equation[2]

$$(\frac{\partial}{\partial t} - 2\alpha\frac{\partial^2}{\partial x^2})\varepsilon A = (\nu_c - \nu)\varepsilon A - \frac{2}{9\alpha}|\varepsilon A|^2\varepsilon A, \tag{11}$$

which has a uniform equilibrium solution $\varepsilon A = [9\alpha(\nu_c - \nu)/2]^{1/2}$.
The uniform solution is given by

$$u = 2|\varepsilon A|\cos(k_c x + \phi), \tag{12}$$

where (10) is to be substituted for εA.

6.   Discussion

The theoretical estimates due to i) cnoidal wave perturbation for
large β, ii) two-mode solution(eq.(8)), and iii) envelope equation
(eq.(12)) are compared with the numerical results in Figs.3 and 4.

Figure 3 shows the equilibrium pulse amplitudes $(a + u_0, u_0)$ for
the strongly dispersive case($\beta = 2.0\times10^{-3}$).   Numerical calculations
were done for various initial conditions: $\cos N\pi x$.   Theoretical estimates due to the cnoidal wave perturbation, two-mode solution, and
envelope equation are shown by dotted, solid, and dot-broken curves.
For the initial condition $\cos\pi x$, the equilibria are 5-7 pulse states.
For $\cos8\pi x$, $\cos10\pi x$, and $\cos12\pi x$, the equilibria are 8, 10, and 12
pulse states.   Note that the cnoidal wave perturbations rather than
the two-mode solutions are good approximations for the numerical results.   Equilibrium amplitudes due to the envelope equation are symmetric with respect to $u = 0$ and are good approximations only around
$\nu \sim \nu_c$.   Pulse amplitudes for $\beta \neq 0$ increase asymmetry as ν decreases.

Figure 4 shows the equilibrium amplitudes of the two-mode solutions and the numerical results for $\beta = 0$.   Numerical calculations
were done for $\cos N\pi x$ with $N = 1,2,4,6,8,10,12$.   For initial conditions $N = 4$ and 8 or $N = 6$ and 12, two-mode equilibria with $N = 8$ or
$N = 12$ were attained eventually.   However, for $N = 1$ and 2, the wave
evolutions were chaotic as is seen in Fig.2.   For the initial condition $\cos\pi x$, chaotic behaviours are observed for $\nu = 0, 1, 2$ but two-
mode equilibria with 10 pulses are observed for $\nu = 3$ and 4.   Therefore we may conclude that for $\beta = 0$, chaotic behaviours occur for the
situations in which many linearly unstable modes are excited and participate in the wave evolutions.   Note that the maximum number of
linearly unstable modes is 14 for $\nu = 0$.

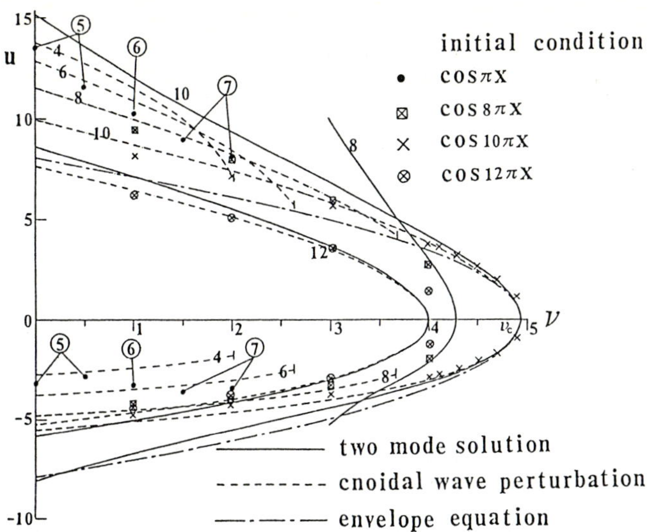

Fig.3 Theoretical estimates of equilibrium amplitudes and numerical results for the strongly dispersive case ($\beta=2.0\times10^{-3}$). Numbers attached to points or curves indicate pulse numbers of equilibrium states.

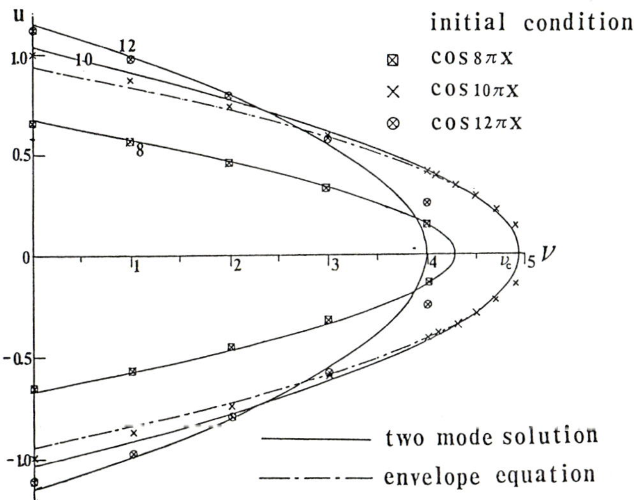

Fig.4 Theoretical and numerical two-mode equilibria for the nondispersive case ($\beta = 0$).

On the contrary, when the initial condition is such as to cause mode selection, the two-mode equilibrium is attained. For the initial condition cos4πx or cos6πx, nonlinearity generates harmonic mode cos 8πx or cos12πx. Because the linear growth rate of the latter is larger than that of 4π or 6π, the generated mode predominates and rather suppresses the initial mode and leads to the two-mode equilibrium balancing with the stable harmonic 16π or 24π. This is the mode selection mechanism due to the nonlinearity.

Summing up the results, we arrive at the conclusions as follows.
For $\beta = 0$, either chaotic behaviours or two-mode equilibria arise
depending on the initial conditions. Even when many linearly unstable modes can participate in wave evolutions, two-mode equilibria can arise due to the mode selection mechanism. For initial conditions from which a number of linearly unstable modes are excited, chaotic behaviours arise. For $\nu \sim \nu_c$, the equilibrium states are nearly sinusoidal and are approximated well by the analytical two-mode solution or the envelope equation. For large β, the equilibrium states involving a number of linearly unstable modes are attained due to the effects of dispersion. Balances between dispersion and nonlinearity lead to an organization of the system with higher equilibrium amplitude for larger dispersion. No mode selection occurs when sufficient dispersion exists. For $\nu \sim \nu_c$, numerical equilibrium states appear to be close to the two-mode solution, since they become nearly sinusoidal as ν approaches ν_c. However, it should be noted that, as soon as ν decreases, periodic solutions lose symmetry and approach cnoidal wave trains involving a number of modes, so that they differ essentially from the two-mode solutions for $\beta = 0$. Detailed balances between dispersive effect and harmonic generation mechanism together with the instability and damping mechanisms in the wave evolution process determine the ultimate equilibrium states. Therefore under the periodic boundary conditions, the final equilibria are initial condition dependent. Solutions in an infinite domain and details about the case of infinitely many degrees of freedom should await further investigations.

The authors would like to thank Prof.T.Tatsumi and Dr.M.Yamada for stimulating discussions.

References

1. P.G.Drazin and W.H.Reid: Hydrodynamic stability(Cambridge Univ. Press, Cambridge, 1981)
2. A.C.Newell and J.A.Whitehead: J.Fluid Mech. **38**,279(1969)
3. J.T.Stuart and R.C.DiPrima: Proc.Roy.Soc.Lond. **A362**,27(1978)
4. M.I.Rabinovich and A.L.Fabrikant: Sov.Phys.JETP **50**,311(1979)
5. K.Nozaki and N.Bekki: Phys.Rev.Lett. **51**,2171(1983)
6. H.T.Moon, P.Huerre, and L.G.Redekopp: Phys.Rev.Lett. **49**,458 (1982), Physica **7D**,135(1983)
7. B.I.Cohen, J.A.Krommes, W.M.Tang, and M.N.Rosenbluth: Nucl. Fusion **16**,971(1976)
8. G.I.Sivashinsky: Physica **4D**,227(1982), Annu.Rev.Fluid Mech. **15**, 179(1983)
9. J.Topper and T.Kawahara: J.Phys.Soc.Jpn. **44**,663(1978)
10. Y.Kuramoto and T.Tsuzuki: Prog.Theor.Phys. **55**,356(1976)
11. T.Yamada and Y.Kuramoto: Prog.Theor.Phys. **56**,681(1976)
12. N.Zabusky and M.D.Kruskal: Phys.Rev.Lett. **15**,240(1965)
13. T.Kawahara: Phys.Rev.Lett. **51**,381(1983)
14. T.Kawahara and S.Toh: Submitted to Phys.Rev.Lett.
15. S.Toh and T.Kawahara: In preparation

Exact Solutions of Two-Dimensional Vortex Systems in Statistical Equilibrium

H.H. Chen, A.C. Ting, and Y.C. Lee

Laboratory for Plasma and Fusion Energy Studies, University of Maryland
College Park, MD 20742, USA

Statistical equilibrium states of two-dimensional vortex systems [1] are described by the solutions of a nonlinear elliptic partial differential equation, the sinh-Poisson equation $\nabla^2 \phi + \lambda^2 \sinh \phi = 0$, with $\phi = 0$ on a rectangular boundary. We present here the first general analytic solutions to such a nonlinear boundary value problem. Some of these solutions were obtained previously only through numerical means. Explicit solutions showing nonlinear superposition are displayed.

A two-dimensional vortex system was studied extensively in the past. Joyce and Montgomery, Pointin and Lundgren studied this problem with statistical mechanics and derived a sinh-Poisson's equation for the stream function [2,3,4]

$$\nabla^2 u + \lambda^2 \sinh u = 0 . \tag{1}$$

This equation also describes the equilibrium states of two-dimensional guiding center plasmas and filamentations of relativistic charged beam systems.[3,5]

Numerical solutions of this equation with vortex structures were revealed previously.[4,5] Physical quantities evaluated for these solutions showed that the state with the largest vortices has the highest entropy and is thus thermodynamically stable.[4]

We present here a method of solving this equation exactly. Even though this is an elliptic boundary value problem, we found no problem in adapting the formalism of inverse scattering, which before had been applied solely to time evolution equations to work for Eq. (1).

It can be checked directly that sinh-Poisson equation (1) with $\lambda^2 = 1$ is the compatibility condition of a linear scattering system.

$$f_x = (\sqrt{E} - \frac{e^{-u}}{16\sqrt{E}})g + (\sqrt{E} - \frac{e^{u}}{16\sqrt{E}})h ,$$

$$g_x = 2(\sqrt{E} - \frac{e^{u}}{16\sqrt{E}})f - \frac{w}{2} g ,$$

$$h_x = 2(\sqrt{E} - \frac{e^{-u}}{16\sqrt{E}})f + \frac{w}{2} h ,$$

$$f_y = i(\sqrt{E} + \frac{e^{-u}}{16\sqrt{E}})g + i(\sqrt{E} + \frac{e^{u}}{16\sqrt{E}})h ,$$

$$g_y = 2i(\sqrt{E} + \frac{e^{u}}{16\sqrt{E}})f - i \frac{w}{2} g ,$$

$$h_y = 2i(\sqrt{E} + \frac{e^{-u}}{16\sqrt{E}})f + i \frac{w}{2} h \tag{2}$$

with $w = i(u_y + iu_x)$.

A finite band solution $u(x,y)$ corresponds to a finite polynomial expansion of f, g, and h, [6,7] that is,

$$f(x,y,E) = \frac{1}{\sqrt{E}} \sum_{j=1}^{N} f_j(x,y)E^j ,$$

$$g(x,y,E) = \sum_{j=0}^{N} g_j(x,y)E^j = \prod_{j=1}^{N} (E-\gamma_j) ,$$

$$h(x,y,E) = \sum_{j=0}^{N} h_j(x,y)E^j .$$

The zero's of the function g are called the auxilliary variables $\gamma_i(x,y)$, $i = 1, \ldots, N$. On the other hand, we see immediately that $P(E) = f^2 - gh$ is also a polynomial

$$P(E) = \prod_{i=1}^{2N} (E - E_i) .$$

The zero's of this polynomial, E_i's, are called the " main spectrum " and are constants.

When these expansions of the eigenfunctions are substituted into Eqs. (2), a set of recursion relations are obtained. From the zeroth order term of the f equations, we have

$$e^{2u} = - g_0/h_0 .$$

Making use of the fact that $g_0 = (-1)^N \prod_{i=1}^{N} \gamma_i$ and $g_0 h_0 = \prod_{i=1}^{2N} E_i$, we get a conversion formula between $u(x,y)$ and $\gamma_i(x,y)$'s .

$$u = \ell n \left(\frac{\prod_{i=1}^{N} \gamma_i}{Q^{1/2}}\right) , \qquad Q = \prod_{j=1}^{2N} E_j .$$

The dependence of γ_i on x and y is given by a set of first order autonomous nonlinear ordinary differential equations. They are obtained by evaluating the derivatives of the function $g(x,y,E)$ at $E = \gamma_i$.

$$\gamma_{ix} = \frac{(\prod_{k\neq i}^{N} \gamma_k/8Q^{1/2} - 2) \sqrt{(\prod_{\ell=0}^{2N} (\gamma_i - E_\ell))}}{\prod_{j\neq i}^{N} (\gamma_i - \gamma_j)} , \tag{3a}$$

$$\gamma_{iy} = i \frac{(\prod_{k\neq i}^{N} \gamma_k/8Q^{1/2} + 2) \sqrt{(\prod_{\ell=0}^{2N} (\gamma_i - E_\ell))}}{\prod_{j\neq i}^{N} (\gamma_i - \gamma_j)} , \tag{3b}$$

where $E_0 = 0$ and $Q^{1/2} = + \left(\prod\limits_{m=1}^{2N} E_m \right)^{1/2}$.

Theorem: Given a set of $2N$ constants E_i, and a set of N variables γ_j which satisfy Eqs. (3), then the function $u(x,y)$ defined by eq. (1) satisfies the sinh-Poisson equation.

Proof: The proof is straightforward and will be omitted here. The upshot of this theorem is that $u(x,y)$ constructed from Eq. (1) and Eq. (3) will always satisfy the sinh-Poisson equation, though the solution may not be real.

For our problem, we are interested in solutions that are real and satisfy the zero boundary condition on a square. These solutions can be extracted from the periodic solutions with odd parity. This in turn imposed symmetry conditions on the main spectrum. They appear in quartets as the following.

$$\left\{ E_i , \ E_i^* , \ \frac{1}{16^2 E_i} , \ \frac{1}{16^2 E_i^*} \right) , \qquad i = 1, \ldots, \frac{N}{2} . \tag{4}$$

The solution of the sinh-Poisson equation has now been reduced to the integration of a set of first order ordinary differential equations with suitable initial conditions. This can easily be implemented with existing numerical techniques and one may consider the job of constructing solutions to the sinh-Poisson equation achieved. However, we shall go further than that by integrating these ordinary differential equations analytically.

The equations of motion contain the Riemann function

$$R(E) = \surd\left(\prod_{i=0}^{2N} (E - E_i) \right) , \qquad E_0 = 0$$

and we shall attempt to linearize the motions of the γ_j's on the two-sheeted Riemann surface generated by this square root function. It is done by transforming γ's into a set of phases ℓ_j's defined as

$$\ell_j = - \sum_{k=1}^{N} \int_{\gamma_0}^{\gamma_k} \sum_{m=1}^{N} C_{jm} d\mu_m \qquad \text{with} \tag{5}$$

$$d\mu_i = \frac{E^{N-i} dE}{R(E)}$$

and C_{jm} are elements of the normalizing matrix of integrals of $d\mu_i$'s along closed contours around the branch cuts of the complex plane. Direct calculations show that

$$\ell_{jx} = (-1)^N \frac{C_{jN}}{8Q^{1/2}} + 2C_{j1} , \tag{6}$$

$$\ell_{jy} = i(-1)^N \frac{C_{jn}}{8Q^{1/2}} - 2iC_{j1} . \tag{7}$$

Therefore,

$$\ell_j(x,y) = k_j x + \omega_j y + \ell_{0j} . \tag{8}$$

All we need to know next is the inversion of the transformation Eq. (5) and thus the dependence of γ_k on x and y. This is known as the classical problem of

Jacobi inversion which has been solved by the introduction of the Riemann theta function

$$\theta(\vec{\ell},\tau) = \sum_{\substack{m_1\cdots m_N \\ = -\infty}}^{\infty} \exp[2\pi i \sum_{i=1}^{N} m_i \ell_i + \pi i \sum_{\substack{i,j \\ = 1}}^{N} m_i \tau_{ij} m_j] \ , \tag{9}$$

where τ_{ij}'s are contour integrals on the Riemann surface.[9] Their values depend only on the main spectrum. The γ_k's turn out to be the zero's of the Riemann theta function, and we have the closed form solution for the sinh–Poisson equation

$$u(x,y) = 2\ \ell n[\theta(\vec{\ell} + \tfrac{1}{2}\ \vec{1},\tau)/\theta(\vec{\ell},\tau)] \ . \tag{10}$$

The remaining hurdle before the actual construction of the solutions is to show that the $u(x,y)$ is a real function. Indeed, with the symmetry of the main spectrum, we can show that the τ matrix is purely imaginary up to integer constants

$$\tau_{ik}^{*} = -\tau_{ik} + J \ , \qquad J : \text{integer}$$

$$\tau_{ii}^{*} = -\tau_{ii} \ ,$$

and the phases are real quantities for real ℓ_{0j}'s. Therefore, both of the Riemann theta functions in Eq.(10) are real and $u(x,y)$ is real. The constants ℓ_{0j}'s should be chosen to insure the odd parity of $u(x,y)$.

The periods of the solution depend on the wave numbers k_j's and ω_j's, which are in turn dependent on the main spectrum.

When $N = 2$, the four E_j's of the main spectrum give a pair of phases ℓ_1 and ℓ_2 where

$$\ell_1 = kx + \omega y + \ell_{01} \ ,$$

$$\ell_2 = kx - \omega y + \ell_{02} \ .$$

Hence, this is a standing wave and the resultant $u(x,y)$ has rectangular boundaries. It is found that when the E_j's lie on the imaginary axis, k and ω are equal and the boundary is square. When they are off the imaginary axis, the solution boundary is a rectangle. The aspect ratio is reversed when the main spectrum goes over to the mirror image position on the other side of the imaginary axis. The amplitude of the solution increases as the E_j's are further away from the inversion circle and approaches the linear limit as the E_j's collapse onto the inversion circle.

When $N = 4$, we can place two quartets of E_j's on the complex plane at mirror image positions about the imaginary axis. The wave numbers for one set of standing wave phases equal the wave numbers of the other set with the x and y components interchanged. When two quartets of E_j's are chosen such that the ratio k/ω is a rational number, the solution $u(x,y)$ has a square boundary, with nodal lines that would have been expected for linear waves with similar superpositions. More quartets can be superposed to arrive at more complex solutions. However, we show here contour plots of examples of solutions for the case of $N = 4$, with different values of k/ω in Fig. (1).

In conclusion, we have succeeded in establishing a method of generating a large class of exact periodic solutions for the sinh–Poisson equation which describe equilibrium states of two–dimensional guiding center plasma or a two–dimensional line vortex systems in fluids. These solutions are parametrized by a set of 2N complex constants and are given in closed form as ratio of Riemann theta functions. The periodic solutions with odd parity would satisfy the zero boundary condition on the square. They exhibit nonlinear superposition.

162

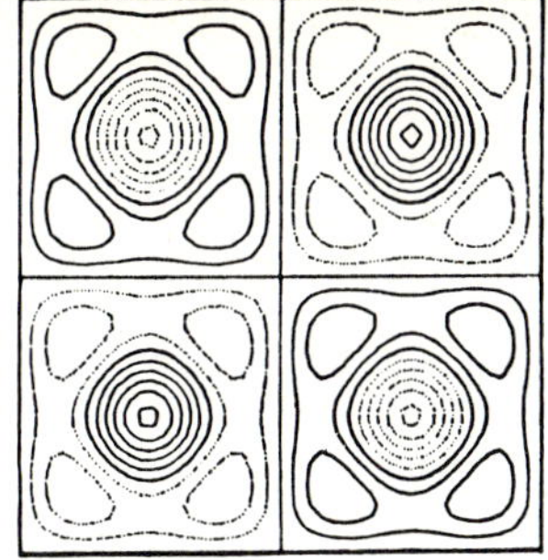

Fig. 1 Contour plots of solutions for the sinh-Poisson equation, with N = 4. the left picture has k/ω = 1/2 and the right one has k/ω = 1/3.

This work was supported by NSF and ONR.

1. L. Onsager, Nuovo Cimento Suppl. $\underline{6}$, 279 (1949). See also C. C. Lin, Proc. Nat. Acad. Sci. USA $\underline{27}$, 570 (1941).

2. G. Joyce, D. Montgomery, J. Plasma Phys. $\underline{10}$, Pt. 1, 107-121 (1973)

3. D. Montgomery, G. Joyce, Phys. Fluids, $\underline{17}$, 1139-1145 (1974)

4. Y.B. Pointin, T.S. Lundgren, Phys. Fluids $\underline{19}$, 1459-1470 (1976)

5. D.L. Book, S. Fisher, B.E. McDonald, Phys. Rev. Lett. $\underline{34}$, 4-8 (1975)

6. M.G. Forest, D.W. McLaughlin, J. Math Phys. $\underline{23}$, 1248-1277 (1982)

7. V.O. Kozel, V.P. Kotlyrarov, Dokl. A. N. Ukr. SSR, Ser. A $\underline{10}$, 878-881 (1976)

8. M.J. Ablowitz, D.J. Kaup, A.C. Newell, H. Segur, Stud. Appl. Math., $\underline{53}$, 249-315 (1974)

9. For application of Riemann complex theory to periodic problems, see <u>Theory of Nonlinear Lattices</u>, by Morikazu Toda, Springer Verlag (1981)

10. J. Zagrodzinski, J. Math. Phys. $\underline{24}$, 46-52 (1983)

Regular and Chaotic Motion of Two-Dimensional Point Vortices

Y. Kimura and H. Hasimoto

Department of Physics, Faculty of Science, University of Tokyo, Hongo, Bunkyo
Tokyo 113, Japan

1 INTRODUCTION

A point vortex is a model of a real vortex with δ-function-type vorticity
configuration in space. The equations of motion for N-point vortices in an
unbounded region are given as follows:

$$\frac{dZ_j}{dt} = \frac{1}{2\pi i} \sum_{m=1}^{N}{}' \frac{\Gamma_m}{\bar{Z}_j - \bar{Z}_m} \qquad (j = 1,2,\cdots,N) \qquad (1)$$

where Z_j ($= x_j+iy_j$) is a position of j-th vortex in the complex Z-plane, and Γ_j is
its strength. The bar on the variable means that we take complex conjugate, and
the prime denotes that we omit the singular terms j=m from the sum. Equation (1)
can be written down in real form with pairs of variables (x_j,y_j) (j=1,$\cdots$,N) as:

$$\frac{dX_j}{dt} = \frac{-1}{2\pi} \sum_{m=1}^{N}{}' \frac{\Gamma_m (Y_j - Y_m)}{r_{jm}^2} \qquad (j = 1,2,\cdots,N) \qquad (2)$$

$$\frac{dY_j}{dt} = \frac{1}{2\pi} \sum_{m=1}^{N}{}' \frac{\Gamma_m (X_j - X_m)}{r_{jm}^2} \qquad (j = 1,2,\cdots,N) \qquad (3)$$

where $r_{jm}^2 = (x_j-x_m)^2 + (y_j-y_m)^2$.

Point vortices have in general the following properties.
1) One vortex yields a circular flow whose velocity is inversely proportional to
the distance from the vortex.
2) Two vortices with the same sense rotate around two circles with a fixed common
centroid between them, and if the strengths are the same, they are on the same
circle.
3) Two vortices with opposite senses move parallel on two circles with their
external centroid as the center, and if the strengths are the same, they make a
vortex-pair and go straight.
4) The system of point vortices can be formulated as a Hamiltonian system, whose
integrability is an interesting problem.

We should note that the following equations which are similar to (1) except for
the absence of the complex conjugate on the right hand side are perfectly
integrable,

$$\frac{dZ_j}{dt} = \sum_{m=1}^{N}{}' \frac{g}{Z_j - Z_m} \qquad (j = 1,2,\cdots,N) \qquad . \qquad (4)$$

Equations(4) are known as the pole type solution of Burger's equation which can be
reduced to the heat equation through Hopf-Cole transformation, and thus perfectly
integrable. [6] [7]

<u>2 INTEGRABILITY of the EQUATION</u>

In this section, we discuss the integrability of the system from two points of view. One of them is the existence of boundaries or external flows, and the other is the conservative quantities (i.e. integrals) of the motion corresponding to the symmetry of the system.

Equation (1) is for the case in which vortices are in an unbounded region. If there is a boundary or an external flow, however, we must put additional terms to (1) in order to satisfy the boundary conditions or flow effects.

2-1 Existence of a boundary or an external flow

Ex.1 Circle [1]

When vortices are in a circular domain, we should consider the contribution of image vortices. Let us denote a vortex with the strength at z as (z,Γ). Then the image vortex is described as $(a^2/\bar{z},-\Gamma)$, where a is a radius of the circle and the equations become: (Fig.1)

$$\frac{dZ_j}{dt} = \frac{1}{2\pi i}\sum_{m=1}^{N}{}' \frac{\Gamma_m}{\bar{Z}_j - \bar{Z}_m} + \frac{1}{2\pi i}\sum_{m=1}^{N} \frac{-\Gamma_m}{\bar{Z}_j - (\frac{a^2}{Z_m})} \qquad (5)$$

$$(j = 1,2,\cdots,N)$$

We can reduce more complicated ones to this case by suitable comformal mappings. For example, the equations of motion for the vortices in a simply connected domain are derived from the following Hamiltonian:[1] [2]

$$H = -\frac{1}{4\pi}\sum_{j=1}^{N}\Gamma_j^2 \log \frac{g'(Z_j)}{1 - g_j\bar{g}_j} - \frac{1}{4\pi}\sum_{j\neq k}\Gamma_j\Gamma_k \log \frac{g_j - g_k}{1 - g_j g_k} \qquad (6)$$

by the canonical equation,

$$\Gamma_j\bar{Z}_j = 2i\frac{\partial H}{\partial Z_j} \qquad (j = 1,2,\cdots,N) \qquad (7)$$

where g(z) is an analytic function which maps a given boundary in the Z-plane onto a unit circle in the ζ-plane, and $g_j = g(z_j)$.

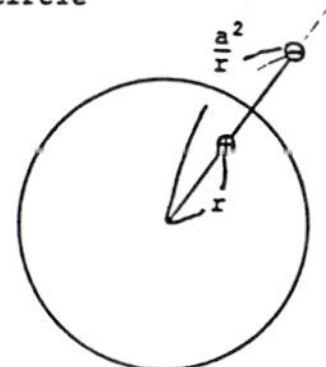

Fig. 1 A vortex in a circle.

Ex.2 Simple shear flow

Suppose that there exists a shear flow shown in Fig.2 as an external flow. In this case vortices are driven with additional velocity in the direction of the x axis and we have:(Fig.2)

$$\frac{dZ_j}{dt} = \frac{1}{2\pi i}\sum_{m=1}^{N}{}' \frac{\Gamma_m}{\bar{Z}_j - \bar{Z}_m} + \alpha\, \mathrm{Im}\, Z_j \qquad (8)$$

$$(j = 1,2,\cdots,N)$$

where Im z_j means imaginary part of z_j and α is a shear rate.

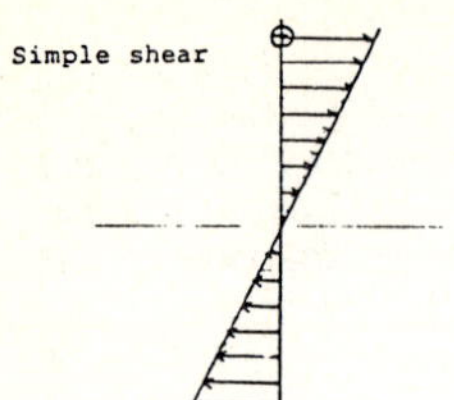

Fig. 2   A vortex in a simple shear flow,

Ex. 3 Circular rotation

When vortices are located in a circular rotational flow,(shown in Fig.3), the velocity of them is the combination of the interaction between vortices and rigid rotation:(Fig.3)

$$\frac{dZ_j}{dt} = \frac{1}{2\pi i} \sum_{m=1}^{N}{}' \frac{\Gamma_m}{\overline{Z}_j - \overline{Z}_m} + i\beta Z_j \qquad (9)$$
$$(j = 1,2,\cdots,N)$$

where i is an imaginary unit and β is the rotatonal speed.

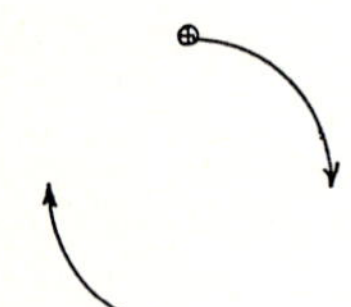

Fig. 3   A vortex in a circular flow.

2-2 Conservative quantites of the motion

The system described by (1) is known to have the following three quantities as integrals.

1) Energy

$$E = -\frac{1}{4\pi} \sum_{\substack{(i\neq j)}} \sum \Gamma_i \Gamma_j \log r_{ij} \qquad (10)$$

2) Moment of Inertia (moment of second order)

$$I = \sum_{i=1}^{N} \Gamma_i \, |z_i|^2 \qquad (11)$$

which is due to rotational symmetry of the system.

3) Centroid

$$G = \sum_{i=1}^{N} \Gamma_i z_i / \sum_{i=1}^{N} \Gamma_i \equiv G_x + iG_y \quad \text{where} \qquad (12)$$

$$G_x = \sum_{i=1}^{N} \Gamma_i x_i / \sum_{i=1}^{N} \Gamma_i \qquad (13)$$

$$G_y = \sum_{i=1}^{N} \Gamma_i y_i / \sum_{i=1}^{N} \Gamma_i \qquad (14)$$

which are due to the translational symmetry of the system.

These quantities are conserved when requisite symmetries are satisfied. If some of them are lost because of the symmetry breaking, the freedom of the system becomes greater and chaotic behavior appears owing to the non-integrability of the system. The integrability for several conditions are listed on Table 1. For example, in a case where three vortices are located in an unbounded region, the number of equations is 6 while there are 4 integrals as mentioned. Thus 6-4=2 is the number of effective unknowns. When the number of unknowns is less than 3, then the system is integrable.

Table 1 Integrability of several conditions.
(triangle means that in high symmetry it is sometimes integrable).

	conditions	numbers of equations	conservatives	integrability
1	Two vortices (without boundaries and external flow)	4	E, I, G_x, G_y	◯
2	Three vortices (without boundaries and external flow)	6	E, I, G_x, G_y	◯
3	Two vortices (in a circle)	4	E, I	◯
4	Three vortices (in a circle)	6	E, I	✕ (△)
5	Two vortices (in a domain without rotational symmetry) ex. Two vortices in a semi circle.	4	E	✕ (△)
6	Two vortices (in a steady flow with translational symmetry in a certain direction)	4	$E, G_x,$(or G_y)	◯

In the following sections we exhibit examples of the regular (integrable) motion and the chaotic (not integrable in general) motion of vortices.

3 SIMILAR SOLUTIONS of THREE POINT VORTICES [3] [4] [5]

As we have mentioned in the previous section, the system of three vortices in an unbounded region is perfectly integrable. Among the solutions we can see several interesting phenomena as similar solutions of motion which are obtainable even when N>3.

Let us denote z_j as a product of a space part K_j and a common temporal part $f(t)$:

$$z_j = K_j f(t) \tag{15}$$

where both K_j and f are complex variables. By substituting this expression into (1) we have two equations for K_j and f respectively:

$$c\dot{K}_j = - \sum_{m=1}^{N} \frac{\kappa_m}{\overline{K}_j - \overline{K}_m} \qquad \kappa_m \equiv \frac{\Gamma_m}{2\pi i} \qquad (16)$$

$$\dot{f}\,\overline{f} = c \equiv A + iB \qquad (17)$$

We first consider the temporal part. As f is a complex variable, we assume that $f = r\exp(i\phi)$ and substitute it into (17) to have following two equations for r and ϕ:

$$r\,\dot{r} = A \qquad (18)$$

$$r^2\,\dot{\phi} = B \qquad (19)$$

These equations can be easily integrated and we have

$$f = \sqrt{2At + \alpha}\,\exp\left[\,i\,\{\frac{B}{2A}\,\log\,(2At + \alpha) + \beta\,\}\right] \qquad (20)$$

Equation (19) shows that, corresponding to the sign of A, f increases with time (A>0), rotates rigidly (A=0), and decreases with time (A<0).

Next for the space part. We take N=3, i.e. the three body problem of vortices. The equations for K_j are explicitly given as follows:

$$c\dot{K}_1 = - \left(\frac{\kappa_2}{\overline{K}_1 - \overline{K}_2} + \frac{\kappa_3}{\overline{K}_1 - \overline{K}_3} \right) \qquad (21)$$

$$c\dot{K}_2 = - \left(\frac{\kappa_1}{\overline{K}_2 - \overline{K}_1} + \frac{\kappa_3}{\overline{K}_2 - \overline{K}_3} \right) \qquad (22)$$

$$c\dot{K}_3 = - \left(\frac{\kappa_1}{\overline{K}_3 - \overline{K}_1} + \frac{\kappa_3}{\overline{K}_3 - \overline{K}_2} \right) \qquad (23)$$

+ complex conjugate of the above equations.

We solve them by the following process.
(1) eliminate $\overline{K}_1$, $\overline{K}_2$, $\overline{K}_3$. (2) denote $x = K_1 - K_2$ and $y = K_2 - K_3$, and express the equations by means of x and y. (3) make a polynomial and factorize it.   Finally we have:

$$(\kappa_1\kappa_2 + \kappa_2\kappa_3 + \kappa_3\kappa_1)(X^2 + X + 1)\{(\kappa_1 + \kappa_2)X^3 + (2\kappa_1 + \kappa_2)X^2$$
$$- (\kappa_2 + 2\kappa_3)X - (\kappa_2 + \kappa_3)\} = 0 \qquad (24)$$

Each term has its own meaning respectively.

The vanishing of the first factor corresponds to the spiralling collapse or expansion of three vortices $A \neq 0$, as is easily seen by introducing (15) and (20) into (10). Otherwise we have $A = 0$, i.e. $r = $ const. In this case the roots of the second factor $X = \omega$, ω^2 where ω is the cubic root of 1, correspond to regular triangle solutions. Therefore, no matter what the strengths are, three vortices located in regular triangle shape rotate without changing the shape.(Fig.4,5) Final factor yields:

$$(\kappa_1 + \kappa_2)X^3 + (2\kappa_1 + \kappa_2)X^2 - (\kappa_2 + 2\kappa_3)X - (\kappa_2 + \kappa_3) = 0 \qquad (25)$$

As this equation is cubic, we can solve it analytically by means of Cardano's method. The details have not been worked out yet, but the real roots are known to correspond to a straight line configuration which rotate without changing the geometry.

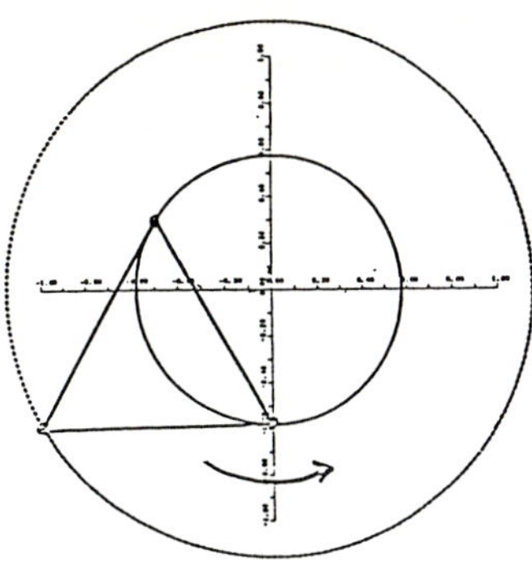

Fig. 4

Regular triangle solution

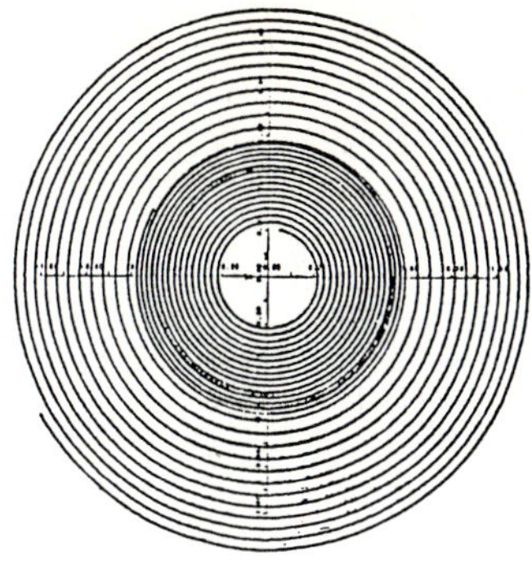

Fig. 5

Contraction case when $(\kappa_1,\kappa_2,\kappa_3) = (2,2,-1)$

These results have an analogy with particular solutions obtained by Lagrange in Celestial mechanics.

4 TWO POINT VORTICES in a SEMICIRCLE [1]

Referring to Table 1, two vortices in a domain without rotational symmetry may reveal chaotic behavior because the integrals, except for energy, are lost.

Simulation of the motion of vortices in a semicircle is realized by putting two image vortices with respect to the diameter, and four, outside the circle. For simplicity, we assume a unit circle and now the strength of the vortices are the same (=1) but senses are opposite.

As a usual step in investigating nonintegrable systems, we first seek for fixed points. There are two fixed points, one of which is stable ($r_1=r_2=(\sqrt{17}-4)^{1/4}$, $\theta_1=\pi/4, \theta_2=3\pi/4$) and the other   unstable ($r_1=0.75264$, $r_2=0.25578$, $\theta_1=\theta_2=\pi/2$). Then we add perturbations to the stable fixed point. Fig6 is an energy contour map when one vortex is located at the original fixed point. Roughly speaking we have four types of contour lines in Fig.6. (1) elliptic-like line around the other fixed point. (2) winding line outside the elliptic-like line. (3) the line along the outer boundary. (4) circular line near the fixed point. These lines are shown in Fig.6 as solid lines. According to these four types, we find four types of orbits of vortices. Of course we should note that if we shift one vortex from the fixed point as an initial condition, both begin to move, so Fig. 6   shows an instant energy cotour map. However, the rough shape of the map is unchanged and the four types of contours always remain.

We exihibit computer simulations for several initial conditions corresponding to the four types of contour lines respectively: Beside those we put the power spectrum of the time development of one variable.

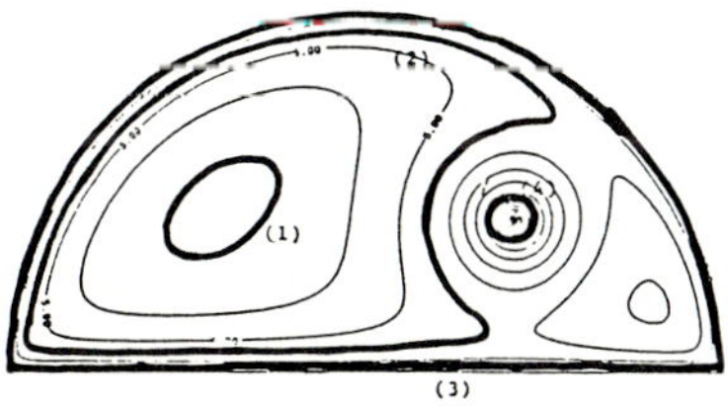

Fig. 6

Energy contour map. There are four types of lines

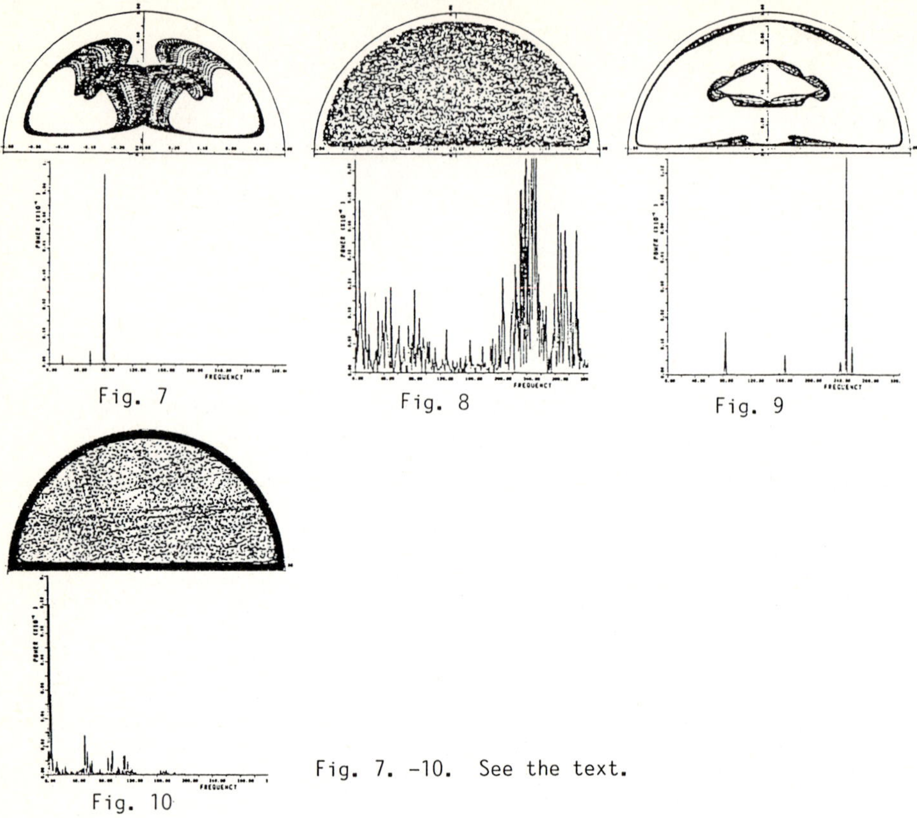

Fig. 7

Fig. 8

Fig. 9

Fig. 10

Fig. 7. -10.   See the text.

The doubly periodic motion is seen in Fig. 7 and in its power spectrum, corresponding to the elliptic-like line. When a shift is made, the orbits become gradually complicated. (Fig.7) Finally they become entirely messy, (Fig. 8) corresponding to the winding lines. But if we shift one vortex further, it moves along the outer boundary, and the motion is periodic again.(Fig.9) Thus the appearance changes as regular-chaos-regular owing to the initial shift from the fixed point. The final figure shows a different kind of chaotic behavior.(Fig. 10) It is obtained when two vortices are put very close together in the semicircle. They go straight, then collide with the boundary and separate. Both move along the boundary and meet to pair again somewhere. In this meet-separate-meet-... process trajectories cover the whole area in a random way. It may be called the billiard-type chaos.

5 CONCLUSION

In relation to their integrability, 2-dimensional point vortices reveal various aspects as regular and chaotic motions. The detail of mechanisms of vortex motion as a singularity, for example, full aspects of collision, or motion in a complex time plane, is a future work.

1.   H. Hasimoto, K. Ishii, Y. Kimura and M. Sakiyama: proc. IUTAM Symp. on Turbulence and Chaotic Phenomena in Fluids, Kyoto, 1983
2.   C. C. Lin: Proc. Nat. Acad. Sci. USA 27, (1941), pp570-575
3.   J. L. Synge: Can. J. Math. 1, (1949), pp257-270
4.   H. Aref: Phys. Fluids 22, (1979), pp393-400
5.   E. A. Novikov and Yu. B. Sedov: Sov. Phys. JETP 50,(1979), pp297-301
6.   D. V. Choodnovsky and G. V. Choodnovsky: IL NUOVO CIMENTO B 40 pp339-353
7.   Y. Ishimori: Prog. Theor. Phys. 72(1), 1984, pp33-37

Explode-Decay Solitons

Akira Nakamura

Physics Laboratory, Osaka University of Foreign Studies, Aomadani, Minoo City
Osaka 562, Japan

1. Origin of Explode-Decay Solitons

In the study of so-called integrable nonlinear systems, the importance
of the role played by solitons has been widely recognized.   Solitons
are the localized nonlinear waves which (i) represent the elementary
mode of excitations of the system under consideration and (ii) have
beautiful superposition properties in spite of the nonlinearity of the
system.
   So far the study of soliton phenomena has been mainly done in the
1+1d (=1 space plus 1 time dimensional) systems. The first big extension of the soliton theory to the 2+1d systems was done in the cylindrical KdV equation [1],

$$u_t + 6uu_x + u_{xxx} + u/(2t) = 0. \tag{1}$$

Suppose, for example, that you throw a stone into a quiet pond.
When the stone lands in the water, you have ring form (cylindrical)
waves going outwards.   Equation (1) describes such waves observed in
the radial direction (in this case x should be read as radius r).
Historically this equation was first   considered in the plasma
physics, where the time reversed process of the above has been investigated. As the ring form wave goes inwards, the amplitude of the wave
increases which physically leads to high density plasma excitations.

   It has been discovered that the cylindrical KdV equation actually
has stable localized waves or solitons, and theoretically has all the
beautiful structures of soliton systems such as the inverse scattering
scheme, Bäcklund transformation, Hirota bilinear form, the infinite
number of conservation laws and so on.   But of course there is one
simple and big difference between cylindrical KdV soliton and other
solitons in the ordinary 1+1d integrable systems.   Namely in the
cylindrical KdV case, the wave form of soliton changes with time
whereas ordinary soliton (such as single soliton of 1+1d KdV equation)
is the propagating wave with constant speed and constant wave form.
Noting this, from now on we call the nonlinear wave exemplified by
the cylindrical KdV soliton as "explode-decay" soliton or in short
"ed-soliton."   (Sometimes we also call ed-soliton as "ripplon.")
This is the origin of the new type solitons of explode-decay mode.

2. Explode-Decay Solitons in 2+1d Integrable Systems

Naturally one becomes curious about whether this rather new mode
wave can exist in other nonlinear systems or not.   Thus one is led
to the search for similar ed-solitons. The 2+1d KdV (=KP) equation,

$$(u_t + 6uu_x + u_{xxx})_x + 3a^2 u_{yy} = 0, \quad a = \text{constant}, \tag{2}$$

was known to be related to the cylindrical KdV equation and consequently
has similar ed-solitons [2].

In the 1+1d soliton case, if certain characteristics were found
in the KdV equation, then the same also happened to be the case
in other soliton equations such as the nonlinear Schrödinger (=NLS)
equation, Toda lattice equation and so on.

Inspired by this fact, the present author investigated 2+1d NLS
equation [3],

$$iu_t - bu_{xx} + cu_{yy} + du^*uu - 2wu = 0, \quad bw_{xx} + cw_{yy} - bd(u^*u)_{xx} = 0,$$

$$b,c,d = \text{constants}, \tag{3}$$

and 2+1d Toda equation [4],

$$au_{tt}(t,y,n) + bu_{yy}(t,y,n) - \exp(-u(t,y,n) + u(t,y,n-1))$$

$$+ \exp(-u(t,y,n+1) + u(t,y,n)) = 0, \quad a,b = \text{constants}. \tag{4}$$

Equation (3) describes the two space dimensional water wave under
gravity and reduces to the usual 1+1d cubic NLS equation when both
u and w do not depend on y. Similarly (4) reduces to the usual
1+1d Toda lattice describing the motion of lattice molecules when
u does not depend on y. It has been found that (3) has ed-solitons
which are expressed by Hermite functions, H_n, Airy functions, Ai, and
Bessel functions of order 1/4, $J_{1/4}$ [3]. Equation (4) has ed-solitons
expressed by Bessel function of order n, J_n [4].

In the three above examples, namely 2+1d KdV, 2+1d NLS, and 2+1d
Toda equations, it has been discovered that ordinary type N-solution
solutions exist. Here "ordinary type soliton" means waves with constant
propagating speed and constant wave form, and mathematically expressed
by secant hyperbolic or exponential function.

We have found that in these three representative 2+1d soliton
equations, not only N-ordinary soliton but also N'-ed-soliton and
even N-ordinary-soliton-N'-ed-soliton solutions exist which are exact
nonlinear superposition state of N ordinary solitons and N' ed-solitons
[3], [4], [5]. Following this, the present author has presented the
following conjectures about ed-solitons [3];

(i) If a 2+1d nonlinear equation has N-(ordinary)-soliton solutions
     then it also has N-ed-soliton solutions and N-ordinary-soliton-
     N'-ed-soliton solutions.
(ii) These ed-solitons are mathematically expressed by special
     functions such as the Bessel function. (5)

3. Explode-Decay Solitons in 2+1d Non-Integrable Systems

Next we consider 2+1d system where only 1-soliton is known to date.
For this example,we consider a2+1d cubic NLS equation of the form [6],

$$iu_t + bu_{xx} + cu_{yy} + du^*uu = 0, \quad b,c,d = \text{constants}. \tag{6}$$

We have found that this equation also has 1-ed-soliton solution.
Clearly (6) includes 1+1d cubic NLS equation as the special choice of
coefficient c=0. Thus in the sense of an essentially 1+1d system, (6)
has N-soliton solutions. In the same way, we can check that it has
N-ed-soliton solutions. This can be shown by the following transformation which gives a new solution of (6) from the arbitrary old solution

of the same equation.   The transformation is written as follows [6];

$$u(x,y,t)=(1/t)U(X,Y,T)\exp(ix^2/(4bt)+iy^2/(4ct)),$$

$$X=x/t, \quad Y=y/t, \quad T=-1/t. \tag{7}$$

One can directly check that if U satisfies (6) with u,x,y,t replaced by U,X,Y,T, then u satisfies (6) itself.   If we apply transformation (7) to the known essentially 1+1d N-soliton solution of (6), then we have the essentially 1+1d N-ed-soliton solution of (6). In the case of (6), even in the limited sense of an essentially 1+1d wave, we do not know at present whether an N-ordinary-soliton-N'-ed-soliton solution exists or not.

4. Explode-Decay Solitons in 1+1d Systems

As shown above, ed-soliton originally appeared in the 2+1d nonlinear systems. But it is of use in searching similar mode in 1+1d systems. At present, in the 1+1d case, we know only certain special cases.

The integro-differential equation,

$$u_t+u_{xxx}-(3uHu_x)_x-(u^3)_x=0, \quad Hu(x)=(P/\pi)\int_{-\infty}^{\infty} u(x')/(x'-x)dx', \tag{8}$$

has an ordinary algebraic soliton and ed-soliton expressed by Airy function Ai [7]. In this case, the equation itself has been derived not through direct physical consideration but in terms of certain modification of the Benjamin-Ono equation which describes the deep water wave.

We have another example of the classical Boussinesq equation written as

$$u_t=((1+u)v+a^2v_{xx})_x, \quad v_t=(u+v^2/2)_x, \quad a=\text{constant}. \tag{9}$$

This equation describes the water wave where certain effects of a higher order than the case of usual KdV equation are taken into account. Recently we have found that in the case of $a^2=-1$, (9) has ed-soliton solutions expressed by Hermite functions [8].

5. List of Explicit Forms of Some Simple Explode-Decay Solitons

In the following, for the above mentioned equations we give the explicit form of some simple ed-soliton solutions. In the following list, we adopt the notation such that the solution (2-1), (3-1), ... corresponds to the solution of the equation (2), (3), ... respectively.

$$u-(2\log f)_{xx}, \quad f=1+s^2(12t)^{-2/3}\int^x dx' Ai^2(z),$$

$$z=x'(12t)^{-1/3}+(y/a)^2(12t)^{-4/3}, \quad s\text{-arbitrary constant}. \tag{2-1}$$

$$u=g/f, \quad w=(2b\log f)_{xx}, \quad g=st^{-1}\exp(ix^2/(-4bt)+iy^2/(4ct)),$$

$$f=1+(1/8)s*sd((x+x_i t)^2/(bt^2)+(y+y_i t)^2/(ct^2)),$$

$$s,x_i,y_i=\text{arbitrary constants}. \tag{3-1}$$

$$u = -\log f(t,y,n+1) + \log f(t,y,n),$$

$$f(t,y,n) = 1 + s^2 \sum_{m=n}^{\infty} J_m^2 ((t^2/a + y^2/b)^{1/2}),$$

$$s = \text{arbitrary constant.} \tag{4-1}$$

$$u = i(\log f^*/f)_x, \quad f = Ai(x(12t)^{-1/3} - is^2 |12t|^{1/6} \text{sgn } t),$$

$$s = \text{arbitrary constant.} \tag{8-1}$$

$$u = -1 - 2(\log ff^*)_{xx}, \quad v = -2i(\log f/f^*)_x, \quad f = x^2 + 2it. \tag{9-1}$$

6.   References

1 S. Maxon and J. Viecelli: Phys. Fluids 17 , 1614 (1974)
2 R.S. Johnson and S. Thompson: Phys. Lett. 66A , 279 (1978)
3 A. Nakamura: J. Phys. Soc. Jpn: 52 , 3713 (1983)
4 A. Nakamura: J. Phys. Soc. Jpn: 52 , 380 (1983)
5 A. Nakamura: J. Phys. Soc. Jpn: 51 , 19 (1982)
6 A. Nakamura: J. Phys. Soc. Jpn: 50 , 2469 (1981)
7 A. Nakamura: J. Phys. Soc. Jpn: 51 , 2057 (1982)
8 R. Hirota and A. Nakamura: preprint (1984)

Part IV

Solitons in Condensed Matter Physics

A Field Theorist's View of Conducting Polymers:
Solitons in Polyacetylene and Related Systems

David K. Campbell

Center for Nonlinear Studies and Theoretical Division, Los Alamos National Lab.
Los Alamos, NM 87545, USA

Introduction

From the perspective of a theoretical physicist, one of the most exciting recent examples of serendipitous interaction between apparently unconnected fields has been the recognition that relativistic field theory models, originally concocted by high energy physicists as theoretical "laboratories" to explore certain features of quantum field theory, can actually be directly applied to real condensed matter systems. The best-known example of this "field theory connection" is polyacetylene -- $(CH)_x$ -- a quasi-one dimensional organic polymer whose exotic properties have stimulated considerable interest in the past several years [1].

In this necessarily brief discussion, I will try to survey, from a field theorist's perspective, the role of nonlinear excitations -- solitons -- in quasi-one-dimensional conducting polymers, including polyacetylene and related systems. This deliberately unconventional -- if not heretical -- view serves, if nothing else, to stress the broad applicability of the important concept of solitons and the importance of transferring insights gained from one nonlinear system to others. After a brief introduction to the background chemistry, I will present a simple microscopic physical model [2-6], involving the coupling of electrons and phonons, for polyacetylene. I next illustrate that, in the static, continuum limit [6], this model is equivalent [7] to a familiar model relativistic field theory [8,9]; this represents a first connection to field theory. This equivalence, together with the prior knowledge of solutions to the field theory [9], permits a ready analytic discussion of the nonlinear excitations -- kinks and polarons, in trans-$(CH)_x$, polarons and "multi" polarons in cis-$(CH)_x$ -- expected in the real materials. Returning to our simple microscopic physical model, I show that adding the effect of electron-electron repulsive interactions to the discrete version of the model leads to a Hamiltonian of the general type considered in several recent simulations of low-dimensional field theories in elementary particles and condensed matter physics [10-14]; this is a second field theory connection. I review some of the quantum Monte Carlo results [12-14] for these systems and close by indicating a number of areas of continuing interest.

Background Chemistry

As the chemical formula $(CH)_x$ suggests, polyacetylene consists of a large grouping of C (carbon)-H (hydrogen) units. The material occurs in two isomeric forms: trans-$(CH)_x$ and cis-$(CH)_x$. The schematic chemical structure of trans-$(CH)_x$ is shown in Fig. 1 and that of cis-$(CH)_x$ in Fig. 2.

The idealized (infinite chain) structure shown in Fig. 1 summarizes conveniently a number of features contained in the physical model for trans-$(CH)_x$ which we will introduce later. First, the structure shows that the compound contains a carbon backbone consisting of alternating single and double bonds. Hence the fundamental unit which repeats contains two carbons and two hydrogens, and the chain is thus said to be "dimerized". Note that the double bonds are physically shorter than the single bonds, and thus the schematic structure indicates that a uniform bond length state is unfavored, relative to bond alternation. Second, taken literally as drawn, the structures suggest that polyacetylene is a quasi-one dimensional material, with the only significant physical dimension being

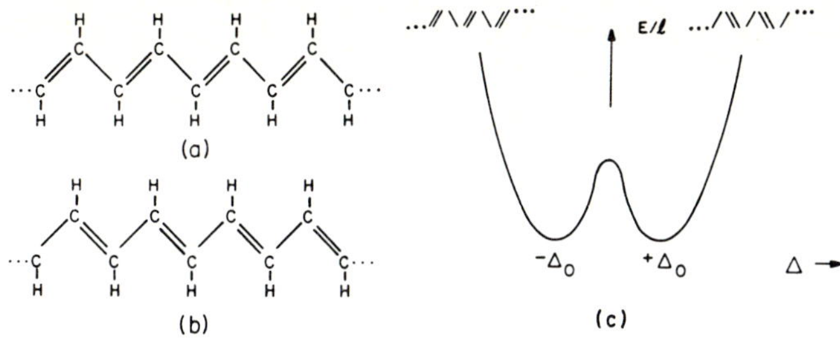

Fig. 1.   For <u>trans</u>-(CH)$_x$: the two degenerate bond alternation patterns [(a) and (b)]; (c) the energy per unit length versus bond alternation (or band gap)

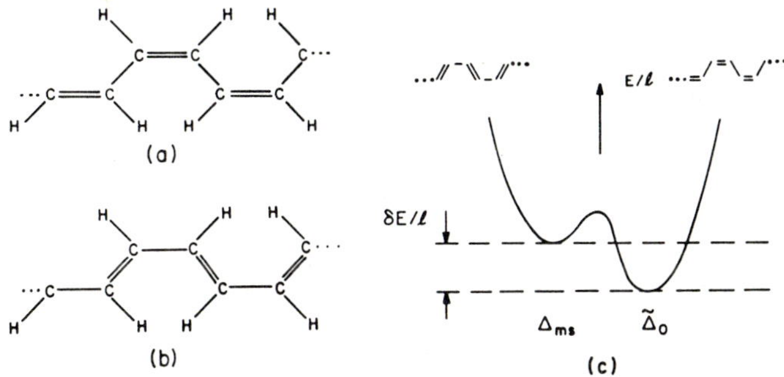

Fig. 2.   For <u>cis</u>-(CH)$_x$: the two bond alternation patterns for (a) the ground state (b) the "metastable" state; (c) the energy per unit length versus bond alternation (or band gap)

along the carbon backbone. Third, the difference between the two structures shown in Fig. 1 is simply whether, for a given carbon atom, the double bond lies to the left or to the right, since all bonds are at the same angle to the backbone axis. For an infinite chain, it seems intuitively clear that this should make no differ-ence to the energy of the system. We shall see below this this is indeed the case and that the ground state of <u>trans</u>-(CH)$_x$, as indicated in Fig. 1c, is two-fold degenerate. In contrast, for <u>cis</u>-(CH)$_x$, as indicated in Fig. 2c, the two distinct bond alternation patters are <u>non</u>-degenerate in energy. The unique ground state of <u>cis</u>-(CH)$_x$ will have profound implications for the allowed exci-tations.

Theoretical Models of <u>trans</u>-(CH)$_x$

To model microscopically the <u>trans</u>-(CH)$_x$ chain shown in Fig. 1, one must describe the coupled motions of the lattice backbone of C-H units and the electrons that can in principle move along the chain. Of the four valence electrons per carbon, three form relatively deeply bound σ-molecular orbitals in the (CH)$_x$ polymer and thus can, at least in the first instance, be treated as nondynamic. Thus the problem reduces to describing the coupled motion of the (C-H) lattice and the single (π-orbital) electron per carbon that, heuristically speaking, "determines where the double bond goes". The explicit Hamiltonian for the discrete system has the form [5]

$$H = \frac{1}{2} M \Sigma_i \, \dot{u}_i^2 + \frac{1}{2} K \, \Sigma_i (u_i - u_{i+1})^2$$

$$+ \Sigma_{i,s} [t_o - \alpha(u_i - u_{i+1})] \times (c^{\dagger}_{i,s} c_{i+1,s} + c^{\dagger}_{i+1,s} c_{i,s}) . \tag{1}$$

177

Here u_i describes the displacement from equilibrium of the (CH) units along the chain, $c_{i,s}$ ($c_{i,s}^\dagger$) annihilates (creates) a π-electron of spin $s(= \pm\tfrac{1}{2})$ at site i, and α represents the coupling between electrons and phonons, i.e., the change in the electron hopping term caused by a change of the relative displacement of adjacent (CH) units. The harmonic strain term proportional to K models the effect of the σ-bonds.

Although the lattice spacing -- call it "a" -- between adjacent (CH) units does not appear explicitly in (1), it plays a crucial role in the discrete theory in that it determines the size in momentum space of the Brillouin zones. To see the implications of this explicitly, let us consider the form of the single π electron spectrum in the lattice model. Figure 3 shows this spectrum for two particular choices of the phonon "coordinates" u_i. In Fig. 3a, we plot the gapless spectrum corresponding to the hypothetical chain with all (CH) units at their equilibrium positions, so $u_i = 0$. In Fig. 3b, we show the spectrum for the perfectly dimerized chain, corresponding to the chemical structure shown in Fig. 1, with $u_n = (-1)^n u_o$. That the true ground state of H in (1) <u>does</u> correspond to the dimerized configuration, $u_n = (-1)^n u_o$, can be shown by <u>direct</u> analytic calculation; this result reflects the well-known Peierls instability [15] of one-dimensional coupled electron-phonon systems. To go beyond this result in the discrete model requires numerical studies, and hence to gain further analytic insight we will turn to the <u>continuum</u> model of <u>trans</u>-$(CH)_x$, which can be derived as the limit, for a $\to$ 0, of the <u>Hamiltonian in (1)</u> [3,6]. We refer to the literature for this derivation [3,6], and note here only that in the continuum model the lattice coordinates u_i are replaced by the bond alternation/band gap order parameter, $\Delta(x)$ -- proportional to $(u_i - u_{i+1})$ and conventionally having dimensions of energy -- and the electron creation and annihilation operators are replaced by a two component field, $\psi_s(x) = (\psi_s^{(1)+}, \psi_s^{(2)+})$. The two components reflect the two distinct "parts" ($\pm k_F$) of the fermi surface, as shown in Fig. 3. The resulting continuum Hamiltonian is

$$\tilde{H} = \Sigma_s \int dx \{ \frac{w_Q^2}{2g^2} \Delta^2(x) + \psi_s^+(x)[iv_F\sigma_2 \frac{\partial}{\partial x} + \Delta(x)\sigma_3]\psi_s(x)\} \tag{2}$$

where $w_Q^2/2g^2$ is the net effective electron-phonon coupling constant, σ_i is the ith Pauli matrix, and v_F is the fermi velocity (in units with $\hbar = 1$, $v_F = Wa/2$). For reasons of later convenience, we have not used the standard basis for the electron wave functions; serious readers will have no difficulty making the transformation necessary for comparison with the literature [6,7]. In deriving (2), the lattice kinetic energy -- which would lead to a term proportional to $\dot{\Delta}^2(x)$ -- has been explicitly ignored. Obviously, this will have no effect on the <u>static</u> solutions we discuss below, but it is very significant for on-going studies <u>involving</u> the dynamics of solitons in $(CH)_x$.

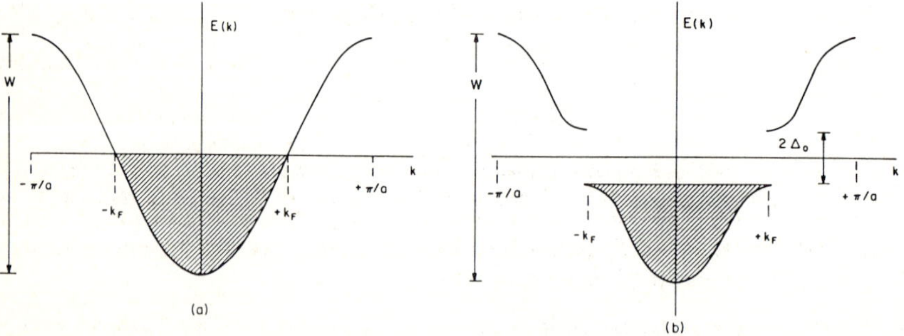

Fig. 3. The single (π) electron spectrum in <u>trans</u>-$(CH)_x$ in (a) the hypothetical uniform (undimerized) case and (b) the actual dimerized ground state. The shading indicates filled electron states. W is the full band width ($\cong$ 10 eV) and $2\Delta_o$ is the full band gap ($\cong$ 1.4 eV). Since there is precisely one "active" electron per carbon, the band is half-filled, so $k_F = \pi/2a$, where the lattice spacing a $\cong$ 1.22 Å

178

The field equations that follow from the variation of $\tilde{H}$ are for the single particle electron wave functions

$$\varepsilon_n \psi_{n,s}^{(1)} = v_F \frac{\partial \psi_{n,s}^{(2)}}{\partial x} + \Delta \psi_{n,s}^{(1)} \tag{3a}$$

$$\varepsilon_n \psi_{n,s}^{(2)} = -v_F \frac{\partial \psi_{n,s}^{(1)}}{\partial x} - \Delta \psi_{n,s}^{(2)} \tag{3b}$$

and for the "self-consistent" band alternation/bond gap parameter

$$\Delta(x) = -g^2 (\omega_Q^2)^{-1} \sum_{n,s}' (|\psi_{n,s}^{(1)}|^2 - |\psi_{n,s}^{(2)}|^2) \ . \tag{3c}$$

The prime on the summation symbol in (3c) indicates that the sum is over all occupied electron states.

How do Eqs. (3) express the conclusions of our discussion of the chemical structure of trans-$(CH)_x$? Recall that we anticipated that the configurational energy as a function of (constant) Δ would look like Fig. 1c, so that a uniform bond length -- equivalently, a gapless electronic spectrum -- state ($\Delta = 0$) is unstable (the "Peierls instability" [15]) and the ground state is two-fold degenerate. Using the explicit plane wave solutions to (3) appropriate for constant Δ, one can show that all these conclusions are indeed correct [3,6,7], that $E(\Delta)$ varies as $\Delta^2 \ln\Delta^2$, and that the gap equation determines Δ_o self-consistently via

$$\Delta_0 = g^2 (2\omega_Q^2)^{-1} \frac{1}{2\pi} \int_{-K}^{K} \frac{dk}{(k^2 v_F^2 + \Delta_0^2)^{\frac{1}{2}}} \tag{4}$$

Note the crucial result that the cut-off wave vector, K, is chosen such that the energy at the cut-off corresponds to the correct π-electron band width (W, as shown in Fig. 3) in $(CH)_x$. Thus $W = 2(K^2 v_F^2 + \Delta_o^2)^{\frac{1}{2}} \cong 2Kv_F$, since $\Delta_o \ll Kv_F$. The need for this cut-off comes solely from the linearization of the spectrum about k_F made in going to the continuum limit; the true spectrum of the discrete model, as shown in Fig. 3, automatically covers a finite range. Solving (4) for $Kv_F \gg \Delta_o$ gives

$$\Delta_0 = W \exp(-\lambda^{-1}) \quad \text{with} \quad \lambda = 2g^2 (\pi v_F \omega_Q^2)^{-1} \ , \tag{5}$$

where the parameters have all been previously introduced (see Fig. 3). The electron spectrum for $\Delta = \Delta_o$ consists of a valence band [$\varepsilon(k) = -(k^2 v_F^2 + \Delta_o^2)^{\frac{1}{2}}$] and a conduction band [$\varepsilon(k) = +(k^2 v_F^2 + \Delta_o^2)^{\frac{1}{2}}$] separated by a gap of $2\Delta_o$. These subbands are the continuum versions of those shown in Fig. 3b for the lattice model.

The Field Theory Connection
The pseudo-relativistic or "Dirac-like" structure of Eq. 3(a) and (b) provides the first hint of a possible relation between the continuum theory of trans-$(CH)_x$ and a model relativistic field theory. In fact, this relation is a precise equivalence: Eqs. (3) are identical [7] to the static, semi-classical equations of the $N = 2$ Gross-Neveu model, a well-studied relativistic field theory in one space dimension and time [8,9].

To explain this result as briefly as possible, we recall that, for arbitrary N, the Gross-Neveu model can be described by the Lagrangian density [8,9]

$$\mathcal{L}(x) = \sum_{\alpha=1}^{N} \bar{\Psi}^\alpha(x) \{i\gamma_\mu \frac{\partial}{\partial x_\mu} - g_{GN}\sigma(x)\} \Psi^\alpha(x) - \tfrac{1}{2}\sigma^2(x) \ . \tag{6}$$

Here γ_μ ($\mu = 0,1$) are the Dirac γ matrices in two dimensions; we follow the usual convention with $\gamma_0 = \sigma_3$ and $\gamma_1 = i\sigma_1$. Recall that $\bar{\Psi} = \Psi^\dagger \gamma_0$. For each α, $\Psi^\dagger$ is a two component Dirac spinor, $(\Psi_1^\dagger, \Psi_2^\dagger)$. The internal SU(N) symmetry index $\alpha = \{1,\dots,N\}$ labels the "particle type". Apart from this internal symmetry, it possesses a discrete "chiral" symmetry, in that it is invariant under $\sigma \to -\sigma$, $\Psi \to \gamma_5 \Psi$, with $\gamma_5 \equiv \gamma_0 \gamma_1$. The static semi-classical equations corresponding to (6) can be shown to be a Dirac equation for single fermion wave functions and a self-consistency equation for $\sigma(x)$. Specifically, defining stationary solutions by $\Psi^\alpha(x,t) \equiv \exp(-i\varepsilon_n t) \, \Psi^\alpha(n;x)$, the Dirac equation becomes

$$(\varepsilon_n \gamma_0 + i\gamma_1 \frac{\partial}{\partial x} - g_{GN}\sigma(x)) \, \Psi^\alpha(n;x) = 0 \, , \tag{7}$$

independently for each α. By making the transformations $g_{GN}\sigma \to \Delta$, $x \to x/k_F$ and performing some straightforward algebra, (7) can be seen to be precisely equivalent to the electron equations -- 3(a) and 3(b) -- for $\underline{\text{trans}}$-(CH)$_x$. For the Gross-Neveu model, the correct form of the self-consistency equation for $\sigma(x)$ can be shown to be [8,9]

$$Z(\Lambda)\sigma(x) = -g_{GN} \sum_{\alpha,n} \bar{\Psi}^\alpha(n;x)\Psi^\alpha(n;x) \, . \tag{8}$$

$$\varepsilon_n < 0$$

In (8), $Z(\Lambda)$ represents the ultraviolet renormalization -- Λ is the ultraviolet momentum cut-off -- necessary to cancel the divergence of the infinite (as $\Lambda \to \infty$) sum over $\bar{\Psi}\Psi$. This sum is over all $\Psi^\alpha(n;x)$ satisfying (7) with energy less than zero. Clearly this sum over the "negative energy sea" is analogous to the sum over the valence band in (CH)$_x$. Further, and very importantly, the sum over α in (8) becomes just like the spin sum in (CH)$_x$ $\underline{\text{if we}}$ $\underline{\text{choose}}$ $\underline{N = 2}$. The explicit form of $Z(\Lambda)$ is

$$Z(\Lambda) = \frac{g_{GN}^2}{\pi} N \int_0^\Lambda \frac{dk}{(k^2+m^2)^{\frac{1}{2}}} = \frac{g_{GN}^2}{\pi} N \, \ell n\{\frac{\Lambda + (\Lambda^2 + m^2)^{\frac{1}{2}}}{m}\} \cong \frac{g_{GN}^2}{\pi} N \, \ell n \, \frac{2\Lambda}{m} \tag{9}$$

where $m \equiv g_{GN}\sigma_0$. Although both the structure and interpretation of (8) are obviously directly analogous to the gap equation in (CH)$_x$, establishing an exact equivalence at first appears to present a problem for the renormalization factor $Z(\Lambda)$ in (8) seems to have no counterpart in (3c). Actually, this reflects merely a (very interesting) difference of interpretation between the solid state and field theory applications. Recall that in the continuum solid state model, because of the linearization of the electron spectrum, we had to cut-off the gap equation integral (3c) at some wave vector corresponding to the full band width, W. Thus the cut-off is directly related to a physical quantity, and there is no need for an "infinite" renormalization. Said another way, the relation between Δ_0 and W [Eq. (3c)] is just the requirement that the self-consistency condition (3c) for Δ_0 be solved with a "renormalization constant" equal to 1! We can achieve a comparable situation in the field theory context by choosing σ_0 such that

$$m \equiv g_{GN}\sigma_0 = 2\Lambda \exp(-\pi/Ng_{GN}^2) \, . \tag{10}$$

Using this result, from (9) it follows that $Z(\Lambda) = 1$. Although this choice might appear unconventional in a field theory analysis, it is in fact completely equivalent to the standard prescription that renormalizes the theory by requiring the second derivative of the effective potential to be unity at an arbitrary subtraction point, σ_s [7-9].

Choosing constants appropriately, the $\underline{\text{formal}}$ equivalence between the continuum electron phonon equations in $\underline{\text{trans}}$-$\overline{(CH)}_x$ and the static, semi-classical Gross-Neveu model is thus established. There remains the important difference in interpretation, however, since both Δ_o and W are measurable in (CH)$_x$ whereas in the field theory, only m is measurable; Λ is an unphysical cut-off introduced to make the theory finite. Let us pursue this difference, as it illustrates how the (CH)$_x$ gap Eq. (3c) reflects three important concepts developed in recent field theory studies and embodied in Eq. (10): "dynamical symmetry breaking", "dimensional transmutation", and "asymptotic freedom". First, as noted above, the original Gross-Neveu Lagrangian possesses a "chiral symmetry" and appears to describe massless fermions. Nonetheless, the actual ground state contains a filled Dirac sea of massive fermions and is not invariant under $\sigma \to -\sigma$, $\Psi \to \gamma_5\Psi$. This is "dynamical spontaneous symmetry breaking" [8,16]. Second, for fixed N, the original Lagrangian contains only one parameter: the dimensionless coupling constant, g_{GN}. In the final, renormalized theory, the one physical parameter, as shown in (10), is the dimensional quantity the fermion mass. This is the phenomenon of "dimensional transmutation" [16]. Third, (10) correctly reflects the "asymptotic freedom" [17] of the Gross-Neveu model, for if one insists on a fixed m -- to define the same theory -- as $\Lambda \to \infty$, then the coupling constant, g_{GN}^2, must go to zero like one over the logarithm of Λ. It is amusing and instructive to see all three of these concepts emerging, albeit with altered interpretation, in Eq. (5), which describes a real material.

With this discussion of the general structure of the equations for trans-(CH)$_x$ and of the ground state solution, there remains only one point to stress before turning to the nonlinear excitations: this is the role of the internal symmetry index, N, in the Gross-Neveu model, which clearly takes the value of N = 2 in trans-(CH)$_x$. Are there other values of N for which the field theory can describe "real", quasi-one-dimensional systems? In fact, the answer is yes, but lack of space precludes a detailed discussion. The pure carbon polymer "carbyne", $(C \equiv C)_x$, can at the static, semi-classical level be shown to be related to as N = 4 Gross-Neveu model [18]. Further, certain charge transfer salts, can by virtue of a large effective Coulomb repulsion that prevents electrons of opposite spins from occupying the same site, be shown to be related to the N = 1 Gross-Neveu system [19]. For details, interested readers are referred to the literature [18,19].

Excitations in Polyacetylene and Related Systems
One obvious (and linear) excitation from the ground state of (CH)$_x$ or any related system is a particle-hole pair, in which an electron is promoted from the valence to the conduction band. The minimum energy for this excitation is $E_{ph}^{min} = 2\Delta_o$. More important and more interesting, in the context of trans-(CH)$_x$, are the nonlinear, soliton-like excitations: "kinks" and "polarons". Thus, let us discuss these in some detail. The existence of kink solutions to (3) is a deep consequence of the two-fold degeneracy of the trans-(CH)$_x$ ground state. As sketched in Fig. 4a, the kink soliton solution interpolates between the two degenerate ground states and has the explicit form [3,6,7]

$$\Delta_K = +\Delta_o \tanh x/\xi_o , \qquad (11)$$

with $\xi_o \equiv v_F/\Delta_o$. In addition, there is an anti-kink, $\bar{K}$, with $\Delta_{\bar{K}}(x) = -\Delta_K(x)$. The single electron spectrum associated with a kink (or anti-kink) consists of extended (modified plane wave) states in the conduction and valence bands plus an additional, localized "mid-gap" state with $\varepsilon_o = 0$. The explicit form of the mid gap electron wave function for the kink is [3,6,7]

$$\psi_o^{(1)} = \psi_o^{(2)} = N_o \operatorname{sech} x/\xi_o , \qquad (12)$$

with $N_o = \frac{1}{2} \xi_o^{-\frac{1}{2}}$. The kink excitation and mid-gap electron distribution are shown in Fig. 4(a).

For either K or $\bar{K}$, the "mid-gap" state can be occupied by 0, 1, or 2 electrons, leading to the localized excitations with bizarre spin/charge assignments. Explicitly, one finds that the neutral kink (K^o) has a single electron in the

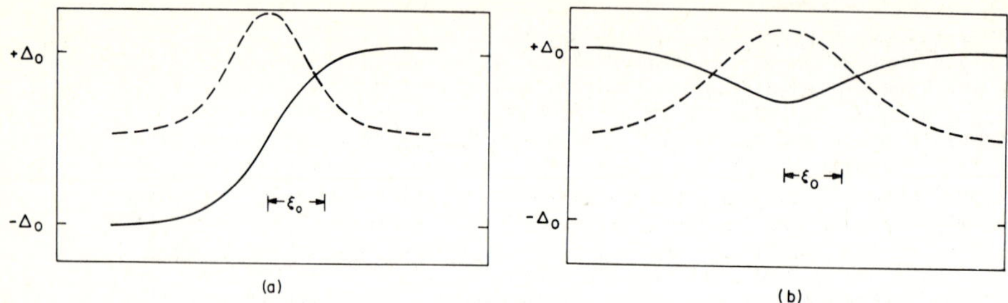

Fig. 4. A plot for <u>trans</u>-(CH)$_x$ of the bond alternation order parameter, $\Delta(x)$ (solid line) and the charge density, $\psi_o^\dagger(x)\psi_o(x)$, for the localized electronic state for (a) the "kink" soliton; (b) the "polaron"

state ε_o; since all states in the valence band remain spin-paired, the spin of the K^o is 1/2 (2,3,5,6)! Similar arguments show that K^+, which has no electrons in ε_o, and K^-, which has two electrons in ε_o, both have spin zero. These results seem to violate conventional solid state spin/charge relations which follow quite generally from the fact that the basic charge-carrying object, the electron, has spin ½ and charge (-e). Nonetheless, recent theoretical studies [20] have confirmed that this effect -- which is related to the concept of "fractional" charge [19,21] -- is expected, and the search for experimental confirmation of this prediction is on.

The energy of a single kink (or anti-kink) of any charge is $E_K = E_{\bar{K}} = 2\Delta_o/\pi$. However, <u>topological</u> constraints require that kinks be produced in $K\bar{K}$ pairs. Hence the minimum energy state involving kinks which can be excited from the ground state is a single $K\bar{K}$ pair with energy $4\Delta_o/\pi$. In a conventional semiconductor, the lowest-lying excitation would be a particle-hole pair, with minimum energy $E_{ph}^{min} = 2\Delta_o$. Thus one has the exciting theoretical prediction that the elementary excitations from the ground state in <u>trans</u>-(CH)$_x$ are the exotic "kink" solitons.

In addition to considering excitations from the ground state which conserve the total electron number, one is interested in predicting what excitations exist when single electrons are added to, or taken from, the ground state. This is particularly relevant for understanding the technologically important "doping" process, for the standard dopants either remove or add electrons to the undoped (CH)$_x$ polymer chains. For definiteness, let us consider what happens when a single electron is added to an infinite chain of <u>trans</u>-(CH)$_x$. If one tried to accommodate this electron in the mid-gap state associated with a kink, since kinks must be produced in $K\bar{K}$ pairs, the minimum energy for adding a single electron would be $4\Delta_o/\pi$. Thus it would be energetically more favorable simply to put the electron in the lowest state in the conduction band, since this costs an energy Δ_o. In fact, as we illustrate in detail below, the most energetically favorable state available to a single electron is the "polaron" [7,22-25] to which we now turn our attention.

Unlike kinks, polarons represent a localized deviation from one of the degenerate ground states of <u>trans</u>-(CH)$_x$, as illustrated in Fig. (4b). The explicit form of the polaron solution can be written in the revealing form (cf. Eq. (11))

$$\Delta_P(x) = \Delta_o - \kappa_o v_F\{\tanh \kappa_o v_F(x+x_o) - \tanh \kappa_o(x-x_o)\} \tag{13}$$

where $\tanh 2\kappa_o x_o = \kappa_o v_F/\Delta_o$. The single electron states for this form of Δ_P again include extended states in the conduction and valence bands, which are plane waves "phase-shifted" by the polaron "potential". The explicit forms are available in the literature [7]. In addition, there are <u>two</u> localized electronic states, with energies symmetrically placed at $\varepsilon_\pm = \pm w_o$, where $w_o = (\Delta_o^2 - \kappa_o^2 v_F^2)^{\frac{1}{2}}$. The electronic wave functions for these localized states are, for $\varepsilon_+ = \mp w_o$

182

$$\psi_+^{(1)} = N_+[\text{sech } \kappa_o(x+x_o)+\text{sech } \kappa_o(x-x_o)] \tag{14a}$$

$$\text{and } \psi_+^{(2)} = N_+[-\text{sech } \kappa_o(x+x_o)+\text{sech } \kappa_o(x-x_o)] \tag{14b}$$

where $N_+ = (\kappa_o/8)^{\frac{1}{2}}$. For $\varepsilon_- = -\omega_o$, $\psi_-^{(1)} = \psi_+^{(2)}$ and $\psi_-^{(2)} = \psi_+^{(1)}$.

Interestingly, the polaron configuration for $\Delta(x)$ -- Eq. (11) and the associated electron wave functions -- satisfy the electron part -- Eqs. (3a) and (3b) -- of the continuum equations for any $\kappa_o v_F$ in the allowed range $0 \leq \kappa_o v_F \leq \Delta_o$. It is the self-consistent gap equation -- (3c) -- that determines the specific value of $\kappa_o v_F$ for an actual solution to the coupled equations. Unsurprisingly, the nature of the solution depends on the occupation numbers -- call them n_+ and n_- -- of the gap levels $\varepsilon_\pm = \pm\omega_o$. Using some aspects of soliton theory [7,23,24] one can in effect convert Eq. (3c) to an algebraic minimization problem. Apart from simplifying the problem technically, this is very appealing intuitively, for the quantity being minimized is essentially the <u>energy</u> of the full interacting electron-phonon system. One finds that

$$E_P^{trans}(n_+,n_-,\kappa_o) = (n_+-n_-+2)\omega_o + \frac{4}{\pi} \kappa_o v_F$$
$$- \frac{4}{\pi} \omega_o \tan^{-1} (\kappa_o v_F/\omega_o) . \tag{15}$$

With $\kappa_o v_F = \Delta \sin\theta$ and $\omega_o = \Delta \cos\theta$ -- possible since $\omega_o^2 + (\kappa_o v_F)^2 = \Delta^2$ -- one can show that E_P^{trans} is minimized for $\theta = \theta(n_+,n_-) = (n_+-n_-+2)\pi/4$. From this result it follows that for a stable, localized polaron solution the electrons must be distributed in one of two configurations:

(1) $n_+ = 1, n_- = 2$, the "electron polaron" state, with $Q = -e$ and $S = \frac{1}{2}$;

or (2) $n_+ = 0, n_- = 1$, the "hole polaron" state, with $Q = +e$ and $S = \frac{1}{2}$.

Thus, unlike kinks, polarons have conventional spin/charge relations. For either electron or hole polarons, $\theta = \pi/4$, so $\kappa_o v_F = \omega_o = \Delta_o/\sqrt{2}$ and $E_P^{trans} = 2\sqrt{2} \Delta_o/\pi$. This is less than Δ_o -- and obviously also less than $4\Delta_o/\pi$, the minimum $K\bar{K}$ pair energy -- and hence the polaron is the <u>lowest energy excitation available</u> to a <u>single electron</u> added to a <u>trans</u>-$(CH)_x$ chain.

Using Eqs. (13) and (15) we can also answer the natural question "What happens if we try to form a "bipolaron" by adding a <u>second</u> electron (or hole) to a polaron state?" For any of the electron-electron $\overline{(n_+ = n_- = 2)}$, hole-hole $(n_+ = n_- = 0)$, or electron-hole $(n_+ = n_- = 1)$ configurations, E_P^{trans} is minimized for $\theta = \pi/2$, so that $\kappa_o v_F = \Delta_o$, $\omega_o = 0$ and from Eq. (4.9), $x_o \to \infty$. Thus these putative "bipolarons" in $\underline{trans}$-$(CH)_x$ actually correspond to an infinitely separated $(x_o \to \infty)$ $K\bar{K}$ pair, as Eq. (13) would suggest. Hence when a single electron is added to a <u>trans</u>-$(CH)_x$ chain, it should form a polaron, whereas when many excess electrons are present on a single chain, it becomes energetically favorable to form many $K\bar{K}$ pairs. A consequence is that the response and transport properties of lightly-doped -- average of $\leq$ one electron (or hole) per chain -- <u>trans</u>-$(CH)_x$ should be controlled by polarons, whereas more heavily doped samples should reflect the kink dominance.

From the perspective of the nonlinear excitations we are discussing, the most important difference between the <u>trans</u> and <u>cis</u> isomers of $(CH)_x$ is that for <u>cis</u>-$(CH)_x$ there is a unique, non-degenerate ground state, as shown in Fig. (2c). As we have indicated previously, an immediate consequence is that since there are <u>not</u> two degenerate ground states for kink solitons to connect, there are no kink soliton solutions in <u>cis</u>-$(CH)_x$! Similarly, in a wide class of other quasi-one-dimensional conducting polymers (poly-diacetylene, poly-paraphenylene, polypyrrole) with non-degenerate ground states, there can be no kink solitons. Since this point is so crucial, let us discuss it from an intuitive viewpoint before presenting the detailed mathematical model.

What would happen if one tried to create a kink/anti-kink pair from the ground state in cis-(CH)$_x$? Suppose that $\delta E/\ell$ (see Fig. (2c)) is small -- as it is in cis-(CH)$_x$ -- so that the metastable and ground state have almost the same energy per unit length. One could then imagine constructing a configuration that interpolated (an "anti-kink") between $\tilde{\Delta}_0$ (the cis-ground state) and Δ_{ms} (the metastable state), remained in Δ_{ms} for a spatial distance d, and then interpolated (a "kink") between Δ_{ms} and $\tilde{\Delta}_0^{ms}$. The form of Δ would look like (13) with d = $2x_0$. For finite d, the configuration would have finite energy, but since $\delta E/\ell > 0$, for d $\to \infty$, its energy would become infinite like $(\delta E/\ell)\cdot d$, and hence an independent, "free" $K\bar{K}$ pair cannot exist. In this sense, one can say that the putative kinks in cis-(CH)$_x$ are "confined". Of course, referring to (13), we see that permanently confined $K\bar{K}$ pairs have the same structure for Δ as polarons. Thus, in cis-(CH)$_x$, only polaron-type nonlinear excitations will exist.

To make this precise, it is very useful to study an explicit model of cis-(CH)$_x$ [23] which is a natural extension of the microscopic electron-phonon Hamiltonian proposed for trans-(CH)$_x$. One assumes that the gap parameter can be written as $\tilde{\Delta}(x) = \Delta_i(x) + \Delta_e$, where the intrinsic gap $\Delta_i(x)$ is sensitive to the electron feedback (as for trans) but Δ_e is a constant, extrinsic component. This Ansatz can be motivated [23,26] in terms of effects arising from molecular orbitals other than the π-orbital explicitly included in (2).

The adiabatic mean field Hamiltonian for cis-(CH)$_x$ then becomes [23,27-28]

$$H_{cis} = \Sigma_s \int dx \{ \frac{\omega_Q^2}{2g^2} \Delta_i^2(x) + \psi^+(x)(iv_F\sigma_2 \frac{\partial}{\partial x} + (\Delta_i(x)+\Delta_e)\sigma_3) \, \psi(x) \} \; . \tag{16}$$

Although in the interest of simplicity we have not indicated this explicitly, it is important to realize that all the physical parameters in the model of cis-(CH)$_x$ -- lattice spacing, effective electron-phonon coupling constant, band width, fermi velocity, band gap -- can have different values from those in trans. Comparing (2) and (16), we see that with the replacement $\Delta(x) \to \tilde{\Delta}(x) = \Delta_i(x) + \Delta_e$, the electron equations for trans can be converted to those for cis. Thus, the structure of the electron spectrum and the eigenfunctions will be the same. The gap equation is, however, modified to read

$$\Delta_i(x) = \tilde{\Delta}(x) - \Delta_e = -g^2(\omega_Q^2)^{-1} \, \Sigma'_{n,s} (|\psi_{n,s}^{(1)}|^2 - |\psi_{n,s}^{(2)}|^2) \; , \tag{17}$$

which leads to different constraints on solutions and on bound state eigenvalues. Specifically, the ground state has the unique value $\tilde{\Delta}_0$ determined from

$$\tilde{\Delta}_0 - \Delta_e = (g^2/\omega_Q^2) \frac{1}{\pi} \int_{-K}^{K} \frac{dk \, \tilde{\Delta}_0}{(k^2v_F^2+\tilde{\Delta}_0^2)^{\frac{1}{2}}} \tag{18}$$

which leads to $\Delta_e = \lambda\tilde{\Delta}_0 \, \ell n \, \tilde{\Delta}_0/\Delta_0$.

In confirmation of our earlier arguments, it can be shown directly that there are no kink solutions when the modified gap equation, (17), is considered. For polaron configurations, one finds that exactly the same functional form as that in (13) does satisfy the cis equations provided κ_0 is chosen appropriately. Again using soliton techniques [7,23,27] to convert the gap equation to an algebraic minimization problem one finds

$$E_P^{cis}(n_+,n_-,\kappa_0) = (n_+-n_-+2)\omega_0 + \frac{4}{\pi} \kappa_0 v_f - \frac{4}{\pi} \omega_0 \tan^{-1}(\kappa_0 v_f/\omega_0)$$

$$+ \frac{4}{\pi} \tilde{\Delta}_0 \, \gamma[\tanh^{-1} (\frac{\kappa_0 v_f}{\tilde{\Delta}_0}) - \frac{\kappa_0 v_f}{\tilde{\Delta}_0}] \tag{19}$$

where $\gamma = \Delta_e/\lambda\tilde{\Delta}_o$ is a measure of the strength of the soliton "confinement" energy. Introducing θ such that $\kappa_o v_f = \tilde{\Delta}_o \sin\theta$ and $w_o = \tilde{\Delta}_o \cos\theta$, one finds that E_p^{cis} is minimized for solutions to the equation

$$\theta + \gamma \tan\theta = (n_+ - n_- + 2)\pi/4 \; ,$$

$$0 \le \theta \le \pi/2 \; . \tag{20}$$

In the limit of zero extrinsic gap ($\Delta_e = 0$) $\gamma = 0$, and as expected (20) reduces to the result for trans-(CH)$_x$. For $\gamma > 0$, (20) -- unlike the corresponding trans equation -- has solutions for all combinations of n_+ and n_-. These are polaron and multipolaron states. Note that as the effective polaron occupation number, $N \equiv n_+ - n_- + 2$, increases, both the polaron width ($2 x_o$) and depth increase, as one would expect intuitively. Note also that the qualitative nature of these multipolaron states varies substantially. The (e-e) bipolaron ($n_+ = n_- = 2$), for example, can only have spin zero and has charge -2 relative to the ground state. In contrast, the (e-h) bipolaron can exist in either singlet (s = 0) or triplet (s = 1) forms and, with charge Q = 0 relative to the ground state, could more properly be called an "exciton". Recently, considerable experimental interest has focused on the existence of bipolarons in polypyrrole and polythiophene [29,30].

Finally, let me refer the serious reader to the literature for details of the panoply of nonlinear excitations in several related systems, such as the pure carbon polymer (-(C≡C)-$_x$) [18], the "diatomic" polymer (-(A=B)-$_x$) [31,32], and the charge transfer salt corresponding to the N = 1 Gross-Neveu theory [19].

Electron-Electron Interactions and Quantum Monte Carlo Simulations

The importance of understanding the combined effects of electron-phonon interactions -- as modeled by Eq. (1) -- and electron-electron interactions in real quasi-one-dimensional materials has in recent years become increasingly clear. To model the short-range electron-electron interactions, one adds to the discrete Hamiltonian (1) "Hubbard" terms of the form

$$\Delta H = \frac{U}{2} \sum_{i,s} n_{i,s} n_{i,-s} + V \sum_i n_i n_{i+1} \tag{21}$$

where $n_{i,s} \equiv c_{i,s}^+ c_{i,s}$ and $n_i = \sum_s n_{i,s}$ and U(> 0) and V ($0 < V < \frac{1}{2}U$) represent on-site and nearest neighbor electron-electron repulsion. The full "Peierls-Hubbard" Hamiltonian is then $\bar{H} = H + \Delta H$. Again, given limitations of space, I must refer to the literature for details [10-14] and at present have only a small fraction of the results. For the Peierls-Hubbard systems, analytic solutions are, in general, not possible, and numerical techniques are thus necessary. Here I focus on the recent quantum Monte Carlo studies [10-14], which have provided essentially exact results for intermediate size -- between twenty and thirty (CH) units -- polyacetylene chains and rings. These studies provide another, different illustration of the field theory connection, as the "Peierls-Hubbard" model (for the half-filled band) is essentially identical to the low-dimensional models studied in field theory, and the Monte Carlo techniques are also identical [10,12-14].

One of the crucial questions regarding "Peierls-Hubbard" systems has been whether the dimerization that exists for U = V = 0 persists in the presence of the electron-electron interactions. Several studies have now established that dimerization indeed does persist out to quite large (U >> 4t$_o$) values of U [11-14,33]. In Fig. 5 I plot the magnitude of the difference in energy $|\Delta E_d|$, versus U, between an undimerized system and a perfectly dimerized system -- that is, a system having sequential transfer integrals that alternate between $t_\pm \equiv t_o(1\pm2\delta)$ -- for $\delta = 0.1$. The system shown is a ring of 32 sites. Note that $|\Delta E_d|$ peaks at $U \cong 4t_o$ and remains large throughout the entire plotted range [33].

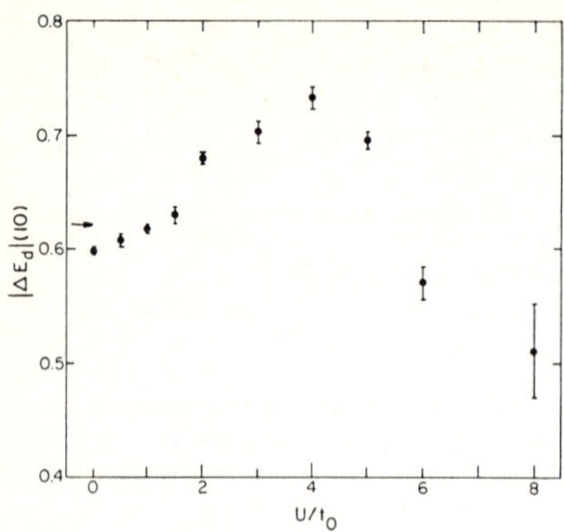

Fig. 5. The modulus of the energy difference per site, $\left|\Delta E_d\right|$, between the undimerized ring and the ring with dimerization $\delta = 0.1$ for $N = 32$ versus U. Both are units of t_o

Another essential question in "Peierls-Hubbard" systems is the effect of electron-electron interactions on the energies of solitons. One measure of these effects is the energy difference, for an open, odd chain, between a perfectly dimerized configuration and a "single site soliton", for which the bond alternation reverses about the central site. For $U = V = 0$, the soliton is known to have lower energy [34]. In Fig. 6 I plot $\left|\Delta E_s\right|$ the magnitude of the energy differences between the kink soliton -- called S in the figure -- and the dimer versus U for several values of $V(< \tfrac{1}{2}U)$. Both neutral (D^o, S^o) and positively charged (D^+, S^+) systems are shown. Several important features of these results should be noted [14,33]. First, the ground state of an odd chain is a soliton for both charged and neutral systems even for nonzero U and V. Second, the degeneracy between S^o and S^+ is destroyed for nonzero U and V. Third, although it is not indicated on Fig. 6b, which shows only S^+, when $V_{1,21} = V$ the results show that $S^+ = S^-$, to within the Monte Carlo errors [33]. Fourth, the neutral soliton stabilization energy <u>decreases</u> continuously with U (for $V = 0$) and with U-V (for $V \neq 0$). The charged soliton stabilization energy remains relatively flat as U is increased for $V = 0$, while for both U and $V > 0$ the charged soliton can be more stable than the pure Peierls $(U = 0 = V)$ case. This very strong stabilization of the charged soliton may be responsible for its apparent ready formation in doping experiments.

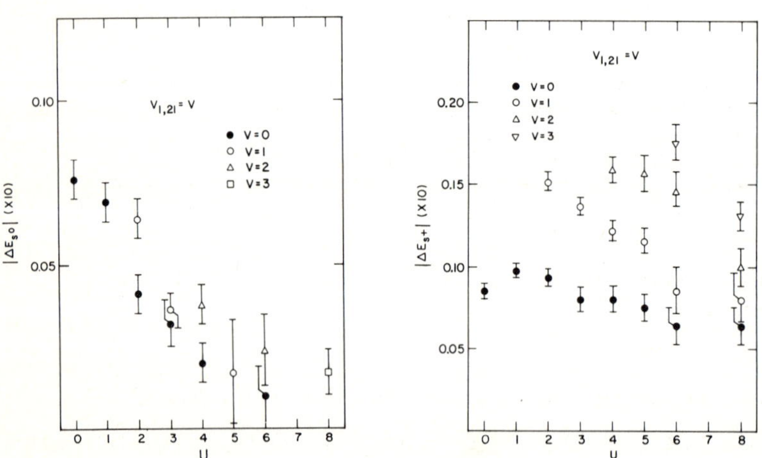

Fig. 6. The magnitude of the energy difference between a "single site" soliton and a dimer for $N = 21$: (a, on left) neutral case; (b, on right) charged case

186

Conclusions
In this very brief survey I have attempted to sketch a field theorist's view of
polyacetylene and related systems. Even given this limited perspective, I have
had to omit many important results, most notably those having to do with time-
dependent adiabatic mean-field studies [22,35,36], the question of fractional
charge [20,21], and the relation between the quantum chemists' approach [11,37]
and that of the physicists. I can only hope that this peek into the fascinating
world of real, quasi-one-dimensional materials will stimulate interested readers
to study further this elegant and important interdisciplinary area.

Acknowledgments
Since this article is a survey, albeit from a limited perspective, it naturally
reflects the work of many individuals and groups. I am particularly grateful
to my collaborators -- Alan Bishop, Tom DeGrand, Klaus Fesser, Baruch Horovitz,
Peter Lomdahl, Sumit Mazumdar, Simon Phillpot, and Michael Rice -- for their
many shared insights.

References

1. For reviews and collections of relevant articles see Proceedings of the
    International Conference on Synthetic Metals, Abano Terme, Italy, June
    1984 (Liq. Cryst. Mol. Cryst., to be published); Proceedings of the
    International Conference on Synthetic Conductors and Superconductors
    in Low Dimensions, Les Arcs, France, December 1982, J. Phys. Colloq.
    C3, No. 6, 44 (1983); Proceedings of the International Conference on
    Low-Dimensional Conductors, Boulder, Colorado, August 1981, Mol. Cryst.
    Liq. Cryst. 77, 1981; Physics in One Dimension, eds. J. Bernasconi and
    T. Schneider (Springer Verlag, 1981); A. J. Heeger and A. G. MacDiarmid,
    in The Physics and Chemistry of Low Dimensional Solids, ed. L. Alcâcer
    (Reidel, 1980), pp. 353-391; S. Etemad, A. J. Heeger, and A. G.
    MacDiarmid, Ann. Rev. Phys. Chem. 33 443-469 (1982); D. Baeriswyl, G.
    Harbeke, H. Kiess, and W. Meyer, Chapter 7 in Electronic Properties
    of Polymers, eds. J. Mort and G. Pfister (Wiley, 1982).

2. A. Kotani, J. Phys. Soc. Japan 42, 408 and 416 (1977).

3. S. A. Brazovskii, JETP Lett. 28, 606 (1978) (trans. of Pisma ZhETF 28, 656
    (1978)); Sov. Phys. JETP 51, 342 (1980) (trans. of ZhETF 78, 677 (1980)).

4. M. J. Rice, Phys. Lett. 71A, 152 (1979); M. J. Rice and J. Timonen, Phys.
    Lett. 73A, 368 (1979); E. J. Mele and M. J. Rice, Chemica Scripta 17, 21
    (1981).

5. W. P. Su, J. R. Schrieffer, and A. J. Heeger, Phys. Rev. Lett. 42, 1698
    (1979); Phys. Rev. B22, 2099 (1980).

6. H. Takayama, Y. R. Lin-Liu, and K. Maki, Phys. Rev. B 21, 2388 (1980);
    J. A. Krumhansl, B. Horovitz, and A. J. Heeger, Solid State Commun.
    34, 945 (1980); B. Horovitz, Solid State Commun. 34, 61 (1980) and
    Phys. Rev. Lett. 46, 742 (1981).

7. D. K. Campbell and A. R. Bishop, Phys. Rev. B24, 4859 (1981) and Nuc.
    Phys. B200[FS4], 297 (1982)

8. D. J. Gross and A. Neveu, Phys. Rev. D 10, 3235 (1974).

9. R. F. Dashen, B. Hasslacher, and A. Neveu, Phys. Rev. D 12, 2443 (1975).

10. J. E. Hirsch and E. Fradkin, Phys. Rev. Lett. 49 402 (1982); Phys. Rev. B
    27, 1680 and 4302 (1983).

11. S. Mazumdar and S. N. Dixit, Phys. Rev. Lett. 51, 292 (1983); Phys. Rev.
    B29, (1984).

12. J. E. Hirsch, Phys. Rev. Lett. 51, 296 (1983).

13. J. E. Hirsch and M. Grabowski, Phys. Rev. Lett. 52, 1713 (1984).

14. D. K. Campbell, T. A. DeGrand and S. Mazumdar, Phys. Rev. Lett. 52, 1717 (1984).

15. R. E. Peierls, Quantum Theory of Solids (Clarendon Press, Oxford, 1955) p. 108; D. Allender, J. W. Bray, and J. Boudreau, Phys. Rev. B 9, 119 (1974).

16. S. Coleman and E. Weinberg, Phys. Rev. D 7, 1888 (1973).

17. D. J. Gross and F. Wilzcek, Phys. Rev. Lett. 30, 1343 (1973); Phys. Rev. D 8, 3633 (1973); and H. D. Politzer, Phys. Rev. Lett. 30, 1346 (1973).

18. M. J. Rice, A. R. Bishop, and D. K. Campbell, Phys. Rev. Lett. 51, 2136 (1983).

19. M. J. Rice and E. J. Mele, Phys. Rev. B25, 1339 (1982).

20. S. Kivelson and J. R. Schrieffer, Phys. Rev. B25, 6447 (1982).

21. R. Jackiw and C. Rebbi, Phys. Rev. D13, 3398 (1976); R. Jackiw and J. R. Schrieffer, Nuc. Phys. B190[F53], 253 (1981); W. P. Su and J. R. Schrieffer, Phys. Rev. Lett. 46, 738 (1981).

22. W. P. Su and J. R. Schrieffer, Proc. Nat. Acad. Sci. 77, 5526 (Physics) (1980).

23. S. A. Brazovskii and N. N. Kirova, JETP Lett. 33, 4 (1981) (trans. of Pisma ZhETF 33, 6 (1981)).

24. I. V. Krive and A. S. Razhavskii, JETP Lett. 31, 610 (1981) (trans. of Pisma ZhETF 31, 647 (1980)).

25. J. L. Bredas, R. R. Chance, and R. Silbey, Mol. Cryst. Liq. Cryst. 77, 319 (1981).

26. M. J. Rice, Phys. Rev. Lett. 37, 36 (1976).

27. A. R. Bishop and D. K. Campbell, in Nonlinear Problems: Present and Future, eds. A. R. Bishop, D. K. Campbell, and B. Nicolaenko (North Holland, 1982) p. 195.

28. I. V. Krive and A. S. Rozhavskii, Sov. J. Low Temp. Phys. 7, 449 (1981) (trans. of Fiz. Nizk. Temp. 7, 921 (1981)).

29. J. C. Scott, J.-L. Bredas, J. M. Kaufman, P. Pfluger, G. B. Street, and K. Yakushi, contribution to Abano Terme Proceedings, ref. 1.

30. J. H. Kaufman, G. B. Street, J. C. Scott, N. Colaneri, T.-C. Chung, A. J. Heeger, and F. Wudl, contribution to Abano Terme Proceedings, ref. 1.

31. M. J. Rice and E. J. Mele, Phys. Rev. Lett. 49, 1455 (1982).

32. D. K. Campbell, Phys. Rev. Lett. 50, 865 (1983).

33. D. K. Campbell, T. A. DeGrand, and S. Mazumdar, to be published.

34. W. P. Su, Solid State Commun. 35, 899 (1980).

35. A. R. Bishop, D. K. Campbell, P. S. Lomdahl, B. Horovitz, and S. R. Phillpot, Phys. Rev. Lett. 52, 671 (1984).

36. F. Guinea, to be published.

37. Z. G. Soos and S. Ramasesha, Phys. Rev. Lett. 51, 2374 (1983); J. Chem. Phys., to be published.

Soliton, Polaron, Phonons in Polyacetylene and Their Interactions

Y. Wada

Department of Physics, University of Tokyo, Bunkyo-ku, Tokyo 113, Japan

1. Introduction

Dynamical properties of solitons in trans-polyacetylene have attracted much interest recently. They have been studied mainly by three types of experiments.
Electron spin resonance experiments have measured temperature dependences of the
halfwidth at the half-maximum and the electron-lattice relaxation rate. The
second type has been the dynamic nuclear polarization experiment. It has measured NMR enhancement by pumping microwave power near the electronic Larmor
frequency. As the magnetic field is swept through the ESR line, we can see a
symmetric pattern if unpaired spins are mobile (Overhauser effect), or an
asymmetric pattern if they are immobile (solid state effect). The third type
has been the NMR, particularly, the temperature and field dependences of the spin-
lattice relaxation time. Although there have been many discussions about the
observed results, it would seem agreed upon that there are mobile unpaired spins
and that they are more mobile at higher temperatures. It is concluded by motional
narrowing of the ESR line shape and the Overhauser effect.

It is quite important to study theoretically the motion of the mobile
unpaired spins - solitons - to understand the observed data. At low temperatures,
it would be modulated mainly by phonons which have small excitation energies. In
the ϕ^4 model, a soliton was shown to undergo a Brownian-like motion [1] by nonlinear interactions with the thermally excited phonons. The diffusion constant of
the soliton was proportional to T^2 at the classical low temperature limit. Quantum correction modified the temperature dependence to an activation type [2].

MAKI [3] discussed the diffusive motion of solitons in trans-polyacetylene,
regarding the system of solitons as a Boltzmann-type gas with mean free paths
limited by phonons which are linearly coupled to solitons. He used the model of
TAKAYAMA, LIN-LIU, and MAKI (TLM) [4], which is a continuum version of that of SU,
SCHRIEFFER, and HEEGER [5]. Since the mean free path becomes shorter as the
temperature increases, the soliton becomes less mobile.

It would thus be interesting to apply the method of [1] to the TLM model. We
must first know the structure of phonons when there is a soliton. Then we can find
interactions between the phonons and the soliton, which would give rise to the
diffusive motion of the latter.

Small oscillations around the soliton solution of the TLM model have been investigated [6]. Self-consistency condition is imposed between the phonon order parameter and the electron wave function for the small time dependent, oscillating components. It gives an eigenvalue equation for the linear modes around the soliton
solution in the form of an integral equation, an adiabatic approximation being used
for the slowly varying phonon motion. It is well approximated by a matrix eigenvalue problem and solved numerically.

Our results are very different from those obtained by NAKAHARA and MAKI [7].
We find three localized modes: Goldstone mode which gives rise to the soliton
translation, amplitude oscillation mode of the soliton, and the new third mode
close to, but smaller than, the extended phonon energies. Phase shifts of the

189

extended phonons are obtained. They have two remarkable properties. In the long
wavelength limit, the phase shift approaches 3π instead of 2π in accordance with
the three localized modes. Moreover, it is found to be parity independent, which
means that the effective potential for the phonons is reflectionless. This property
is commonly seen in soliton-bearing systems. But, it is first shown by the
present work in polyacetylene.

It has been argued that polarons would play important roles in the doping and
optical absorption of polyacetylene. ONODERA [8] will discuss two polaron problem
to clarify the mechanism of polaron pair decay into soliton - antisoliton pairs.
At very low doping levels, however, motion of a single polaron would be modulated
mainly by phonons before finding another polaron to form a pair. Thus, it would
be important to study the structure of phonons in the presence of a polaron to
investigate the interactions between them. The linear mode analysis is also
carried out around the polaron solution in the TLM model [9]. We find five localized
phonon modes. One of them, with a vanishing frequency, corresponds to the Goldstone
mode. Phase shift of the extended phonons approaches 5π in the long wavelength
limit. Again it is parity independent.

2. Linear Mode Analyses

The Hamiltonian in the TLM model is given in a form

$$H = (1/2g^2)\int dx[\dot{\Delta}^2(x,t) + \omega_Q^2\Delta^2(x,t)]$$

$$+ \sum_s \int dx \psi_s^\dagger(x,t)[-iv_F\sigma_3\partial_x + \sigma_1\Delta(x,t)]\psi_s(x,t) \quad . \tag{1}$$

Here, ω_Q is the bare optical phonon frequency. The phonon order parameter $\Delta(x,t)$
is defined by

$$\Delta(x,t) = g(M/a)^{1/2}\tilde{y}(x,t) \quad , \tag{2}$$

with g the electron-phonon coupling constant, M the mass of a CH unit, a the
lattice constant and $\tilde{y}(x,t)$ the staggered lattice displacement which is the
continuum version of $(-1)^n$ times the displacement of the n-th CH unit along the
chain direction, with $x = na$. The electronic wavefunctions $\psi_s^\dagger$ and ψ_s with the
spin index s are spinors, the first and   second components representing the
right- and left-going waves respectively, σ_1 and σ_3 the Pauli matrices and v_F the
Fermi velocity.

The selfconsistency condition is given by the coupled equations

$$i\dot{\psi}_{n,s}(x,t) = [-iv_F\sigma_3\partial_x + \sigma_1\Delta(x,t)]\psi_{n,s}(x,t) \quad ,$$

$$\ddot{\Delta}(x,t) + \omega_Q^2\Delta(x,t) = -g^2\sum_{n,s}{}' \psi_{n,s}^\dagger(x,t)\sigma_1\psi_{n,s}(x,t) \quad , \tag{3}$$

where the suffix n implies the electronic state index and the prime indicates the
sum over the occupied states. The adiabatic approximation has been used, since
the occupancy of the electronic states has been assumed to be invariant. As we
are interested in the frequencies of   linear modes which are much smaller than
the gap in the elctronic spectrum $2\Delta_0$, this assumption applies. Static solutions
of (3) were known for a soliton [4] and for a polaron [10-11]. They are denoted
by $\Delta_0(x)$ and $\psi_n^{(0)}(x)$ with ε_n corresponding electronic eigenvalue. Introducing
the small deviations $\delta\Delta$ and $\delta\psi_n$ by

$$\Delta(x,t) = \Delta_0(x) + \delta\Delta(x,t) \quad , \tag{4}$$

$$\psi_n(x,t) = \exp(-i\varepsilon_n t)[\psi_n^{(0)}(x) + \delta\psi_n(x,t)] \quad , \tag{5}$$

we linearize (3) with respect to the deviations. The time derivative of $\delta\psi_n$ is neglected, using the adiabatic approximation, since the dominant time dependence is taken account of by the first factor in (5). Elimination of $\delta\psi_n$ gives the integral equation for the Fourier transform of $\delta\Delta(x,t)$

$$(1 - \frac{\Omega^2}{\omega_Q^2})\,\delta\Delta(x,\Omega) = \frac{\lambda}{2}\int dy\, K(x,y)\delta\Delta(y,\Omega) \quad , \tag{6}$$

where $\lambda = g^2/\pi v_F\omega_Q^2$ is the dimensionless coupling constant, and the kernel $K(x,y)$ can be obtained, using the explicit form of $\psi_n^{(0)}$

$$K(x,y) = -4\pi v_F\cdot\mathrm{Re}\,{\sum_{n,s}}'\,{\sum_m}''\,(\psi_n^{(0)\dagger}(x)\sigma_1\psi_m^{(0)}(x))\cdot$$

$$\cdot(\psi_m^{(0)\dagger}(y)\sigma_1\psi_n^{(0)}(y))/(\varepsilon_n-\varepsilon_m) \quad . \tag{7}$$

It is composed of a linear combination of contributions from pairs of the electronic states, n occupied and m unoccupied. We find that the kernel does not depend on the occupancy of the midgap states. Thus, the structure of phonons with a neutral soliton is the same with that of a charged soliton. Similarly, the phonons around a polaron do not change by the polaron charge.

Introducing the soliton width $\xi=v_F/\Delta_0$, we use ξ and ξ^{-1} as the units for the coordinate x and the wave number k, respectively. The kernel $K(x,y)$ is actually given in a form of double integral over two wave numbers k and k'. The diagonal components $K(x,x)$ are logarithmically divergent for large values of $|k|$ and $|k'|$. It is to counterbalance the first term on the lefthand side of (6). We make use of a cutoff $|k|,\,|k'| < \Lambda$ with $\Lambda = W\xi/2v_F$. Here, W is the electronic band width. In the weak coupling limit $\lambda\to 0$, Λ goes to infinity with the help of the relation $\Delta_0 = W\exp(-1/2\lambda)$.

Dispersion relation of the extended phonons is identical with that of the perfectly dimerized state where there are no solitons or polarons. It was calculated by [7] and takes a value $\omega_0 = (2\lambda)^{1/2}\omega_Q$ in the long wavelength limit.

3. Numerical Results

Using parity relations, satisfied by the kernel K, we reduce (6) into two independent eigenvalue problems

$$(\frac{1}{2\lambda} - \frac{\Omega^2}{\omega_0^2})\,\delta\Delta_\pm(x,\Omega) = \frac{1}{4}\int_0^L dy\, K_\pm(x,y)\,\delta\Delta_\pm(y,\Omega) \quad , \tag{8}$$

with $K_\pm(x,y) = K(x,y) \pm K(x,-y)$. The plus and minus signs represent even and odd functions, respectively. For the numerical work, we truncate the y-integral in (8) at a value L and divide the region $[0, L]$ into N small regions with a mesh dx, reducing (8) to eigenvalue problems of real symmetric matrices. For N, we take 100 and 150 and several values of dx are chosen between 0.05 and 0.25. The cutoff Λ is fixed to π/dx. A typical choice in the weak coupling case, dx = 0.1, gives $\Lambda = 10\pi$ which corresponds to $\lambda = 0.12$. An appropriate value $\lambda = 0.19$ for polyacetylene gives $\Lambda = 7.14$, which would not belong to the weak coupling case, in a rigorous sense, since it can not reproduce the phonon dispersion derived by [7].

In Table 1, the eigenvalues of the localized modes around a soliton are shown with the cutoff $\Lambda = 10\,\pi$ on the first column. Increasing the cutoff as far as $40\,\pi$, we use an extrapolation to obtain the eigenvalues on the second column with no cutoff in the weak coupling limit. We can see the energy of the Goldstone mode is not zero when there is a cutoff. It tends to zero as the cutoff increases. On the next two columns parities of the localized modes and numbers of

Table 1. Eigenvalues, parities, and numbers of nodes of the localized modes around a soliton

| Ω^2/ω_0^2 | | Parity | Number of nodes |
$(\Lambda=10\pi)$	$(1/\Lambda \to 0)$		
0.0108	0.000	even	0
0.723	0.703	odd	1
0.956	0.941	even	2

Table 2. Eigenvalues, parities, and numbers of nodes of the localized modes around a polaron

| Ω^2/ω_0^2 | | Parity | Number of nodes |
$(\Lambda=10\pi)$	$(1/\Lambda \to 0)$		
0.0156	0.000	odd	1
0.424	0.407	even	2
0.738	0.729	even	2
0.865	0.846	odd	3
0.978	0.968	even	4

nodes are shown. There is the third mode with a rather high energy close to the extended phonons. It is expected to be infra-red active. Corresponding results for the localized modes around a polaron are shown in Table 2. The energy of the Goldstone mode again vanishes in the large cutoff limit. Its parity is odd since the polaron has a symmetric form. We find four additional localized modes. It is quite interesting to see that the parity does not alternate as the eigenvalue increases, accordingly, the number of nodes does not increase regularly. In Figs. 1 and 2, we show the explicit shapes of these modes for the soliton and the polaron. These shapes do not depend on the choice of L. This indicates that these modes are actually localized.

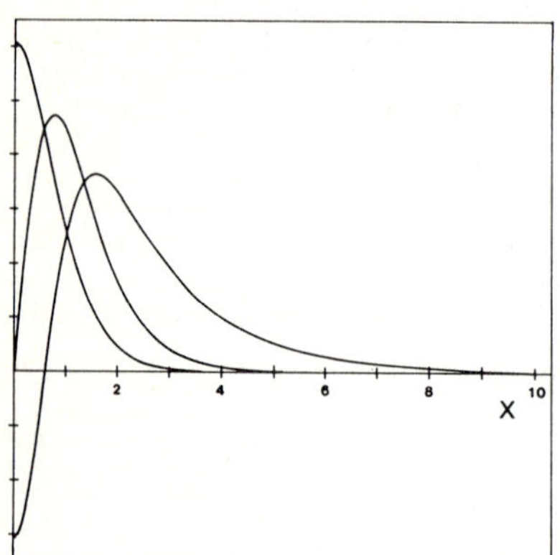

Fig.1 Shapes of the three localized modes around a soliton, the unit of the abscissa being ξ

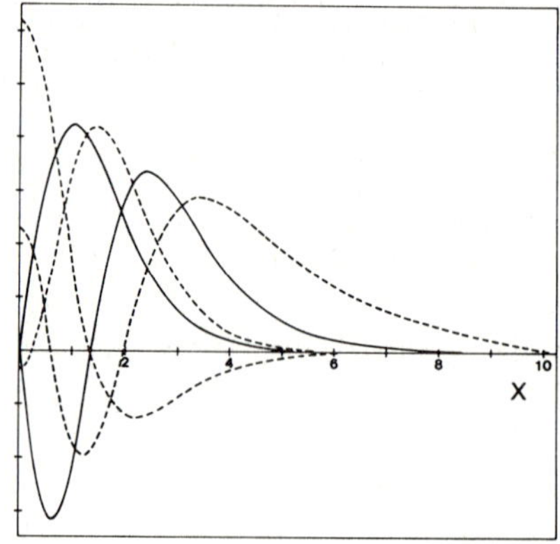

Fig.2 Shapes of the localized modes around a polaron, the three even modes by broken curves and the two odd ones by solid curves, the unit of the abscissa being ξ

Eigenvalues of the extended modes form a continuous spectrum. The eigenfunctions with even or odd parities approach the forms $\cos(kx+\delta_e(k)/2)$ or $\sin(kx+\delta_o(k)/2)$ respectively in the region far from the soliton or polaron, $\delta_e(k)$ and $\delta_o(k)$ being the phase shifts. Since our eigenfunctions might be modified at $x \sim L$ by the artificial truncation of the integral in (8), we use the functional forms in the intermediate region to determine the phase shifts. Both phase shifts turn out to be identical to each other, for the soliton as well as for the polaron. They are shown in Fig. 3. The curve S is for the soliton and the curve P for the polaron. Extrapolations of the curves suggest that the phase shifts seem to reach 3π and 5π at $k = 0$, for the soliton and the polaron respectively. This is consistent with the numbers of the localized modes, according to Levinson's theorem. Phase shifts are always positive. Effective potentials for phonons are attractive. Since the phase shift of the even parity is the

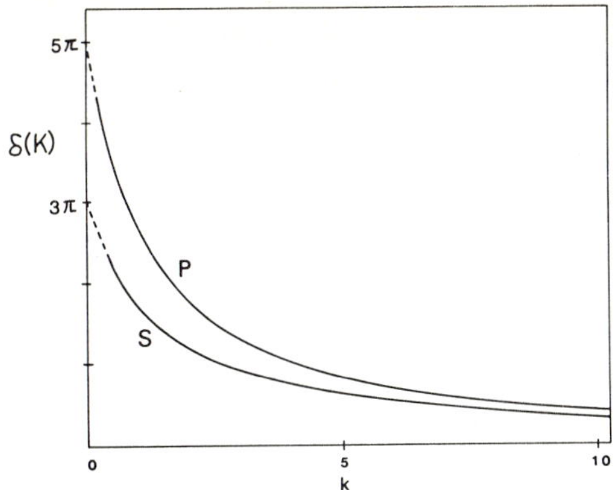

Fig.3 Phase shifts of the
extended modes around
a soliton (curve S) and
a polaron (curve P),
the even and the odd phase
shifts being represented by
the same curves, with the
unit of the abscissa ξ^{-1}

same with that of the odd parity, the effective potentials are reflectionless for
the phonon scattering [12]. Using the phase shifts, we can estimate corrections
to the creation energies of the soliton or the polaron. It turns out that the
creation energy of the soliton is $2\Delta_0/\pi$ - 0.54 ω_0 , while that of the polaron is
$2\sqrt{2}\Delta_0/\pi$- 0.8 ω_0. For the former, the correction was estimated as -0.60 ω_0 by
[7]. The creation energy increases due to the third localized mode.

4. Discussions

We have studied the optical phonon modes in the presence of a single soliton or a
polaron. New, interesting results have been obtained. The soliton generates three
localized modes and the polaron, five. The modes with the smallest frequency have
been identified with the Goldstone modes which give the shift of the location of
the soliton or polaron. The phase shifts of the extended phonons approach 3π or
5π in the respective cases. The phase shifts are parity independent, indicating
that the effective potentials for phonons are reflectionless. They are attractive. The corrections for the creation energies of soliton and polaron, due to the
phonons, are estimated. The creation energy is somewhat larger than the previous
value for the soliton.

It is desirable to observe these structures of phonons experimentally. One
possibility would be the infrared absorption or the Raman scattering. It is to be
discussed by ITO and ONO [13].

The above analyses show that the Goldstone mode does not interact with the
phonon modes in the first order with respect to the deviation from the static
soliton or polaron solutions, since they are both normal modes. But, in the
higher order terms, they interact with each other. These results of the linear mode
analyses are being used to find interactions between a soliton and phonons. Suppose we study a collision between the static soliton and a wave packet of a phonon.
The order parameter and the electronic wavefunctions are expanded, the amplitude
of the incoming phonon being the smallness parameter. When the second order
component of the order parameter is expanded in terms of the linear modes, the
Goldstone mode satisfies a second order differential equation with respect to
time, with a vanishing excitation energy. Ordinarily, this should mean that, if
the time dependence of the Goldstone mode were given   in a form of Fourier
transform over the frequency, the integrand would have a pole of order two at zero
frequency. We find, however, that nonlinear interaction has a form such that it
gives a matrix element which vanishes linearly at the zero frequency. The pole
becomes simple, which means that the soliton - phonon collision, in the second
order, gives rise to a shift of the soliton location and not to a momentum transfer between them. This was the case, we found, in the ϕ^4 system. The soliton would
undergo a random walk, since the collisions with the thermally excited phonons
take place randomly. The step of the random walk would be proportional to a quadratic form of the phonon amplitudes, since the interactions are nonlinear. Since

the diffusion constant of the soliton is proportional to the square of the step,
it is proportional to a quartic form. It would thus be proportional to T^2 at
classical low temperatures.

In the ϕ^4 system, it has been known that the phonons can transfer the momentum
to the soliton in the fourth order processes.   The transferred momentum would be
dissipated by viscosity by the phonons themselves. This is to be discussed by
OGATA and WADA [14]. The solitons and phonons in polyacetylene may have similar
property. The temperature dependence of the soliton diffusion constant would be
modified by this fourth order effect at higher temperatures.

References

1. Y. Wada and J. R. Schrieffer: Phys. Rev. B18, 3879 (1978)
2. Y. Wada: J. Phys. Soc. Japan 51, 2735 (1982)
3. K. Maki: Phys. Rev. B26, 2181, 2187, 4539 (1982)
4. H. Takayama, Y. R. Lin-Liu, and K. Maki: Phys. Rev. B21, 2388 (1980)
5. W. P. Su, J. R. Schrieffer, and A. J. Heeger: Phys. Rev. Letters 42, 1698
   (1979); Phys. Rev. B22, 2099 (1980); B28, 1138(E) (1983)
6. H. Ito, A. Terai, Y. Ono, and Y. Wada: J. Phys. Soc. Japan, in press (1984)
7. M. Nakahara and K. Maki: Phys. Rev. B25, 7789 (1982)
8. Y. Onodera: The current Proc. 7th Kyoto Summer Institute (1984)
9. A. Terai, H. Ito, Y. Ono, and Y. Wada: to be submitted to J. Phys. Soc. Japan
10. D. K. Campbell and A. R. Bishop: Nucl. Phys. B200, 297 (1982)
11. S. A. Brazovskii and N. N. Kirova: Zh. Exsp. Teor. Fiz. Pis'ma Red. 33, 6
    (1981); JETP Lett. 33, 4 (1981)
12. J. H. Eberly: Am. J. Phys. 33, 771 (1965)
13. H. Ito and Y. Ono: to be presented in this Summer Institute
14. M. Ogata and Y. Wada: J. Phys. Soc. Japan, in press (1984); and to be
    presented in this Summer Institute

Soliton-like Excitations and Their Interactions in the Continuum Model of Polyacetylene

Y. Onodera

Physics Department, Tokyo Metropolitan University, 2-1-1 Fukazawa, Setagaya-ku Tokyo 158, Japan

1. Introduction

Polyacetylene is a long-chain polyene synthesized by acetylene polymerization. As a consequence of its one-dimensional geometrical structure and its conjugate double bond of π electrons, any local modification in the geometrical configuration of the chain exerts a strong influence on the electronic energy levels, often giving rise to bound levels. The electrons, in turn, favor local distortions of the chain. In short, polyacetylene is a nonlinear system, where electrons and atomic displacements are strongly coupled.

The simple bond-alternating structure of *trans*-polyacetylene admits the two ground-state configurations shown in Fig.1. A domain-wall-like transition region, which connects the two configurations of Fig.1 is called a kink soliton (or simply, soliton). Since the introduction of this soliton concept [1,2], *trans*-polyacetylene has been extensively studied. Various striking properties—electric, magnetic, and optic—of *trans*-polyacetylene are now interpreted mostly in terms of the soliton model [3]. The microscopic theory of SU, SCHRIEFFER, and HEEGER [1] and its continuum version [4] form the theoretical basis of such a soliton model.

Fig.1 The two ground states of *trans*-polyacetylene

In addition to the kink soliton, the possibility of another kind of soliton—polaron soliton—was raised recently [5-9]. Polarons are generated by doping with acceptors or donors. Figure 2 shows the hole polaron generated by acceptor doping; the hole (vacant π electron) on the double bond in the middle of the figure relaxes the bond. That distortion of the chain creates a localized electronic level and traps the hole. The total entity of the hole and the surrounding distortion of the chain is called the polaron. In contrast to the tremendous amount of experiments on the kink solitons, evidence of the polarons are sparse [10]. It is believed [11] that the polarons(P) generated by doping decay, in pairs, into kink(K)-antikink($\bar{\text{K}}$) pairs,

$$P + P \rightarrow K + \bar{K}. \tag{1}$$

Fig.2 Hole polaron in *trans*-polyacetylene

Relatively rapid reaction will readily kill the polarons; only the
resulting charged kinks are observed in experiments.

The above-mentioned two kinds of solitary excitations—kinks and
polarons—perhaps deserve being called solitons in view of the strong
nonlinearity of the system. However, because the term soliton was
coined to imply stability on interaction with another soliton, it is
intriguing to see the interactions between the kinks and the polarons.
The interaction between two kinks on a single *trans*-polyacetylene chain
has already been studied [12] and I will consider below the interaction between two polarons [13] and a polaron and a kink coexisting on
a single chain [14]. The result will clarify the reaction process (1).

2. Continuum Model of *trans*-polyacetylene

A simplified model of *trans*-polyacetylene, which treats the π electrons
in the Hückel approximation and neglects the Coulomb interactions between electrons, was put forth by SU et al.   The characteristic length
appearing in this model is

$$\xi = v_F / \Delta_0 \, ,$$

where v_F denotes $\hbar$ times the Fermi velocity and $2\Delta_0$ is the band-gap
energy (resulting from the dimerization). Since the magnitude of ξ
is about 7 CH units [1], the kinks and the polarons are extended over
many carbon atoms (rather than localized on a single atom as shown in
Fig.2), which justifies continuum approximation to the original model.

In the continuum description, the dimerization parameter $\Delta(x)$ represents, apart from a constant multiplier, the staggered displacement
$(-1)^n u_n$ of carbon atoms along the chain direction. The π-electron
field is now described by a two-component spinor $\Psi(x)$, whose upper and
lower components correspond to the even and odd sites of the dimerized
chain. The Hamiltonian in the continuum limit becomes [4]

$$H = \int \Delta(x)^2 \, dx \, / \, \pi \lambda v_F + \int dx \, \Psi^\dagger(x) \, [\, -iv_F \sigma_y \frac{d}{dx} + \Delta(x) \, \sigma_x \,] \, \Psi(x). \quad (2)$$

The first term represents the elastic deformation energy and λ is the
dimensionless coupling constant. The second term consists of the electron kinetic energy and the electron-lattice interaction energy. σ_y
and σ_x are the Pauli matrices related to the two components of the
spinor. The electron-spin index is omitted for simplicity.

Exact static solutions to the above continuum model have been obtained in the adiabatic approximation. We first fix the displacement
$\Delta(x)$ and solve the one-electron eigenvalue problem,

$$[\, - v_F \frac{d}{dx} + \Delta(x) \,] \, g_i(x) = \varepsilon_i \, f_i(x),$$

$$\quad (3)$$

$$[\quad v_F \frac{d}{dx} + \Delta(x) \,] \, f_i(x) = \varepsilon_i \, g_i(x),$$

where $f_i(x)$ and $g_i(x)$ are one-electron wave functions on even and odd
sites. Total energy of the system (the adiabatic potential) is given by

$$E[\Delta(x)] = \int \Delta(x)^2 \, dx \, / \, \pi \lambda v_F + \sum_i \varepsilon_i [\Delta(x)] \quad (4)$$

as a functional of $\Delta(x)$. The summation on the right hand side is
done over the occupied one-electron levels.

The ground state is described by the constant dimerization param-
eter $\Delta(x) = \pm \Delta_0$, the double sign corresponding to the two configurations

of Fig.1. The electron energy spectra become

$$\varepsilon(k) = \pm (\Delta_0^2 + v_F^2 k^2)^{\frac{1}{2}},$$

and there occurs a band gap $2\Delta_0$ at $k=0$, which makes polyacetylene an insulator.

3. Kinks and Polarons

Both kinks and polarons should be stable excitations that minimize the energy (4). To find the expressions of $\Delta(x)$ for such excitations, we focus our attention on the scattering data of the eigenvalue problem (3), which, according to the inverse scattering theory [15,16], will uniquely prescribe the "potential" $\Delta(x)$. The energy (4) can be explicitly written in terms of the scattering data, and its extremization with respect to the reflection coefficients tells us that $\Delta(x)$ should be reflectionless [8]. Thus, the expressions for $\Delta(x)$ can be constructed by solving the inverse scattering problem. Strikingly enough, the above analysis had been done [17] long before the advent of the microscopic SSH model for polyacetylene [1].

$\Delta(x)$ of kink (and antikink) excitations are obtained as the reflectionless potential with one bound level in the midgap,

$$\Delta(x) = \pm \Delta_0 \tanh(x/\xi), \tag{5}$$

whereas the polaron distortion [6-9]

$$\Delta(x) = \Delta_0 - \kappa_0 v_F [\tanh(\kappa_0 x + \tfrac{1}{2}\alpha) - \tanh(\kappa_0 x - \tfrac{1}{2}\alpha)], \tag{6}$$

$$\alpha = \tanh^{-1}(\kappa_0 \xi)$$

has two bound levels $\pm\omega_0$, where

$$\omega_0 = \kappa_0 v_F = \Delta_0/\sqrt{2},$$

as shown in Fig.3(a)-(c). The kinks can be either positively charged, neutral, or negatively charged, according as to whether the midgap level is empty, singly occupied, or doubly occupied. As for the polarons, the lower level is singly occupied in the hole polaron shown in Fig.3(c). In the electron polaron, the lower level is doubly occupied and the upper one singly occupied. The displacement pattern (6) is common to both kinds of polarons.

It should be stressed here that these bound levels are real ones that accommodate π electrons, contrary to the abstract, auxiliary ones

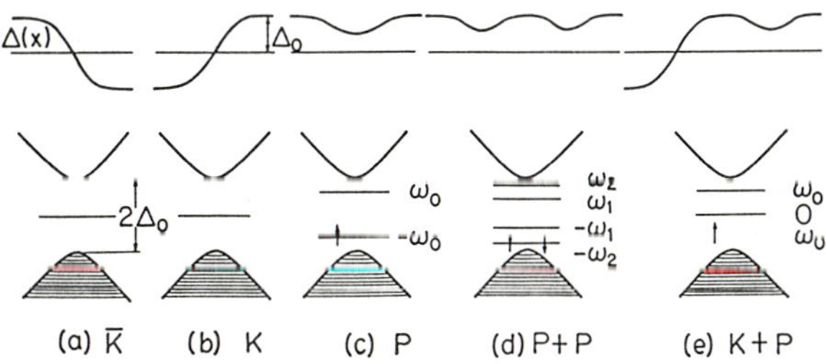

Fig.3 Displacement patterns and electron configurations for positively charged (a) antikink, (b) kink, (c) polaron, (d) two polarons, and (e) kink and polaron

which appear in solving soliton-bearing equations by means of the inverse scattering transform.

4. Two-Polaron Solution

Let us now consider two polarons on a single chain. As shown in Fig. 3(c), a single polaron has two symmetric bound levels. When two such polarons approach each other, the two energy levels are each split into two levels (bonding and antibonding levels) owing to the overlap of wave functions. Therefore, the two-polaron system should have four discrete levels, as shown in Fig.3(d). The system lowers its energy by accommodating the two electrons in the lowest level with energy $-\omega_2$. This means that the two polarons attract each other.

The dimerization function $\Delta(x)$ is constructed by solving the inverse problem of (3) on the requirement that it be a reflectionless potential with four symmetric bound levels. The inverse problem is solved after transforming (3) to decoupled Schrödinger equations, and the result is [13,18]

$$\Delta(x) = \Delta_0 + \frac{v_F(\kappa_1^2-\kappa_2^2)}{\kappa_1\coth(\kappa_1 x-\tfrac{1}{2}\alpha_1) - \kappa_2\tanh(\kappa_2 x-\tfrac{1}{2}\alpha_2)}$$
$$- \frac{v_F(\kappa_1^2-\kappa_2^2)}{\kappa_1\coth(\kappa_1 x+\tfrac{1}{2}\alpha_1) - \kappa_2\tanh(\kappa_2 x+\tfrac{1}{2}\alpha_2)} \quad , \quad \text{where} \tag{7}$$

$$\alpha_j = \tanh^{-1}(\kappa_j\xi) \qquad (\ j=1,2\ ),$$

$$(\kappa_j v_F)^2 + \omega_j^2 = \Delta_0^2 \qquad (\ j=1,2\ ).$$

The above function $\Delta(x)$ contains two independent parameters ω_1 and ω_2. Total energy E of the system (relative to the ground state) calculated from (4) becomes

$$E = \frac{4}{\pi}\sum_{j=1}^{2}[\ \kappa_j v_F + \omega_j\sin^{-1}(\omega_j/\Delta_0)\ ] - 2\omega_2. \tag{8}$$

The last term comes from the two electrons that occupy the lowest bound level. The dependence of this energy E on the two parameters ω_1 and ω_2 is seen from the constant-energy lines drawn in Fig. 4.

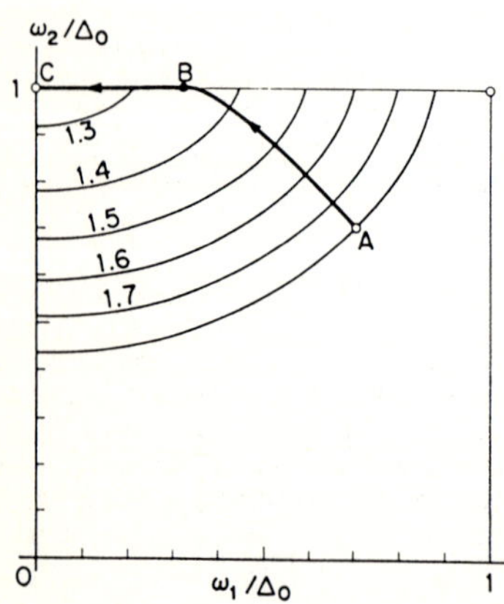

Fig.4 Constant-energy lines for the energy (8) of the two-polaron system.

We can understand the quasistatic interaction between two polarons
from this map. Starting from the point A, where the two polarons are
infinitely separated, the system will descend along the steepest path
(the thick line of Fig.4) on the adiabatic potential surface. The
terminal point of this reaction process is the minimum point C, where
two charged kinks are infinitely separated. The displacement patterns
shown in Fig.5 exhibit the entire process of this reaction: Two separated polarons attract each other with a rather long-range potential
(because of the large extension of polarons), form a single complex
that may be called bipolaron, and then break into a kink-antikink pair.
Thus, the above two-polaron solution describes the reaction process (1).

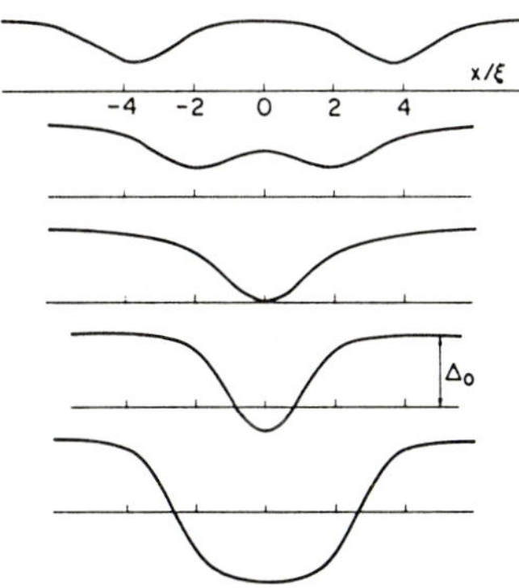

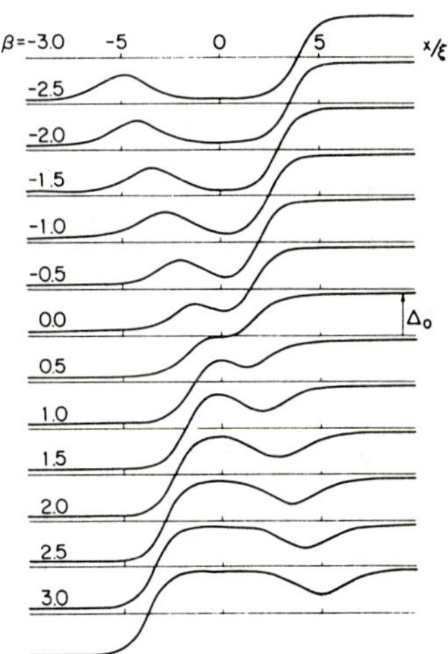

Fig.5 Displacement functions
$\Delta(x)$ for representative five
points on the steepest descent
path of Fig.4. The figure,
from top to bottom, shows the
reaction process (1).

Fig.6 Displacement functions
of a kink-plus-polaron system

5.   Interaction of a Kink and a Polaron

When a kink and a polaron coexist on a single chain, we have the displacement pattern and electron configuration as shown in Fig.3(e).
The function $\Delta(x)$ can be constructed similarly as in the two-polaron
case. Now, $\Delta(x)$ should be the reflectionless potential with three
symmetric bound levels. Figure 6 shows the result [14] for various
values of the parameter β which determines the separation between the
kink and the polaron. This figure has features common to soliton
interactions in general; the displacement pattern becomes considerably
distorted when the two excitations approach each other, and they re-
gain their identities with a phase shift at large separations. An
important difference, however, is that I am considering here static
(or at best, quasistatic) problems. Hence Fig.6 may not be understood
to represent a dynamical collision event.

Calculation of the total energy (4) for this coexistence system
shows that the energy simply remains the sum of the free kink and free
polaron energies, independent of their separation [14]. This means
that the kink and the polaron can pass through each other without attractive or repulsive interactions. In relation to the polaron reaction (1), this free passage implies that even when a kink happens to
exist between two polarons, that kink will exert no disturbance on the
reaction of the two polarons into charged kinks.

<u>6.   Conclusion</u>

I have reviewed the kinks and polarons in *trans*-polyacetylene on the
basis of the continuum model, and then considered interactions between
those soliton-like excitations. The inverse scattering theory plays
a fundamental role in the theoretical description of those stable excitations and their interactions as well. The result obtained clari-
fies the mechanism of polaron reaction (1) and qualitatively accounts
for the observed predominance of charged kinks over polarons in doped
trans-polyacetylene.

<u>References</u>

1. W. P. Su, J. R. Schrieffer, and A. J. Heeger: Phys. Rev. Lett. **42,**
   1698 (1979); and Phys. Rev. B **22**, 2099 (1980); **28**, 1138(E) (1983)
2. M. J. Rice: Phys. Lett. **71A**, 152 (1979)
3. For a review, see, e.g., D. Baeriswyl, G. Harbeke, H. Kiess, and
   W. Meyer: in *Electronic Properties of Polymers*, edited by J. Mort
   and G. Pfister (Wiley, New York, 1982) p.267
4. H. Takayama, Y. R. Lin-Liu, and K. Maki: Phys. Rev. B **21**, 2388
   (1980)
5. W. P. Su and J. R. Schrieffer: Proc. Natl. Acad. Sci. USA **77**,
   5626 (1980)
6. S. A. Brazovskii and N. N. Kirova: JETP Lett. **33**, 4 (1981)
7. D. K. Campbell and A. R. Bishop: Phys. Rev. B **24**, 4859 (1981)
8. D. K. Campbell and A. R. Bishop: Nucl. Phys. B **200**, 297 (1982)
9. A. R. Bishop and D. K. Campbell: in *Nonlinear Problems: Present
   and Future*, edited by A. R. Bishop, D. K. Campbell, and B. Nicolaenko (North-Holland, Amsterdam, 1982) p.195
10. S. Etemad, A. Feldblum, A. G. MacDiarmid, T.-C. Chung, and
    A. J. Heeger: J. Phys.(Paris) **44**, C3-413 (1983)
11. J. H. Kaufman, J. W. Kaufer, A. J. Heeger, R. Kaner, and A. G.
    MacDiarmid: Phys. Rev. B **26**,2327 (1982).
12. Y. R. Lin-Liu and K. Maki: Phys. Rev. B **22**, 5754 (1980)
13. Y. Onodera and S. Okuno: J. Phys. Soc. Jpn **52**, 2478 (1983)
14. S. Okuno and Y. Onodera: J. Phys. Soc. Jpn **52**, 3495 (1983).
15. G. L. Lamb, Jr.: *Elements of Soliton Theory* (Wiley, New York, 1980)
16. I. S. Frolov: Dokl. Akad. Nauk USSR **207**, 44 (1972)
17. R. F. Dashen, B. Hasslacher, and A. Neveu: Phys. Rev. D **12**,2443
    (1975)
18. Y. Onodera: Phys. Rev. B **30,** in press (1984)

Lattice Relaxation Theory of Soliton and Polaron Generation in Polyacetylene

Zhao-bin Su and Lu Yu

Institute of Theoretical Physics, Academia Sinica, Beijing, PR China

I. Introduction

The lattice relaxation theory of multiphonon processes developed by Huang and
Rhys [1] and others [2] in the early 50s in connection with colour centers in
ionic crystals has proved to be very successful in traditional solid state
physics. The key idea in this theory is explicitly to take into account the difference in the symmetry breaking (lattice relaxation) of the initial and the final
states of quantum transitions involving multiphonon processes. We have shown that
the lattice relaxation theory offers a natural and convenient vehicle to treat the
soliton and polaron generation in polyacetylene [3-5] important for some recent
development in this exciting field [6]. We have generalized the original version
of this theory to include the self-consistency of the electronic states with the
lattice configuration as well as the many-electron background effects both of
which are essential for the collective, self-localized excitations such as soliton
and polaron. Our calculations have been published in [4], whereas the connection
with the recent experiments has been discussed in [5]. Since the reference [4] is
not easily accessible, we will outline here the basic idea of the method itself
which may be useful for some other purposes.

II. Basic Formalism

We illustrate the basic idea of the lattice relaxation theory on the example of
nonradiative transitions. The standard continuum version [7] of the SSH [8]
Hamiltonian can be identically rewritten as

$$H = H_0^e + H_0^{ph} + H_{int} \; , \tag{1}$$

$$H_0^e = \int dx [\psi^+(x)(\frac{\hbar v_F}{i} \frac{d}{dx} \tau_3 + 4\alpha u_c(x)\tau_1)\psi(x) + \frac{\rho}{2} \omega_b^2 u_c^2(x)] \; , \tag{2}$$

$$H_0^{ph} = \int dx \{ \dot{u}(x)\dot{u}(x) + \omega_b^2 [u(x) - u_c(x)][u(x) - u_c(x)] \} \; , \tag{3}$$

$$H_{int} = \int dx [4\alpha \, \psi^+(x)\tau_1 \, \psi(x) + \rho \omega_b^2 u_c(x)][u(x) - u_c(x)] \; , \tag{4}$$

where $\psi^+(x)$, $\psi(x)$ are spinor fermion operators, τ_1, τ_3 Pauli matrices, $u(x)$ staggered lattice displacement operator with $u_c(x)$ as its classical counterpart and
$\dot{u}(x)$ as the velocity operator. Here $\rho \equiv M/a$ is the linear density with M as the (CH)
group mass and a the lattice constant, whereas v_F is the Fermi velocity and α the
electron-phonon coupling constant. For simplicity we limit ourselves to the single
frequency model ($\omega_b^2 = 4K/M$, K the spring constant) for the optical phonons. A summation over the spin indices is understood where needed, while the x integration
is taken over the chain length L.

In the Born-Oppenheimer approximation the state vector can be decomposed as

$$| \; > = |e> \otimes |\tilde{n}> \; , \tag{5}$$

with $|e\rangle$ diagonalizing the electron Hamiltonian and $|\tilde{n}\rangle$ -- the phonon Hamiltonian of shifted origin. Also, the diagonal matrix elements of H_{int} should vanish for both $|e\rangle$ and $|\tilde{n}\rangle$. The nonradiative decay rate is then determined by the matrix element of H_{int} between the initial and the final states.

It is essential to note that the total Hamiltonian is split into interacting and noninteracting parts differently in the initial and the final states in accordance with their own lattice relaxation $u_c(x)$ and $v_c(x)$. Therefore, they belong to different complete sets. Nevertheless, the result depends only on the transition amplitude and not on the overlap integral.

In the low temperature ($\hbar\omega_b \lesssim k_B T$) and strong coupling ($|S\hbar\omega_b - W_{if}|/S\hbar\omega_b \lesssim 1$) limit the probability of nonradiative decay is given by

$$W = \frac{\pi\lambda v_F}{4}\left(\frac{2\pi}{S}\right)^{1/2}\exp\left\{-\frac{(W_{if} - S\hbar\omega_b)^2}{2S\hbar^2\omega_b^2}\right\}\sum_\nu|\langle e_f|K_\nu|e_i\rangle|^2 \, , \quad \text{where} \tag{6}$$

$$W_{if} \equiv E_i - E_f \, , \tag{7}$$

$$\lambda = 32\alpha/\pi\rho\omega_b^2 \, \hbar \, v_F \, , \tag{8}$$

$$K_\nu \equiv \int dx \, \psi^+(x)\tau_1\psi(x)\xi_\nu(x) + \pi\omega_b^2 \, u_\nu^c/4\alpha \, , \tag{9}$$

$$S \equiv \frac{\rho\omega_b^2}{2\hbar}\int dx[u_c(x) - v_c(x)]^2 \tag{10}$$

with $\xi_\nu(x)$ as eigenmodes and u_ν^c as the expansion coefficients for the lattice relaxation $u_c(x)$. Here the Huang-Rhys factor S is the average number of phonons needed to bring the initial lattice configuration into the final one.

It is interesting to note that (6) can be derived in the quasi-classical approximation as a zero point motion activated process. Within the single frequency model we can replace the continuum of the lattice displacement field $u(x)$ by a single coordinate Q. The energy diagram is schematically shown in Fig.1. The energies of the initial and the final states are given by

$$E_i = W_i + \frac{1}{2}\omega^2 Q^2 \, , \tag{11}$$

$$E_f = W_f + \frac{1}{2}\omega^2(Q - \Delta_{if})^2 \, , \tag{12}$$

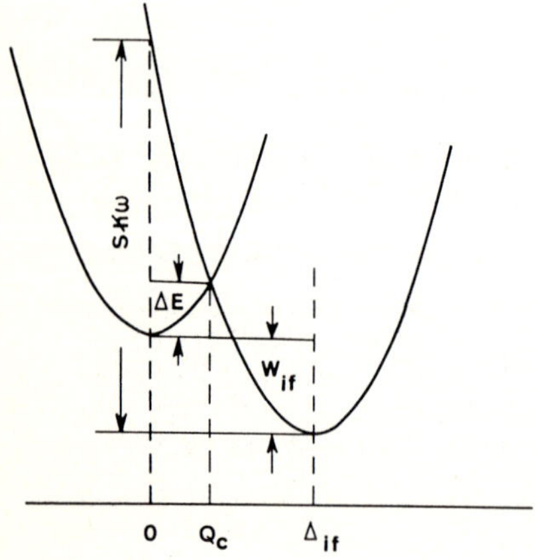

Fig.1. Schematic diagram for nonradiative decay.

where Δ_{if} is the difference in the lattice relaxation. The crossing point Q_c is given by

$$Q_c = E_1/\omega^2 \Delta_{if} \; , \tag{13}$$

while the adiabatic change of the energy turns out to be

$$E_i(t) - E_f(t) = E_m \cos(\omega t) - E_1 \; . \tag{14}$$

where

$$E_1 \equiv S\hbar\omega - W_{if} = \frac{1}{2}\omega^2 \Delta_{if}^2 - W_{if} \; , \tag{15}$$

$$E_m \equiv \omega^2 \Delta_{if} Q_m \; . \tag{16}$$

The transition probability in this case was calculated by LANDAU in the early 30s in connection with the predissociation of the molecule as [9]

$$P = \frac{2V^2\omega}{\hbar v |F_f - F_i|} \; , \tag{17}$$

where V is the electronic transition amplitude, v the velocity of nucleus, F_i and F_f the forces acting on the nucleus in the initial and the final states at the crossing point. Equation (17) can be rewritten as

$$P = \frac{2V}{\hbar E_1} \frac{(\Delta E)^{1/2}}{(E_Q - \Delta E)^{1/2}} \; , \qquad \text{where} \tag{18}$$

$$E_Q = \frac{1}{2}\omega^2 Q_m^2 \; , \tag{19}$$

$$\Delta E = \frac{1}{2}\omega^2 Q_c^2 \; . \tag{20}$$

From a purely classical point of view, the system sitting in the initial state $|i\rangle$ cannot change into state $|f\rangle$, but the transition can take place when the zero point motion is taken into account by the ground state oscillator wave function

$$f(E_Q) = \frac{2}{\hbar\omega} \exp(-\frac{2E_Q}{\hbar\omega}) \; . \tag{21}$$

Averaging (18) over (21) will lead to (6) as stated before.

III. Summary of the Main Results

We have calculated explicitly the probability of nonradiative decay of an e-h pair into a soliton pair and that of an electron (hole) into a polaron. This probability turns out to be of order $10^{13} \mathrm{sec}^{-1}$ in consistency with the numerical calculations of SU and SCHRIEFFER [10] using the same set of parameters.

We have also calculated the cross section of soliton pair photoproduction and compared it with the photoconductivity data [3,4] as well as with the excitation profile of the photoinduced infrared absorption [5] in good agreement. Moreover, we have derived from the symmetry argument a selection rule which forbids the direct process of photoproduction for neutral soliton pairs. However, the neutral pairs are allowed in the indirect process of first exciting an e-h pair and its subsequent decay. This way we can estimate the branching ratio of photoproduced neutral to charged soliton pairs in consistency with the photoinduced ESR experiments [3-5].

IV. Tunneling versus Perturbation

As seen from section II, the nonradiative decay is a kind of tunneling process.
One would expect to use the WKB type approach (or instanton technique in modern terminology), but we have treated it by a simple golden rule calculation in the strong
coupling limit. This is nothing new. The reduction of a tunneling problem to a
simple perturbation in certain limiting cases has been known for a long time.

ZENER also, derived in the early 30s, a formula for the probability of molecule predissociation as [9]

$$W = 2 \, e^{-P/2} \, (1 - e^{-P/2}) \, . \tag{22}$$

It is easy to see that if $P \ll 1$, or the splitting of the energy levels

$$\Delta E \sim V \tag{23}$$

is much less than the linewidth due to the transition

$$\Gamma \sim \hbar v |F_i - F_f|/V \, , \tag{24}$$

(22) reduces to (17). This means that in such cases tunneling processes can be
treated perturbatively. It is well known that there are many problems connected
with tunneling type processes, e.g., hopping conductivity, scattering of electrons
on atoms in very intensive laser fields, macroscopic tunneling, nuclear fission and
fate of false vacuum, etc. It is interesting to find out whether there are any
limiting cases for these problems which can be treated perturbatively.

References

1.   K. Huang and A. Rhys, Proc. Roy. Soc. A204, 406 (1950).
2.   See, e.g., K. Huang, Progr. in Phys. (Nanjing, China) 1, 31 (1981).
3.   Z.B. Su and L. Yu, Phys. Rev. B27, 5199 (1983); Errutum B29, 2309 (1984).
4.   SU Zhao-bin and YU Lu, Commun. in Theor. Phys. (Beijing, China) 2, 1203, 1323,
     1341 (1983).
5.   Z.B. Su and L. Yu, in Proceedings of ICSM 84, see Ref.6.
6.   Proceedings of the International Conference on Physics and Chemistry of Low-
     Dimensional Synthetic Metals, to be published in Mol. Cryst. and Liq. Cryst.
7.   H. Takayama, Y.R. Lin-Liu and K. Maki, Phys. Rev. B24, 2388 (1980).
8.   W.P. Su, J.R. Schrieffer and A.J. Heeger, Phys. Rev. Lett. 42, 1698 (1979).
9.   L.D. Landau and E.M. Lifshitz, *Quantum Mechanics*, 3rd ed. (Pergamon, Oxford,
     1977), §90.
10. W.P. Su and J.R. Schrieffer, Proc. Natl. Acad. Sci. (U.S.A.) 77, 5526 (1980).

Breathers in the PPP Model of Polyacetylene

Masaki Sasai and Hideo Fukutome

Department of Physics, Kyoto University, Kyoto 606, Japan

In the Pariser-Parr-Pople(PPP) model of polyacetylene with the Coulomb interaction, charged solitons with opposite charges and neutral solitons with opposite spins, respectively, have attractive interactions and may make breathers of two kinds, the zwitterionic and diradical ones. We numerically simulate those breathers in the adiabatic approximation. They have different spreadings, energies and oscillation periods but have similar oscillation patterns.

Recent experiments about photoinduced absorption of polyacetylene [1]∿[6] have suggested that charged solitons are photogenerated and some of them may be in a self-pinned state. Bishop et al. [7] have shown, utilizing the Su-Schrieffer-Heeger (SSH) model [8], that an electron-hole pair relaxes to a pair of running away charged solitons, thereby leaving a breather in between them. The breather may explain some of the photoinduced intragap absorption bands and the self-pinning of photogenerated solitons. We also have shown [9], utilizing the PPP model[10]∿[13], that charged solitons with opposite charges may make a breather. In the SSH model, neglecting the Coulomb interaction, charged and neutral solitons have essentially the same structure, so that only one kind of breather is possible. In the PPP model, on the other hand, charged and neutral solitons have different structures owing to their different Coulomb interaction dependences[12]. We shall show that breathers of two kinds may be possible in the PPP model, one consisting of charged solitons with opposite charges, the zwitterionic breather, and another consisting of neutral solitons with opposite spins, the diradical breather.

In the PPP model, charged and neutral solitons have π electron clouds with alternating charge density d_n (CD) and spin density s_n (SD), respectively, which arise from the Coulomb correlation of π electrons [12]. When the tails of the π electron clouds of a soliton (S) and an antisoliton ($\bar{S}$) are overlapped, the $S + \bar{S}$ pair turns out to interact. If the phases of the CD or SD modulations in S and $\bar{S}$ are the same , then the π electron cloud in the overlapped region is enhanced owing to the constructive interference of the CD or SD modulations giving an attractive interaction between S and $\bar{S}$ owing to an increase of the correlation stabilization energy. The zwitterionic pairs $S^+ + \bar{S}^-$ and $S^- + \bar{S}^+$ and the diradical pairs, $S^0_\uparrow + \bar{S}^0_\downarrow$ and $S^0_\downarrow + \bar{S}^0_\uparrow$, have attractive interactions due to this mechanism where the superscript represents the charge and the subscript the spin. The pairs $S^+ + \bar{S}^+$, $S^- + S^-$, $S^0_\uparrow + \bar{S}^0_\uparrow$ and $S^0_\downarrow + \bar{S}^0_\downarrow$, on the other hand, have repulsive interactions due to the destructive interferences of the CD or SD modulations.

When the π electron clouds in an attractive $S + \bar{S}$ pair are overlapped, the lattice inside the pair turns out to have no static equilibrium geometry and begins to move. We numerically followed

the lattice motion and the accompanying change of the π electronic
structure in the adiabatic approximation utilizing the PPP model
formulated in [10]. We integrated the equation $M\ddot{u}_n = F_n$ of lattice
motion by the Runge-Kutta-Gill method with the time interval
$\Delta t = 2.5 \times 10^{-16}$ sec. where u_n is the displacement of n-th CH group
along the chain direction, M is the mass of a CH group and F_n is
the Hellmann-Feynman force. The SCF solution of the UHF equation
for π electrons was obtained at every instance with a rather severe
convergence condition, the root mean square values of the changes
of the electron density and bond order per iteration cycle being
smaller than 2×10^{-4}, which is necessary to avoid the artifactory loss
of energy due to the error of the SCF energy. For the sake of economy
we adopted the PP parametrization [10] for the parameters
of the PPP model which gives much narrower soliton spreadings and
consequently a much faster convergence of the SCF procedure than
another parametrization, the SOK one, which gives good agreement
with experiments [11]~[13]; thus the present calculation is not
intended to compare with experiments. We used, as the effective
Coulomb interaction, the screened Ohno potential [11],[12] with the
dielectric and exponential screening parameters $1/\varepsilon = 0.5$ and
$\alpha = 1.5$. All the calculations were carried out in a chain with 62
carbons and with the periodic boundary condition.

We show in Fig.1a and b the initial lattice structures in terms
of the order parameters $\rho_n^A = (-1)^n (r_n - r_{n-1})/2r'$ for antisymmetric
distortions and $\rho_n^A = (r_n + r_{n-1})/2r - 1$ for symmetric distortions,
where r_n is the n-th bond length and r' and r are the bond
alternation and average bond length in the regular lattice, and in
Fig.1a' and b' the initial CD and SD SCF structures of the
zwitterionic and diradical breathers, respectively. The diradical
breather has much narrower spreading than the zwitterionic one due
to the much narrower spreading of the neutral soliton than the
charged one.

We show in Fig.2a and b the time dependences of the potential
energies V for lattice motions, which are the sums of the π electron
and elastic energies, and the CD and SD at the centers of the
zwitterionic and diradical breathers, respectively, which start to
oscillate from the initial structures shown in Fig. 1. We show in

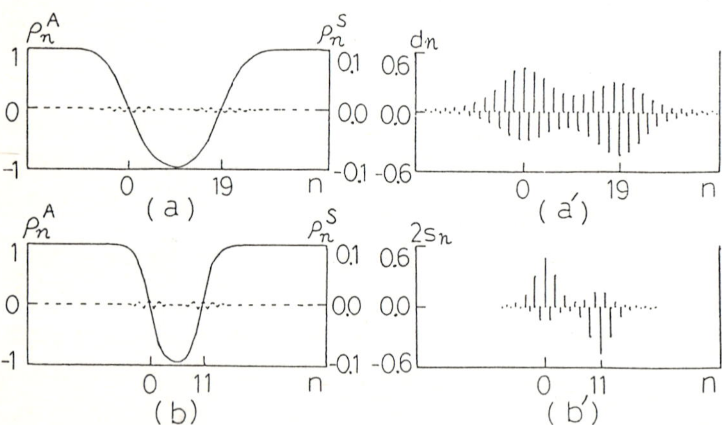

Fig. 1. The initial lattice and CD or SD structures of the
zwitterionic (a),(a') and diradical (b),(b') breathers.
ρ_n^A (—) and ρ_n^S (---) are shown in (a) and (b). The centers of S
and $\bar{S}$ are indicated on the abscissa.

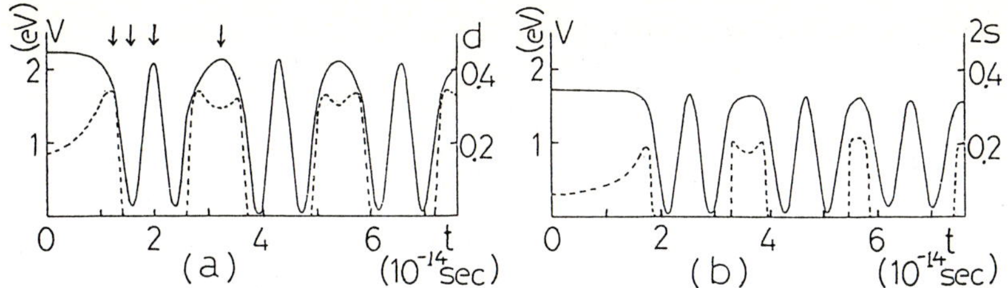

Fig. 2. The oscillation patterns of the zwitterionic (a) and
diradical (b) breathers. The potential energy (-) and the CD at
n=9 (a) or the SD at n=5 (b) (---) are shown.

Fig.3a $\sim$ d the lattice structures of the zwitterionic breather at
the times indicated by the arrows in Fig.2a. We also show in
Fig.3a' and d' the CD structures at the times indicated by the
first and last arrows in Fig.2a. At the other two times no CD
exists in the breather.

From Figs.1 $\sim$ 3 we can see that the oscillation of the
zwitterionic breather proceeds as follows. The initial state of
the breather is $S^+ + S^-$ (Fig.1a and a'). First, the reversed bond
alternation at the central region is relaxed (Fig.3a) and the CD
clouds of S^+ and S^- collapse (Fig.3a'), so that the CD at the
center increases (Fig.2a). Second, the lattice becomes almost

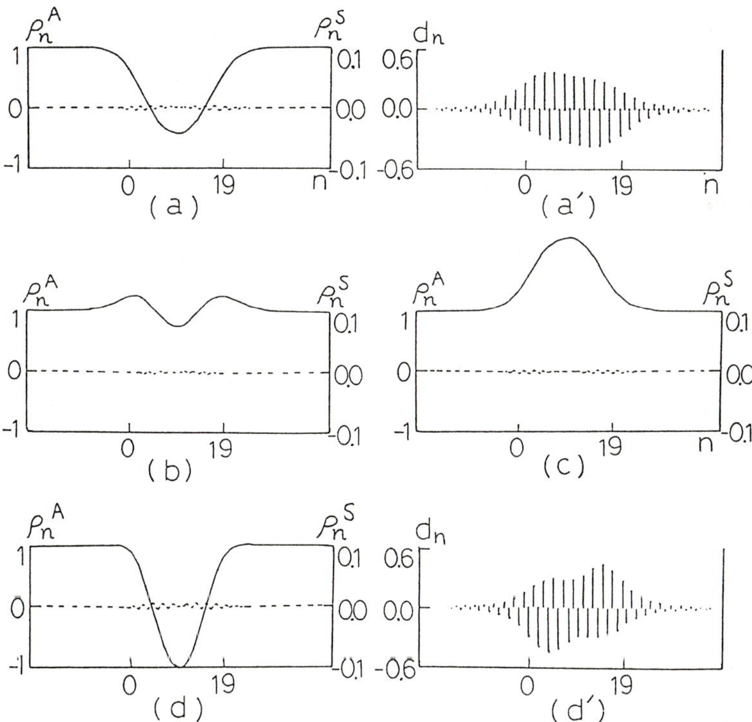

Fig. 3. The lattice (a) $\sim$ (d) and CD (a'),(d') structures of the
zwitterionic breather at the times indicated by the arrows in Fig.
2a. No CD exists in (b) and (c). ρ_n^A (-) and ρ_n^S (---) are shown in
(a) $\sim$ (d).

regular though a small distortion remains (Fig.3b), so that the CD
cloud disappears and the potential energy reaches the first minimum
(Fig.2a). Third, the lattice makes a localized overshoot of the
bond alternation, namely the lattice has a more enhanced
antisymmetric distortion than the regular bond alternated lattice
in the region of the breather (Fig.3c) and the potential energy
increases and reaches the first maximum (Fig.2a). Fourth, the
overshoot of the bond alternation relaxes and the lattice again
becomes almost regular, giving the second minimum of the potential
energy (Fig.2a). Fifth, the sense of the bond alternation at the
central region again reverses (Fig.3d) and the CD reappears and
shows a tendency to split into the pair $S^- + S^+$ (Fig.3d') giving
the second broad maximum of the potential energy (Fig.2a). The
next cycle repeats the above steps with the reverted charge
arrangement. Hence, the one period of the breather oscillation is
twice that of the above cycle. During the oscillation, little symmetric lattice distortion is excited.

The diradical breather has narrower spreading (Fig.1) and
smaller energy (Fig.2) than the zwitterionic breather. Its
oscillation pattern, however, is very similar to that of the
zwitterionic breather though its oscillation period is a little
shorter (Fig.2b). The interval between the first and second
potential energy minima is almost the same in the two kinds of
breather. The difference in the oscillation periods comes from the
difference in the duration of the broad potential energy peak with
the SD or CD (Fig.2).

The zwitterionic breather has a large oscillating electric
dipole moment, so that it strongly interacts with the electromagnetic
field. Its oscillation frequency is in the far infrared region.
Therefore, the photoinduced far infrared absorption observed in
trans-polyacetylene [4] may be due to zwitterionic breathers.

The oscillations of both breathers in the time range of the
present calculation are still not stationary. The duration of the
broad potential energy peaks tends to be shorter for later time (Fig.2).
This tendency is larger in the diradical breather. This might be
due to the growth of a small but non-negligible symmetric distortion
with time in it. Extension of the computation to a longer time
interval is economically difficult because it needs a very long cpu
time. Therefore, unfortunately, we cannot infer any estimation of
the lifetimes of the breathers from the present calculation.

References

[1] J. Orenstein and G.L. Baker, Phys. Rev. Lett. 49, 1040 (1980).
[2] J. Orenstein, G.L. Baker and Z. Vardeny, J. Phys.(Paris)
Colloq. 44, C3-407(1983).
[3] Z. Vardeny, J. Orenstein and G.L. Baker, Phys. Rev. Lett. 50.
2032(1983).
[4] G.B. Blanchet, C.R. Fincher, T.C. Chung and A.J. Heeger,Phys.
Rev. Lett. 50, 1938(1983).
[5] Z. Vardeny, J. Strait, D. Moses, T.C. Chung and A.J. Heeger,
Phys. Rev. Lett. 49, 1657(1982).
[6] C.V. Shank, R. Yen, R.L. Fork, J. Orenstien and G.L. Baker,
Phys. Rev. Lett. 49, 1660(1982).
[7] A.R. Bishop, D.K. Campbell, P.S. Lomdahl, B. Horovitz and S.R.
Phillpot, Phys. Rev. Lett. 52, 671(1984).
Synth. Met. 9, 223(1984).
[8] W.P.Su, J.R. Schrieffer and A.J. Heeger, Phys. Rev. Lett. 42,

1698(1979), Phys. Rev. B22, 2099(1980), M.J. Rice, Phys. Lett. 71A, 152(1979).
[9] M. Sasai and H. Fukutome, Synth, Met., 9, 295, 495(1984).
[10] H. Fukutome and M. Sasai, Prog, Theor. Phys. 67, 41(1982).
[11] H. Fukutome and M. Sasai, Prog, Theor. Phys. 69, 1(1983).
[12] H. Fukutome and M. Sasai, Prog, Theor. Phys. 69, 373(1983).
[13] M. Sasai and H. Fukutome, Prog, Theor. Phys. 70, 1471(1983).

On the Nonlinear Excitations in One-Dimensional Uniaxial Anisotropic Heisenberg Ferromagnetic Spin Chain in External Magnetic Fields

M. Lakshmanan[1,2] and M. Daniel[1]

[1]Department of Physics, Bharathidasan University, Tiruchirapalli-620 023, India
[2]Department of Physics, Kyoto University, Kyoto 606, Japan

K. Nakamura

Fukuoka Institute of Technology, Higashi-ku, Fukuoka 811-02, Japan

1. Introduction

The dynamics of the Heisenberg ferromagnetic spin chain is of fundamental importance in soliton theory, as it is a typical model system
which is completely integrable under suitable circumstances and is
endowed with interesting geometrical aspects [1,2]. In particular, the
one-dimensional continuum limit of the classical Heisenberg ferromagnet
with uniaxial anisotropy is a completely integrable solitonic system
and is gauge equivalent to the nonlinear Schrödinger equation (or
generalizations) [3,4]. In this article, we consider the influence of
external magnetic fields on the dynamics of the spin chain.

2. The Evolution Equations

We consider the one-dimensional classical Heisenberg ferromagnetic
spin Hamiltonian with uniaxial anisotropy and external magnetic field
included:

$$H = -J \sum_{\{i,j\}} \vec{S}_i \cdot \vec{S}_j + A\sum_i (S_i^z)^2 - \mu\vec{B}\cdot\sum_i\vec{S}_i \ . \tag{1}$$

In (1), $\vec{S}_i$'s are three component unit vectors interacting only between nearest neighbors, $J>0$ is the exchange integral, A is the anisotropy parameter, $\mu = g\mu_B$ is the gyromagnetic ratio in Bohr magnetons
and $\vec{B} = \vec{B}(t)$ is the external magnetic field which can in general be
time-dependent. Conventionally the z-axis is treated as the easy axis
of magnetisation as well as the chain direction. Then the corresponding
Landau-Lifshitz equation in the continuum limit [5], when the lattice
parameter $a \to 0$, becomes, after a rescaling ($t \to Ja^2t$, $A \to A/Ja^2$, $\mu \to \mu/Ja^2$)

$$\partial\vec{S}(z,t)/\partial t = \vec{S} \times [\vec{S}_{zz} - 2A(\vec{S}\cdot\vec{n})\vec{n} + \mu\vec{B}], \tag{2a}$$

where

$$\vec{S} = (S^x,S^y,S^z), \ \vec{S}^2 = 1, \ \vec{n} = (0,0,1) . \tag{2b}$$

We may distinguish two cases depending on the value of the anisotropy
parameter A, as below:

value of A	Nature of anisotropy	Example
< 0	Easy axis	$CoCl_2\cdot2NC_5H_5$
> 0	Easy plane	$CsNiF_3$.

The anisotropy can be more general, for example, biaxial. We do not
consider their dynamics here. Typically, the orientation of the magnetic field has enough freedom to be directed along different directions
in the spin space. This may be, for example, along the easy axis of

magnetisation (z-axis) — <u>longitudinal field</u>, that is $\vec{B}(t) = (0,0,B^L(t))$; or perpendicular to it — <u>transverse field</u>, that is $\vec{B}(t) = (B^T(t),0,0)$. Here we consider the dynamics of both the systems and show that in the former case the system is exactly integrable and in the latter case the dynamics is more complicated and probably non-integrable.

3. Equations of Motion under Stereographic Projection

It has been found that when the anisotropy is present it is advantageous to study the spin field by stereographically projecting it onto a complex plane [5,6], so that the three coupled evolution eqs. (2a), along with the length constraint (2b), become the evolution equation for a complex field: The unit sphere of spin $\vec{S}^2 = 1$ is projected onto the complex plane

$$\omega(z,t) = (S^X + iS^Y)/(1 + S^Z) \tag{3a}$$

or equivalently

$$S^X(z,t) = (2\mathrm{Re}\ \omega(z,t))/(1+\omega\omega^*),\quad S^Y(z,t) = (2\mathrm{Im}\ \omega(z,t))/(1+\omega\omega^*),$$

$$S^Z(z,t) = (1 - \omega\omega^*)/(1 + \omega\omega^*)\ . \tag{3b}$$

Making use of these variable transformations and their derivatives, and after a detailed analysis, it is possible to obtain the following evolution equations in place of (2) for the longitudinal and transverse field cases separately.

A. <u>Longitudinal field</u>: $\vec{B}(t) = (0,0,B^L(t))$
We have

$$i(1 + \omega\omega^*)\omega_t + (1 + \omega\omega^*)\omega_{zz} - 2\omega^*(\omega_z)^2 + 2A\omega(1 - \omega\omega^*)$$

$$- \mu B^L\omega(1 + \omega\omega^*) = 0,\quad (\omega_t = \partial\omega/\partial t,\ \omega_z = \partial\omega/\partial z,\ \text{etc.}) \tag{4}$$

B. <u>Transverse field case</u>: $\vec{B}(t) = (B^T(t),0,0)$
The equation of motion (2) becomes

$$i(1 + \omega\omega^*)\omega_t + (1 + \omega\omega^*)\omega_{zz} - 2\omega^*(\omega_z)^2 + 2A\omega(1 - \omega\omega^*)$$

$$+ \frac{\mu B^T}{2}(1 + \omega\omega^*)(1 - \omega^2) = 0\ . \tag{5}$$

We will take up for analysis each of the above equations separately in the following.

4. Elementary Excitations in The Longitudinal Field Case

Considering the equation of motion (4), we observe that in addition to space-time translation invariance, it also admits a U(1) gauge symmetry corresponding to $\omega \to \omega e^{i\theta}$, a consequence of which is that $S^Z - (1 - \omega\omega^*)/(1 + \omega\omega^*)$ is a conserved density. Further a time-dependent gauge transformation

$$\omega \longrightarrow \hat{\omega} = \omega\exp[+i\mu\!\int\! B^L(t)dt] \tag{6}$$

allows us to rewrite (4) (after omitting the 'hat') as

$$i(1 + \omega\omega^*)\omega_t + (1 + \omega\omega^*)\omega_{zz} - 2\omega^*(\omega_z)^2 + 2A\omega(1 - \omega\omega^*) = 0\ . \tag{7}$$

Rewriting eq.(7) in terms of the matrix

$$S = 1/(1 + \omega\omega^*) \begin{bmatrix} (1 - \omega\omega^*) & 2\omega^* \\ 2\omega & -(1 - \omega\omega^*) \end{bmatrix} , \quad S^2 = 1, \quad S^\dagger = S \qquad (8)$$

as

$$S_t = (1/2i)[S,S_{zz}] - (A/2i)[S,\sigma^z S\sigma^z], \quad \sigma^z = \begin{bmatrix} 1 & 0 \\ 0 & -1 \end{bmatrix} , \qquad (9)$$

it can be shown that (9) is a Galilean invariant under the transformation

$$S(z,t) = RSR^{-1}(z + \lambda t,t), \quad R \in SU(2), \qquad (10)$$

provided $R(z,t)$ satisfies the linear equations [3]

$$R_z = LR, \quad L = i\lambda S + \nu[\sigma^z,S] \qquad (11a)$$

$$R_t = MR, \quad M = 2i\lambda^2 S + \lambda SS_z + 2\lambda\nu[\sigma^z,S] - i\nu[\sigma^z,SS_z]$$
$$+ 4i\nu^2\{\sigma^z,S\}\sigma^z \qquad (11b)$$

where

$$\nu = i\sqrt{A}/4 \qquad \text{for } A > 0, \text{ easy plane case}$$

$$\nu = \sqrt{|A|}/4 \qquad \text{for } A < 0, \text{ easy axis case}$$

and λ is a parameter. The quantities L and M are indeed the Lax-pairs of (9) or (7) [3,4] in terms of which an inverse scattering analysis can be performed and the elementary excitations are the usual envelope solitons, kinks and spin waves (depending on the sign of A). For example, the typical nonlinear excitation in the easy axis case is

$$\omega(z,t) = [(\gamma_a^2/2)\{2(\Omega^2 - 2Av^2)^{\frac{1}{2}} \cosh^2[\gamma_a^{-1}(z - vt - z_0)]$$
$$+ \Omega - 4A - (\Omega^2 - 2Av^2)^{\frac{1}{2}}\} - 1]^{-\frac{1}{2}}$$
$$\times \exp i[\psi_0 + \Omega t + (v/2)(z - vt - z_0) + \tan^{-1}\{(\gamma_a/|v|)$$
$$\times [\Omega - v^2/2 + (\Omega^2 - 2Av^2)^{\frac{1}{2}}]\cdot\tanh(\gamma_a^{-1}(z - vt - z_0))\}] \qquad (12)$$

and in the easy plane case it is

$$\omega(z,t) = [-1 + 2/\{1 \pm \gamma_p^{-1}\mathrm{sech}\sqrt{2A}\cdot\gamma_p^{-1}(z - vt - z_0)\}]^{\frac{1}{2}}$$
$$\times \exp i[\psi_0 \mp \tan^{-1}\{(\sqrt{2A}/|v|)\ \sinh\sqrt{2A}\gamma_p^{-1}(z - vt - z_0)\}]. \qquad (13)$$

Here $\gamma_a^{-1} = (\Omega - 2A - v^2/4)^{\frac{1}{2}}$ and $\gamma_p^{-1} = (1 - v^2/2A)^{\frac{1}{2}}$.

It is of further interest to note that [1 - 4] there exists gauge equivalence / geometrical equivalence between the isotropic, anisotropic Heisenberg continuum spin chain and the nonlinear Schrödinger equation as per the following diagram.

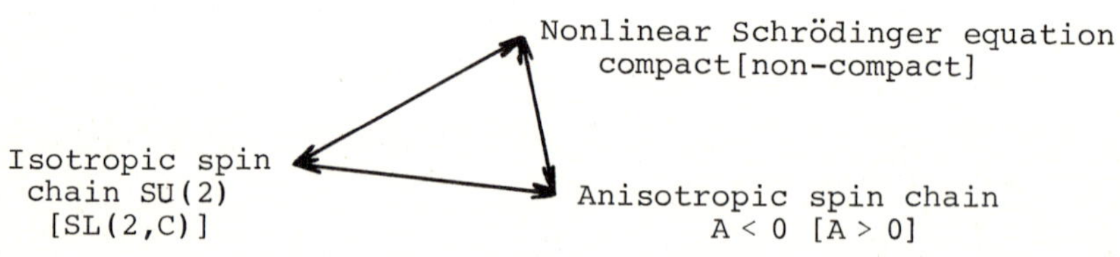

5. The Transverse Field Case

a. Breaking of Symmetries

Considering now the equation of motion (5), we find that it is no longer invariant under the U(1) gauge transformation $\omega \rightarrow \omega \exp i\theta$. As a consequence, $S^z = (1 - \omega\omega^*)/(1 + \omega\omega^*)$ is no longer a conserved quantity. Further, the time-dependent gauge transformation

$$\omega \rightarrow \hat{\omega} = \omega \exp[-i\int B^T(t)dt] \tag{14}$$

leads (5) to the form (after dropping the 'hats')

$$i(1 + \omega\omega^*)\omega_t + [(1 + \omega\omega^*)\omega_{zz} - 2\omega^*(\omega_z)^2] + 2A\omega(1 - \omega\omega^*)$$

$$+ \mu B^T(1+\omega\omega^*)[(1-\omega)\exp(-i\mu\int B^T(t)dt) - \omega^2\exp(i\mu\int B^Tdt)] = 0 \tag{15}$$

so that we have an explicit time-dependent perturbation to the completely integrable anisotropic chain. Due to the breaking of the gauge symmetry, the Galilean symmetry is also seen to be broken as no linear eigenvalue problem for R in (10) could be obtained as far as we know. In this connection, it is well realized that a useful approach to analyse the complete integrability properties of nonlinear partial differential equations is to investigate the Painlevé properties in the vicinity of a movable singular manifold [7]. In the following we show that while the longitudinal field cases(equivalently the pure anisotropic case) possess single-valued solutions in the neighborhood of a singular manifold, the transverse case is not of P-type and thereby showing its non-integrable nature.

b. Painlevé Analysis

Calling

$$\omega = E , \quad \omega^* = G , \tag{16}$$

we rewrite eq.(5) as

$$i(1 + EG)E_t + (1 + EG)E_{zz} - 2GE_z^2 + 2AE(1 - EG)$$

$$+ \mu(B^T/2)(1 + EG)(1 - E^2) = 0 \tag{17a}$$

$$-i(1 + EG)G_t + (1 + EG)G_{zz} - 2EG_z^2 + 2AG(1 - EG)$$

$$+ \mu(B^T/2)(1 + EG)(1 - G^2) = 0 , \tag{17b}$$

where, to be specific, we assume $B^T(t) = B^T(0) = $ constant. In the neighborhood of the singular manifold $\phi(x,t)$, we look for Laurent type expansion,

$$E = \phi^p\Sigma a_r\phi^r \quad , \quad G = \phi^q\Sigma b_r\phi^r \tag{18}$$

giving rise to the following leading orders:

$$[p(p-1) - 2p^2]\phi_z^2 \cdot a_0^2 b_0 \phi^{2p+q-2} - \mu(B^T/2)a_0^3 b_0 \phi^{3p+q} = 0 , \tag{19a}$$

$$[q(q-1) - 2q^2]\phi_z^2 \cdot b_0^2 a_0 \phi^{2q+p-2} - \mu(B^T/2)b_0^2 a_0 \phi^{3q+p} = 0 . \tag{19b}$$

We now discuss the cases $B^T = 0$ and $B^T \neq 0$ separately.

Case 1: Pure anisotropic ($B^T = 0$). From (19), we identify two possible branches of solutions:

Branch 1: $p = -1$, $q = -1$, a_0, b_0: arbitrary $\tag{20a}$

$\underline{\text{Branch 2}}$: p = -1, q = 0,(p=0,q=-1), a_0,b_0: arbitrary. $\hspace{1cm}$ (20b)

Then the coefficients of a_r,b_r in (18) after substituting them in (17) to leading orders give rise to a characteristic resonance determinant proportional to

$\underline{\text{Branch 1}}$: $(r^2 + r)(r^2 + r) = 0 \longrightarrow r = -1,-1,0,0$ $\hspace{1cm}$ (21a)

$\underline{\text{Branch 2}}$: $(r^2 + r) = 0 \longrightarrow r = -1,0$ $\hspace{1.5cm}$ (21b)

and r essentially corresponds to the powers in (18) at which arbitrary functions enter, as confirmed by (20). Thus in fact the Painlevé property is satisfied for the pure anisotropic case, confirming its complete integrability. In fact the resonance values (21) ensure that a Hirota bilinearization is really possible, as noted by Hirota [8].

$\underline{\text{Case 2: Transverse field included}}$ $(B^T \neq 0)$. Here the following branches with the associated leading orders and resonances are available.

$\underline{\text{Branch 1}}$: p = -2, q = -2, $a_0 = b_0 = -(4/\mu B)\phi_z^2$
$$r = -2,-2,-1,-1 \hspace{1cm} \text{(22a)}$$

$\underline{\text{Branch 2}}$: p = -2, q = -2, a_0 or $b_0 = 0$, r = arbitrary $\hspace{1cm}$ (22b)

$\underline{\text{Branch 3}}$: p = -1, q = -1, a_0,b_0: arbitrary, r = -1,-1,0,0 $\hspace{0.5cm}$ (22c)

$\underline{\text{Branch 4}}$: p = -1, q = 0, a_0,b_0: arbitrary, r = -1,0 . $\hspace{1cm}$ (22d)

The existence of r arbitrary in the 2nd branch shows that in the neighborhood of the movable singularity manifold, the solution may be expressible in terms of arbitrary number of arbitrary functions, showing that the solution contains a movable essential singularity manifold and so is of non-Painlevé type and so the underlying system is unlikely to be completely integrable in general.

6. Transverse Field Case: Special Static Solutions and Linear Stability Aspect

Due to the breaking of U(1) symmetry and possible non-invariance under Galilean transformation, no exact wave profiles of the form ω = Aexpi(kz $- \Omega t + \delta$), A,k,Ω,δ: parameters, exist. Neither any other explicit time-dependent solution is known. We therefore try to find possible static configurations which (5) can admit. Even here it does not seem to be feasible to obtain the general static configurations, except for two special configurations. To see these transparently, we first introduce the two real functions P and Q as

$$\omega(z) = P(z) + iQ(z) \hspace{1cm} \text{(23)}$$

and rewrite the static part of (5) in the following form:

$$(1 + P^2 + Q^2)P_{zz} - 2PP_z^2 + 2PQ_z^2 - 4P_zQQ_z + 2AP(1 - P^2 - Q^2)$$
$$+ (\mu B/2)(1 + P^2 + Q^2)(1 - P^2 + Q^2) = 0 , \hspace{1cm} \text{(24a)}$$

$$(1 + P^2 + Q^2)Q_{zz} + 2P_z^2Q - 2QQ_z^2 - 4PP_zQ_z + 2AQ(1 - P^2 - Q^2)$$
$$- \mu B(1 + P^2 + Q^2)PQ = 0, \quad B = B^T(0). \hspace{1cm} \text{(24b)}$$

System (24), considered as a set of dynamical equations for the 'two degrees of freedom' variables P and Q with associated conjugate momenta

214

$\Pi_P = P_z/(1 + P^2 + Q^2)^2$ and $\Pi_Q = Q_z/(1 + P^2 + Q^2)^2$, and z treated as the time parameter, gives rise to an integral of 'motion', the 'Hamiltonian':

$$I = (P_z^2 + Q_z^2)/2(1 + P^2 + Q^2)^2 - A(1 - P^2 - Q^2)^2/4(1 + P^2 + Q^2)^2$$

$$- (\mu B/2)[1 - P/(1 + P^2 + Q^2)] \tag{25}$$

such that $dI/dz = 0 \rightarrow$ (24). A Poincaré surface of section shows that the system (24) is essentially irregular and chaotic. However, two special solutions do exist corresponding to purely in-plane $(S^x - S^y)$ motions and out-of plane $(S^x - S^z)$ motions: (i) $P^2 + Q^2 = \omega\omega^* = 1$ and (ii) ω = Real, $Q = 0$.

(a). <u>In-plane configurations</u>: Using the condition $P^2 + Q^2 = 1$, eqs. (24) degenerate into

$$P_{zz} + PP_z^2/(1 + P^2) + \mu B(1 - P^2) = 0. \tag{26}$$

Eq. (26) is integrable and as a specific limit we obtain

$$P = 1 - 2\,\mathrm{sech}^2(\sqrt{\mu B}\cdot z + \delta), \quad Q = 2\,\mathrm{sech}(\sqrt{\mu B}\cdot z + \delta)\tanh(\sqrt{\mu B}\cdot z + \delta) \tag{27}$$

so that the spin configurations are

$$S^x = 1 - 2\,\mathrm{sech}^2(\sqrt{\mu B}\cdot z + \delta), \quad S^y = 2\,\mathrm{sech}(\sqrt{\mu B}\cdot z + \delta)\tanh(\sqrt{\mu B}\cdot z + \delta),$$

$$S^z = 0 . \tag{28}$$

One might note that (28) is essentially the static counterpart of pulse solutions of the isotropic case $(A = 0)$.

(b). <u>Off-plane configurations</u>: $Q = 0$. Correspondingly, we have

$$P_{zz} - 2PP_z^2/(1 + P^2) + 2AP(1 - P^2)/(1 + P^2) + \mu B(1 - P^2)/2 = 0. \tag{29}$$

Eq. (29) can be integrated exactly, and as a limiting form we obtain

$$P = \frac{2D + (C - D)\,\mathrm{sech}^2[\sqrt{2A + \mu B}\cdot(z/2) + \delta]}{2D - (C + D)\,\mathrm{sech}^2[\sqrt{2A + \mu B}\cdot(z/2) + \delta]} \tag{30}$$

where $C = 1 + [A/(A + \mu B)]$, $D = [(A/(A+\mu B))^2 - 1]^{\frac{1}{2}}$. So

$$S^x = 1 - \frac{2C^2\,\mathrm{sech}^4[\sqrt{2A + \mu B}(z/2) + \delta]}{4D^2\tanh^2[\sqrt{2A+\mu B}(z/2) + \delta] + (C^2+D^2)\,\mathrm{sech}^4[\sqrt{2A+\mu B}(z/2) + \delta]} , \tag{31a}$$

$$S^y = 0, \tag{31b}$$

$$S^z = \frac{-2CD[1+\tanh^2(\sqrt{2A+\mu B}(z/2)+\delta)]\cdot\mathrm{sech}^2[\sqrt{2A+\mu B}\cdot(z/2)+\delta]}{4D^2\tanh^2[\sqrt{2A+\mu B}\cdot(z/2)+\delta]+(C^2+D^2)\,\mathrm{sech}^4[\sqrt{2A+\mu B}(z/2)+\delta]} . \tag{31c}$$

(c). <u>Linear Stability</u>: Treating the above static solutions as unperturbed, we apply an infinitesimal perturbation

$$P(z,t) = P_0(z) + \varepsilon\chi(z)e^{\lambda t}, \quad Q = Q_0(z) + \varepsilon\psi(z)e^{\lambda t} , \tag{32}$$

so that using these in (5) along with (23), we obtain a coupled linear eigenvalue problem with the redefinitions

$$\chi = P_0\alpha + Q_0\beta , \quad \psi = Q_0\alpha - P_0\beta \tag{33}$$

for the solutions (28) as

$$\alpha_{zz} - [\mu B + 2A - 6\mu B \operatorname{sech}^2(\sqrt{\mu B}\cdot z + \delta)]\alpha = -\lambda\beta \ , \tag{34a}$$

$$\beta_{zz} - \mu B[1 - 2\operatorname{sech}^2(\sqrt{\mu B}\cdot z + \delta)]\beta = +\lambda\alpha \ . \tag{34b}$$

When $A = 0$, it can be shown that $\lambda = 0$ is the lowest eigenvalue and linear stability is ensured for the pulse soliton as it should be. When $A \neq 0$, the eigenvalue problem is not exactly solvable. However, a perturbation analysis seems to show that the static solutions (28) are in general linearly unstable. Similar analysis can be carried out for the static modes (31).

7. Perturbation of solitons

Considering eq.(15) for the transverse field case, the term $\mu B^T(1+\omega\omega^*)$ $\times[(1-\omega)\exp(-i\mu\int B^T(t)dt)-\omega^2\exp(i\mu\int B^Tdt)]$ may be considered as a perturbation for the anisotropic soliton, when B is small. The fact that even when B is constant it is a nonautonomous perturbation and that S^z is a non-conserved quantity ensures that perturbation leads to an irregular motion of the soliton. The details are under further investigation.

Acknowledgments

M.L. wishes to thank Professor H.Hasegawa for his hospitality at Kyoto and the Japan Society for Promotion of Science for financial support.

References:

1. M. Lakshmanan, Th.W. Ruijgrok and C.J. Thompson: Physica 84A, 577 (1976); M. Lakshmanan: Phys. Lett. A64, 53 (1977)
2. V.E. Zakharov and L.A. Takhtajan: Theor. Math. Phys. 38, 17 (1979)
3. K. Nakamura and T. Sasada: J. Phys. C15, L915 (1982)
4. A. Kundu and O. Pashaev: J. Phys. C16, L585 (1983)
5. M. Daniel: 'Nonlinear Excitations in the Heisenberg Ferromagnetic Spin Chain', Ph.D. thesis, University of Madras, 1983
6. M. Lakshmanan and M. Daniel: Physica 107A, 533 (1981)
7. J. Weiss, M. Tabor and G. Carnevale: J. Math. Phys. 24, 522 (1983)
8. R. Hirota: J. Phys. Soc. Japan 51, 323 (1982)

Part V

Solitons in Biological Systems

Solitons and Excitons in Quasi-One-Dimensional Systems

Alexander S. Davydov

Institute for Theoretical Physics of the Academy of Sciences of the Ukrainian SSR
SU-252130 Kiev-130, USSR

1. INTRODUCTION

Many processes in living organisms are associated with a space propagation of energy
and electrons along protein molecules. For example, the energy released under hydrolysis of ATP molecules in some cases is transferred along alpha-helical protein
molecules as the vibrational oscillation (amide I) of atoms C=O of peptide groups
(PG's) contained in these molecules. Protein molecules also transport electrons
from donors to acceptors very effectively.

The investigation [2, 4, 10-14] carried out at ITP AN Ukr. SSR have shown that
the above processes of effective transfer are due to particular properties of excited
states of alpha-helical protein molecules in which PG's are arranged as three parallel quasi-periodic chains. PG's have a constant electric dipole moment (3.5 Debye).
Therefore they form a potential well, capable of retaining an extra electron which
goes to a molecule from a donor. The overlap of the wave function of an extra electron and the resonance dipole-dipole interaction of amide I vibrations of the neigh-
bouring PG's lead to the collectivization of the corresponding excitations.

If PG's are fixed in sites na (n is an integer), the collective excitations will
form energy bands of the corresponding quasi-particles (excitons or extra electrons)
with effective mass (m) defined by the resonance dipole-dipole (for amide I vibrations) or exchange (for electrons) interaction D,

$$m = \hbar^2 / 2 a^2 D. \tag{1}$$

In the region of quasi-particle location the interpeptide distances change. Therefore excited states of a protein molecule have a complicated character which is
stipulated by the interaction between particles and local deformations of the chain.

The peculiarities caused by the presence of three parallel chains of PG's in
alpha-helical molecules were investigated in [20, 21-23]. Here, for simplicity, we
restrict ourselves to the consideration of the single chain of PG's in a protein
molecule. The Hamiltonian of an infinite chain of PG's is

$$H = H_p + H_{vib} + H_{int} \tag{2}$$

where

$$H_p = D \sum_n [2 \psi_n - \psi_{n+1} - \psi_{n-1}] \tag{3}$$

is the Hamiltonian of quasi-particles when the energy is counted off from the energy
band bottom. The square modulus of the wave function $\psi_n(t)$ characterizes the probability of location of a quasi-particle on PG of number n: The Hamiltonian of displacements u_n of PG of number n and of mass M from equilibrium position na has the
form

$$H_{vib} = M \sum_n [2^{-1} a^2 (du_n / dt)^2 + V_0^2 U(\rho_n)], \tag{4}$$

$$V_0 = a (\mathscr{x} / M)^{1/2}, \qquad \rho_n = u_n - u_{n+1}.$$

V_0 is the velocity of longitudinal sound waves in the chain with elasticity coefficient $\mathfrak{x}$.

$$U(\rho_n) = 2^{-1} \rho_n^2 + 3^{-1} \gamma\rho_n^3 \tag{5}$$

is the dimensionless potential energy of displacements with allowance for a cubic anharmonicity. The connection of displacements with quasi-particles is characterized by the operator

$$H_{int} = 2^{-1} \alpha \sum_n |\psi_n|^2 (u_{n+1} - u_{n-1}). \tag{6}$$

The Hamiltonian (2) corresponds to the set of finite difference equations

$$(D \, d^2 / d\,\xi^2 + \chi\rho - \varepsilon) \, \phi(\xi) = 0, \tag{7}$$

$$\nu \, d^2 \rho / d\xi^2 + (1 - s^2) \, \rho + \gamma\rho^2 = \chi\phi^2 / MV_0^2, \tag{8}$$

$s=V/V_0$, when the envelope $\phi_n(t)$ of the wave function $\psi_n(t)$ is separated by means of the expression

$$\psi_n(t) = \phi_n(t) \exp \{i \, h^{-1} [m \, a \, n \, V - (\varepsilon - 2^{-1} m \, V^2) \, t]\} \tag{9}$$

and by the transition to the continuum approximation ($n \to x/a$), taking into account the approximate equalities

$$\rho_n \to \rho \, (x) = -a \, du / dx, \qquad \nu = 1 / 12,$$

$$u_{n-1} + u_{n+1} - 2 \, u_n \to a^2 \, (d^2u / dx^2 + \nu \, a^2 \, d^4u / dx^4)$$

in the coopdinate system

$$\xi = (x - x_0 - V \, t) / a, \tag{10}$$

moving with constant velocity V. The value ε determines the energy of quasi-particles in the field of local deformation of the chain. In order to obtain a total excitation energy it is necessary to add the chain deformation energy

$$W = (1/2) \, M \, V_0^2 \int d \, \xi \, [(1 + s^2) \, \rho^2 + \nu \, (d\rho / d\xi)^2 + (2/3) \, \gamma\rho^3 \tag{11}$$

to ε.

2. EXCITONS AND SOLITONS IN HARMONIC APPROXIMATION

We begin by analyzing the solution of the set of (7) and (8) in the absence of anharmonicity ($\gamma=0$) and the displacement dispersion ($\nu=0$). From this there follows follows

$$\rho \, (\xi) = \chi\phi^2 / MV_0^2 \, (1 - s^2) \tag{12}$$

and (7) is reduced to the Schrödinger equation for the envelope

$$D \, d^2 \phi / d\xi^2 + G(s)\phi^3 = \varepsilon\phi, \qquad G(s) = \chi^2 / \mathfrak{x}a^2 \, (1 - s^2). \tag{13}$$

The solution of (13) has been studied in [2, 9-14]. In a very long chain ($l \gg 1$) a particular solution of (13) is $\phi=1/\sqrt{L}$. In this case $\rho(\xi)=G(\xi)=0$ and the function (9) is a plane wave, characterizing the motion of a quasi-particle with the effective mass m. These solutions are valid both at $V<V_0$ and at $V>V_0$. However, at $V<V_0$ there are spatially localized solutions of less energy-SOLITONS which characterize the motion of a quasi-particle coupled with the local chain deformation. They are described by the functions

$$\phi_{sol}(\xi) = (Q / 2)^{1/2} \, sech \, (Q\xi) \tag{14}$$

$$\rho_{sol}(\xi) = \chi\phi^2(\xi) / \text{æ}\, a^2 (1 - s^2) \tag{14a}$$

when

$$Q = m\chi^2 / 2\,\text{æ}\hbar^2 (1 - s^2). \tag{15}$$

Solitons transfer the energy

$$E_{sol} = -\Delta + (1/2) M_{sol} v^2. \tag{16}$$

Here

$$\Delta = \chi^4 / 48\,\text{æ}^2 a^4 D \tag{17}$$

is the energy gain when a quasi-particle is coupled with the chain deformation.

$$M_{sol} = m [1 + M\chi^4 / 6\,\text{æ}^2 \hbar^2 (1 - s^2)] \tag{18}$$

is the effective soliton mass.

The soliton stability is determined by three causative factors: 1) The splitting of a slowly moving soliton into an exciton or electron and a local deformation requires the binding energy (17). 2) A soliton always travels with a velocity less than V_0. Therefore it does not use up energy emitting phonos. 3) When a soliton travels, the displacement of PG's from their equilibrium positions takes place

$$u(\xi) = \chi [1 - \tanh (Q\xi) / a\,\text{æ}(1 - s^2). \tag{19}$$

In order to destroy a soliton and to return the chain of PG's to the ground state one must return PG's to initial positions. This property is called topological stability.

The soliton stability is due to the mutual compensation of dispersion and non-linearity. Exceptional stability of solitons and electrosolitons makes helical protein molecules ideal carriers of the hydrolysis energy of ATP molecules and electrons from donor molecules to acceptors.

Solitons can be excited only by local external events at the edge of the molecular chain. For example, under chemical reactions. This property of solitons is based on their topological stability.

3.   THE SOLITON PHOTODISSOCIATION

The infrared radiation with the wave length $\sim6.10^{-4}$cm corresponding to the energy of vibrations of amide I excites metastable exciton states in proteins. The soliton excitation, as shown by EREMKO and the author [3], is forbidden due to the Frank-Condon principle since it must be accompanied by displacements of heavy PG's at the moment of excitation. However there exists the probability, although slight, of a soliton splitting into an exciton and a local chain deformation under the influence been studied by ERNEMKO [17].

Owing to the Frank-Condon principle, during the optical transition the coordinates state into that of exciton occurs at $k=0$ and the retention of PG's location in the initial soliton state is characterized by the function $\rho(\xi)$. In this case the the initial soliton state characterized by the function $\rho(\xi)$. In this case the transition energy is greater than the binding energy (17) by the value of the chain deformation energy which is $W=2\Delta$. Thus, the energy of the soliton photodissociation is determined by the equality

$$\hbar\omega = 3\Delta = \chi^4 / 16\,\text{æ}^2 a^4 D. \tag{20}$$

Using the parameters $D=7.8$cm$^{-1}=15.5$ j, $\text{æ}=19.5$ neuton/m and $\chi/a=2\Delta=(6.4-8).10^{-11}$

Newton, we can see that the resonance frequency ω corresponds to the radiation wave length $\lambda=4.6-8.8$mm. This value coincides with the range of the wave lengths of electromagnetic radiation which actively interacts with living organisms [1, 7, 18].

Photodissociation of solitons into quickly damping and "smearing" excitons and a local deformation relaxing into a heat motion must greatly depress the energy transport and, hence, the process of vital activity. The probability of soliton dissociation is very small, but with a long radiation its effect may manifest itself. Thus, the "mysterious" resonance character, noted by many scientists, of the influence of radiation of millimeter range waves on living organisms can be explained.

4. THE ELECTROSOLITON PAIRING IN SOFT MOLECULAR CHAINS

If a pair of extra electrons gets into a chain of PG's then as it has been shown by BRIZHIK and the author [8] their "pairing" occurs in the region of a local chain deformation formed by these electrons. Due to the nonlinearity effect the potential well formed by a single electron attracts another one (with an opposite spin) which in its turn deepens the well. The energy gain when two electrosolitons are paired is determined by the value

$$E_{bond}(s) = m \chi^4 (1 - 5s^2) / 4 \hbar^2 \varkappa^2 (1 - s^2) a^2$$

from which it follows that the pairing is energetically productive only at small velocities V.

Coupled electrosolitons moving with constant velocity $V \ll V_0$ have the energy

$$E(V) = m V^2 - m \chi^4 / 3 \varkappa^2 a^2 \hbar^2 + M_{bond} V^2 / 2.$$

The effective mass of paired electrosolitons M_{bond} is

$$M_{bond} = 2 m (1 + 2 M \chi^4 / 3 \hbar^2 \varkappa^3 a^2).$$

It is at least twice as great as the mass of two isolated electrosolitons (18). Consequently with increasing velocity the kinetic energy of the pair increases more rapidly than the sum of energy of each electrosoliton taken separately. This fact explains the violation of the pairing condition at large velocities.

The electron pairing in soft chains is stipulated by the nonlinear connection of electrons with chain inertia deformation. In a nonlinear and inertialess field with self-action there are no bound states.

The effect of electron pairing in a singlet spin state manifests itself in some biological phenomena. So, in the redox reactions a pair of electrons is transferred from one molecule to another. It is known that pairs of electrons rather than single ones are transferred during ATP molecule synthesis in conjugate membranes of mitochondria and chloroplasts [19].

5. SOLITONS IN MOLECULAR SYSTEMS WITH NONLINEAR INTERACTION BETWEEN MOLECULES

The presence of the multiplier $(1-s^2)$ in the denominators of the expression (12), (13), etc. limits the region of the theory applicability. This limitation is due to the fact that PG's are greatly displaced from the equilibrium positions at $s\to1$ which does not agree with the used harmonic approximation. It has been shown by ZOLOTARIUK and the author [5, 15, 16] that taking the anharmonicity of PG displacements into account removes this difficulty. In a particular case of cubic anharmonicity the problem is reduced to the solution of the set of eqs. (7), (8) at $v=0$. Below we present the wave function and the chain energy only for solitons moving with the limiting velocity $V=V_0$. In dimensionless units

$$\tau = \chi / M V_0^2, \qquad \sigma = 2 m a^2 \chi / \hbar^2$$

they are expressed by the formulas

$$\phi(\xi) = \frac{1}{2}\,(3\sigma\,/\,4)^{1/3}\,(\tau\,/\,\gamma)^{1/6}\,\mathrm{sech}^2\,(q\xi), \qquad q = \frac{1}{2}\,(\tau\sigma\,/\,6\gamma)^{1/2},$$

$$\rho(\xi) = (\tau\,/\,\gamma)^{1/3}\,\phi(\xi), \qquad E(V_0) = \frac{1}{2}\,m\,V_0^2 + \frac{\tau}{\gamma}\,[1 - \frac{3}{2}\,(\gamma\sigma\tau^2\,/\,36)^{1/3}].$$

Hence, at any $\gamma{\neq}0$ the energy $E(V_0)$ is finite.

6. SUPERSONIC ACOUSTIC SOLITONS

When there is no connection between amide I vibration and the displacements ($\chi{=}0$) the set of eqs. (7), (8) is divided. The equation (8) is transformed into the Boussinesq equation at $\nu{=}1/12$

$$[\,\frac{1}{12}\,d^2\,/d\xi^2 + (1 - s^2)]\,\rho(\xi) + \gamma\rho^2(\xi) = 0. \tag{21}$$

In the systems admitting structural phase transitions and in soft molecular chains the anharmonicity effects ($\gamma{\neq}0$) are very important and cannot be considered using the perturbation theory methods.
The lowest energy of excitation moving with velocity $V{>}V_0$ corresponds to the exact solution (21) as a solitary wave [2]

$$\rho(\xi) = (q^2\,/\,2\gamma)\,\mathrm{sech}^2\,(q\,\xi), \qquad q^2 = 3\,(s^2 - 1). \tag{22}$$

The excitation described by the bell-type function (22) characterizes the decrease in distances between PG's in the region

$$\Delta\xi = 2\pi q.$$

With decreasing velocity the soliton amplitude decreases and its width increases. In the limit $V{\to}V_0$ the soliton vanishes. Such a soliton is called a supersonic

acoustic soliton. Stabilization of such excitation is due to the mutual compensation of dispersion ($\nu{=}1/12$) and the nonlinearity of intramolecular interactions ($\gamma{\neq}0$)

7. SOLITONS IN SYSTEMS WITH DISPERSION AND ANHARMONICITY

In the previous section it was shown that, allowing for anharmonicity of local deformations of chains and without a dispersion soliton, transferring quasi-particle and local deformations can only move with subsonic velocities. With velocities exceeding that of longitudinal sound in a chain, the local deformation has no time to follow a quasi-particle.
However, taking into account dispersion in nonlinear elastic chains supersonic solitons transferring local contraction can spread. Therefore, one can expect that allowing for dispersion and anharmonicity of the chain interacting with a quasi-particle solitons with supersonic velocities may also appear. Such solutions have been analyzed by ZOLOTARIUK and the author [6] on the basis of the set of equations (7), (8) with the parameters D, χ, γ and ν different from zero.
It was shown that when the approximate equality is satisfied

$$g = (\gamma\,D\,/\,3\nu\chi\,)^{1/2} = 1$$

the set of equations (7), (8) has exact solutions of the form

$$\phi(\xi) = (\mu_0\,/\,2)^{1/2}\,\mathrm{sech}\,(\mu_0\xi),$$

$$\rho(\xi) = 2\mu_0^2\,D\,\chi^{-1}\,\mathrm{sech}^2\,(\mu_0\xi),$$

in which the value μ_0, depending on the velocity $V{=}sV_0$ is a root of the cubic equation

$$4\nu\mu^3 - (1 - s^2)\,\mu - \chi^2\,/\,4\,M\,V_0^2\,D = 0.$$

This equation has only one positive root μ_0 for the fixed value s both at s^2 1 (subsonic soliton) and at $s^2 > 1$ (supersonic soliton).

REFERENCES

1. V.S. Bannikov, S.B. Rojkov: Dokl. Acad. Sci. USSR, 255, no.3, 746 (1980)
2. A.S. Davydov: Theor. Nat. Phys. 40, no.3, 408 (1979)
3. A.S. Davydov, A.A. Eremko: Ukr. Fiz. Zh. 22, no.6, 881 (1977)
4. A.S. Davydov, A.A. Eremko, A.I. Sergeenko: Ukr. Fiz. Zh. 23, no.6, 983 (1978)
5. A.S. Davydov, A.V. Zolotariuk: Preprint, ITP-81-112 R, Kiev (1981)
6. A.S. Davydov, A.V. Zolotariuk: Preprint, ITP-83-117 R, Kiev (1983)
7. N.D. Devyatkov: Usp. Fiz. Hayk. 110, no.3, 452 (1973)
8. L.S. Brizhik, A.S. Davydov: Low. Temp. Phys. 10, no.7, 748 (1984)
9. A.S. Davydov, N.I. Kislukha: Phys. Stat. Sol. (b) 59, 465 (1973)
10. A.S. Davydov: Phys. Stat. Sol. (b) 75, 735 (1976)
11. A.S. Davydov: J. Theor. Biol. 66, 739 (1977)
12. A.S. Davydov: Intern. J. Quant. Chem. 26, 5 (1979)
13. A.S. Davydov: Physica Scripta 20, no.3/4, 387 (1979)
14. A.S. Davydov: Physica 3D, no.1-2, 1 (1981)
15. A.S. Davydov, A.V. Zolotariuk: Phys. Stat. Sol. (b) 115, 115 (1983)
16. A.S. Davydov, A.V. Zolotariuk: Physics Letters 94A, no.1, 49 (1983)
17. A.A. Eremko: Preprint ITP-83-109 E, Kiev (1983)
18. H. Frohlich, F. Kremer, ed.: Coherent Excitation in Biological Systems, Berlin, N.Y., Tokyo, 224 (1983)
19. P.C. Hinke, R.E. Mc Carty: Sci. Amer. 238, no.3, 104 (1978)
20. J.M. Hyman, D.W. Mc Laughlin, A.C. Scott: Physica D 3D, no.1-2, 23 (1981)
21. A.C. Scott: Phys. Rev. A 26, no.1, 578 (1982)
22. A.C. Scott: Phys. Scripta 25, 651 (1982)
23. A.C. Scott: in Structure and Dynamics Nucleic Acids and Protein E. Clementy, R. Sarma, ed.: Adenine Press, N.Y., 389-404 (1983)

Biological Solitons

Alwyn C. Scott

Center for Nonlinear Studies, Los Alamos National Laboratory,
Los Alamos, NM 87545, USA

1. Introduction

A fundamental problem in biochemistry is to understand how metabolic energy is
stored and transported in biological molecules [1,2]. In order to achieve this
understanding, two distinct questions must be addressed. The first is a question
of biophysics: What mechanisms are available for the storage and transport of
biological energy? Only after this first question has been answered in detail
can one approach with confidence the biological question: What mechanisms are
actually employed by biological organisms for the storage and transport of energy?

The aim of this review is to provide some answers to the first question.

2. Davydov's Soliton

An interesting candidate for the transport and storage of metabolic energy is the
amide-I (or CO stretching) vibration in protein. This vibration has a quantum
energy of about 0.2 ev which is appropriate to store or transport the 10 kcal/mole
(.43 ev) of free energy released in the hydrolysis of adenosine triphosphate
(ATP). However a linear theory won't fly. If amide-I vibrational energy is assumed to be localized on one or a few neighboring peptide groups at some time, it
will rapidly disperse and distribute itself uniformly over the molecule. The
cause of dispersion is the dipole-dipole interactions between vibrating and non-
vibrating peptide groups. This interaction requires that initially localized
energy become nonlocalized in about a picosecond, a time that is much too short
for biological significance.

To see this effect in more detail consider the alpha-helix shown in Fig. 1.
An important variable is the probability amplitude $a_{n\alpha}$ for finding an amide-I
vibrational quantum in the peptide group specified by the subscripts n and α.
Thus the probability of a vibrational quantum being at the (n,α) peptide group is
$|a_{n\alpha}|^2$ where n is an index that counts turns of the helix and $\alpha(= 1, 2$ or $3)$
specifies one of the three peptide groups in each turn.

Taking account of dipole-dipole interactions $a_{n\alpha}$ obeys Schrödinger's time
dependent equation

$$i\hbar\dot{a}_{n\alpha} = E_o a_{n\alpha} - J(a_{n+1,\alpha} + a_{n-1,\alpha})$$

$$+ L(a_{n,\alpha+1} + a_{n,\alpha-1}) \tag{1}$$

where E_o is the energy of an amide-I quantum, J is the strength of the longitudinal, nearest neighbor, dipole-dipole interaction and L plays a corresponding
role for lateral interaction. Equation (1) is highly dispersive. Initially
localized vibrational energy would quickly spread longitudinally through J and
laterally through L.

In 1973 DAVYDOV and KISLUKHA [3,4] proposed a nonlinear mechanism that might
prevent energy dispersion in (1). This mechanism involves interaction with longitudinal sound waves or stretching of the hydrogen bonds (see Fig. 1). Without
this nonlinear effect, longitudinal sound waves would be governed by the equation

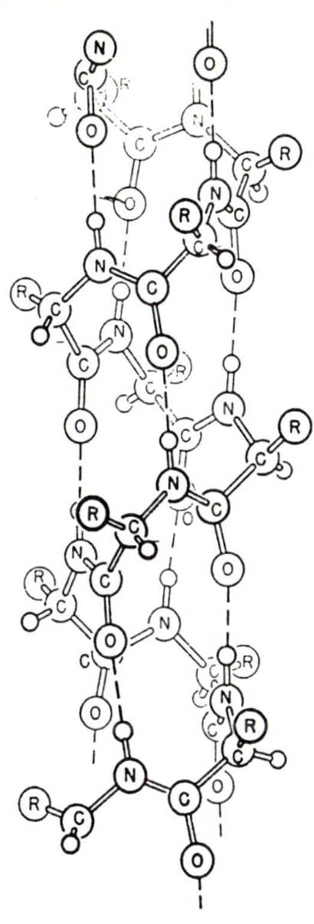

Fig. 1. The alpha helix structure in protein

$$M\ddot{z}_{n\alpha,tt} - K(z_{n+1,\alpha} - 2z_{n\alpha} + z_{n-1,\alpha}) = 0 \qquad\qquad (2)$$

where $z_{n\alpha}$ is the longitudinal displacement of the (n,α) peptide group, M is the mass of a peptide group plus residue, and K is the spring constant of a hydrogen bond.

The specific nonlinear effect considered by Davydov was the effect of stretching a hydrogen bond on the amide-I quantum energy (E_o). If R is the length of the amide's hydrogen bond, this effect can be expressed as the nonlinear parameter

$$\chi = \frac{dE_o}{dR} , \qquad\qquad (3)$$

Since E_o can be measured in joules and R in meters, χ has the units of newtons. Its value has been calculated (using self consistent field methods) as [5]

$$\chi = 3 - 5 \times 10^{-11} \text{ newtons} .$$

It has also been determined experimentally (from a comparison of hydrogen bonded polypeptide crystals with different bond lengths and amide-I energies) as [6]

$$\chi = 6.2 \times 10^{-11} \text{ newtons} .$$

Augmenting (1) and (2) by the interaction expressed in (3), led Davydov to the nonlinear wave system

225

$$i\hbar\dot{a}_{n\alpha} = [E_o + \chi(z_{n+1,\alpha} - z_{n\alpha})]a_{n\alpha}$$

$$- J(a_{n+1,\alpha} + a_{n-1,\alpha}) + L(a_{n,\alpha+1} + a_{n,\alpha-1}) \tag{4a}$$

$$M\ddot{z}_{n\alpha} - K(z_{n+1,\alpha} - 2z_{n\alpha} + z_{n-1,\alpha}) = -\chi(|a_{n\alpha}|^2 - |a_{n-1,\alpha}|^2) . \tag{4b}$$

Note that the only changes from (1) and (2) are to take account of the force due to stretching of the hydrogen bond in the $\chi(z_{n+1,\alpha} - z_{n\alpha})a_{n\alpha}$ term of (4a) and the corresponding source term $-\chi(|a_{n\alpha}|^2 - |a_{n-1,\alpha}|^2)$ in (4b). It is important to observe that each parameter ($\hbar, E_o, \chi, J, L, M$ and K) which appears in (4) has been independently determined. Thus a study of the dynamical behavior permits no parameter adjustment whatsoever, a rather unusual situation in biological science.

A detailed analytical study of (4) has shown that solitons do indeed form [7-9]. If amide-I vibrational energy is localized on one or a few neighboring peptide groups, then the right hand side of (4) is nonzero and acts as a source of longitudinal sound. This longitudinal sound, once created, reacts, through the term $\chi(z_{n+1,\alpha} - z_{n\alpha})$, in (4a) as a potential well to trap the localized amide-I vibrational energy and prevent its dispersion by the effects of dipole-dipole interactions.

Davydov's soliton concept is rather similar to the polaron [10]. For the polaron, localized electronic charge distorts the lattice in its vicinity, lowering its energy and thereby trapping it. For Davydov's soliton, localized amide-I vibrational energy distorts the lattice in its vicinity, lowering its energy and thereby trapping it.

An extended analytical and numerical study of (4) has also shown that a threshold for soliton formation must be considered [11-15]. If a certain amount of amide-I vibrational energy is initially placed at one end of an alpha-helix, the nonlinear parameter χ may or may not be large enough to hold a soliton together. In other words a realistic quantity of amide-I vibrational energy, acting through χ on the right hand side of (4b), may (or may not) create enough longitudinal sound to react, again through χ in (4a), to a degree sufficient to support a soliton. This is a key scientific issue which can be expressed as:

QUESTION: Is the experimentally determined value of χ sufficiently large to hold the amount of energy released in ATP hydrolysis together as a soliton?

To answer this question I have conducted a numerical study of a system of equations similar to (4) but including ten additional dipole-dipole interaction terms in order to avoid underestimating the effects of dispersion [12,15]. The result of these calculations was that the threshold value of χ to achieve soliton formation is about 4.5×10^{-11} newtons. This value compares favorably with the above noted experimental and calculated values for χ; thus the answer to the QUESTION is "Yes".

Let us consider in somewhat greater detail the significance of the dependent variables $a_{n\alpha}$ in (4). In Davydov's theory these are (c-number) expansion coefficients in the wave function

$$|\psi\rangle = \sum_{n\alpha} a_{n\alpha}(t) B_{n\alpha}^{\dagger}|0\rangle \tag{5}$$

where $|0\rangle$ is the vacuum state and $B_{n\alpha}^{\dagger}$ is a creation operator for amide-I bosons. With this interpretation, the probability requirement $\langle\psi|\psi\rangle = 1$ leads to the normalization condition

$$\sum_{n\alpha} |a_{n\alpha}|^2 = 1 \ . \tag{6}$$

Similar equations can be derived from a classical perspective. Transforming the classical coordinate for an harmonic mode (q_i) to the complex amplitude defined by PIERCE [16] as

$$a_i = (m/2\omega_i)^{\frac{1}{2}} (\omega_i q_i + i\dot{q}_i) \tag{7}$$

the classical Hamiltonian becomes

$$H = \sum_i \omega_i |a_i|^2 + \sum_{i,j} f_{ij}(a_i + a_i^*)(a_j + a_j^*)$$
$$+ \sum_{i,j,k} g_{ijk}(a_i + a_i^*)(a_j + a_j^*)(a_k + a_k^*) + \cdots \ . \tag{8}$$

The first term in (8) describes noninteracting modes, the second term describes linear interactions between these modes, and the third term describes nonlinear (anharmonic) interactions between modes. In addition to H, the <u>number</u> defined as

$$N = \sum_i |a_i|^2 \tag{9}$$

is also conserved.

For an harmonic description of amide-I modes assume $\omega_i = \omega_o = E_o/\hbar$. If $f_{ij} \ll \omega_o$, terms of the form $f_{ij}a_i a_j$ and $f_{ij}a_i^* a_j^*$ are small and vary sinusoidally in time. Neglecting these, the first two terms of (8) become

$$H_{LIN} \doteq \sum_i \omega_o |a_i|^2 + \sum_{i,j} f_{ij}(a_i a_j^* + a_i^* a_j) \ . \tag{10}$$

With this approximation one obtains a set of <u>classical</u> equations that are formally identical to (1).

Under quantization the conjugate variables a_i and a_i^* become annihilation and creation operators and N becomes the number operator. This is consistent with the Bohr-Sommerfeld condition $\sum_i \oint a_i^* da_i = iN \sum_i \oint d\theta_i = 2\pi i$ (integer) which, together with the definition of N, implies the normalization condition

$$N = \text{integer} \geq 0 \ . \tag{11}$$

In a quantum analysis the approximation indicated in (10) corresponds to neglecting operators of the form $a_i a_j$ and $a_i^\dagger a_j^\dagger$. The validity of this approximation has been questioned [17,18] and recent studies by TAKENO [19,20] show how these nonresonant terms can be taken into account. Qualitatively different forms for the anharmonicity have been studied by COLLINS [21].

3. Davydov-like solitons in Crystalline Acetanilide

We have seen that the physical parameters of alpha-helix are such that one can expect soliton formation at the level of energy released by ATP hydrolosis. It is now appropriate to ask whether there is any direct experimental evidence for such self-trapped states.

An interesting material to consider is crystalline acetanilide $(CH_3CONHC_6H_5)_x$ or ACN. This is an organic solid in which chains of hydrogen bonded peptide groups run through the crystal in a manner quite similar to the three chains of hydrogen bonded peptide groups seen in Fig. 1. Around 1970 Careri noted that the peptide bond angles and lengths in ACN are almost identical to those in natural

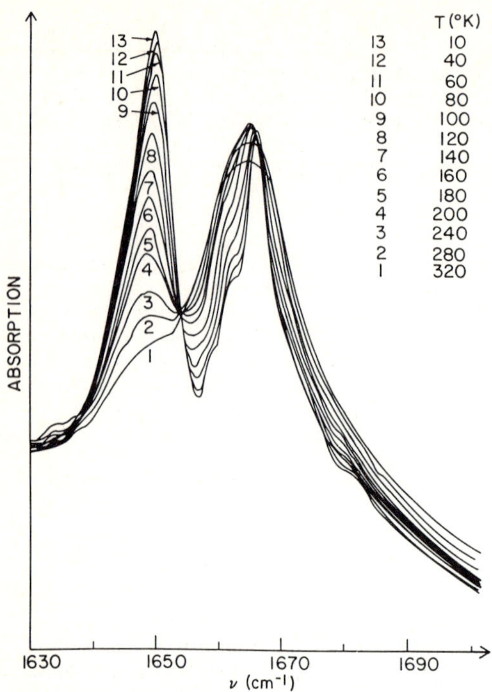

Fig. 2. Infrared absorption spectra of polycrystalline acetanilide. The peak at 1650 cm^{-1} is assigned to soliton absorption (data by E. Gratton [6]).

protein, and he began a systematic spectral study to see whether ACN displayed any unusual properties that might shed light on the dynamical behavior of natural proteins. He soon found an "unconventional" amide-I absorption line at 1650 cm^{-1} that is red shifted from the conventional peak by about 15 wavenumbers [22]. This effect is displayed in Fig. 2.

The identical 1650 cm^{-1} line is observed in Raman scattering; but (since ACN crystal has inversion symmetry) this is forbidden under the assumptions of linear vibrational analysis [23]. Yet N^{15} substitution experiments, polarized Raman measurements and polarized IR absorption measurements clearly identify it as amide-I. Careful measurements of x-ray structure and of specific heat as functions of temperature rule out a first order phase transition and studies of the Raman scattering below 200 cm^{-1} as a function of temperature preclude a second order phase transition. Other considerations make an explanation of the 1650 cm^{-1} band based on Fermi resonance or on localized traps unlikely [6]. Thus the assignment of the 1650 cm^{-1} band to a self-trapped (soliton) state is quite attractive.

In such an assignment the self-trapping is assumed to arise through the interaction of localized amide-I vibrational energy with low frequency optical phonons. This interaction displaces the ground states of the low frequency phonons slightly leading to a Franck-Condon factor

$$F \geq \exp\left(- \frac{\Delta E}{4\hbar\omega}\right) \tag{12}$$

where ΔE is the binding energy of the soliton (15 cm^{-1}) and $\hbar\omega$ is the energy of the low frequency phonon. For an acoustic phonon, as in (4), $\hbar\omega \ll \Delta E$ and $F \ll 1$. Thus, as Davydov has pointed out [8], direct optical absorption is forbidden for solitons that are self-trapped through interaction with acoustic phonons.

228

At non-zero temperatures the probability amplitude for the ground state of a
low frequency phonon is proportional to $[1-\exp(-\hbar\omega/kT)]^{\frac{1}{2}}$. Since the Franck-Condon
factor involves the <u>square</u> of the <u>product</u> of the ground states before and after
the transition, it is proportional to $[1-\exp(-\hbar\omega/kT)]^2$. Fitting this function to
the intensity dependence of the 1650 cm^{-1} peak in Fig. 2 implies $\hbar\omega$ = 130 cm^{-1}
which is close to several optical modes of ACN. Thus the 1650 cm^{-1} peak can be
interpreted as self trapping through interaction with optical phonons for which
the low temperature Franck-Condon factor $F \cong 1$.

4. Solitons in DNA

As a long biopolymer with complex dynamic behavior which is poorly understood, DNA
offers attractive possibilities for nonlinear pulse propagation. In thinking
about Davydov's soliton, we introduced the linear equation describing dispersion
of amide-I vibrational energy (1) and the linear equation for longitudinal sound
waves (2) before considering their augmentation to the nonlinear coupled system
(4). Thus the place to start a study of nonlinear wave dynamics on DNA is with a
firm understanding of the linear behavior. Some progress has recently been made
towards understanding the propagation of linear acoustic waves on DNA based upon
Brillouin scattering measurements of wave speeds [24,25]. This work indicates
rather high longitudinal wave speeds: ~2200 meters/second in dry fiber (A-conformation) and ~1900 meters/second in wet fiber (B-conformation). Extensive normal
mode calculations for both acoustic and optic modes show that the acoustic wave
speeds noted above require long range forces in the A-confirmation and probably
also in the B-conformation [26-28].

The first specific suggestion of soliton states on DNA invoked thermal
effects to describe the breaking of hydrogen bonds between base pairs which, in
turn, was used to explain experimental measurements of hydrogen-deuterium exchange
rates [29]. This idea has been developed in some detail by YOMOSA [30] as the
"dynamic plane base-rotator model" for which chaotic behavior has been investigated by TAKENO and HOMMA [31].

From both theoretical [32] and numerical studies [33], it is clear that the
thermal excitation of kinks and antikinks plays a key role in phase transitions.
Starting with this insight, KRUMHANSL and ALEXANDER [34] are constructing a
dynamical model of DNA with topological solitons for which the state of the system
approaches (say) the A-conformation as x → +∞ and the B-conformation as x → -∞.
Rather than attempting to follow the motions of all atoms, they are isolating a
few significant coordinates for dynamic simulation. Their present effort is
directed toward making this model consistent with the linear description.

SOBELL [35] has proposed for DNA the nontopological structure shown in Fig.
3. Scanning from left to right this figure shows a kink transition from B-conformation into a central region of modulated β alternation in sugar puckering along
the polymer backbone, followed by an antikink transition back to the B-conformation. The β premelted core exhibits a breathing motion which facilitates drug
intercalation and it may provide a nucleation center for RNA polymerase-promoter
recognition. This is a chemical model that is suggested by the physical equations,
but a direct connection has yet to be shown.

Some appreciation for the complex dynamic behavior to be expected in DNA is
obtained from recent molecular dynamics simulations [36]. These followed the
motions of the 754 atoms in the dodecamer (CGCGAATTCGCG)$_2$ and the motions of the
1530 atoms in the 24-bp fragment (A)$_{24}$(T)$_{24}$ for times up to 96 ps. (48,000 time
steps). Motions that could encourage drug intercalation were observed. Ultim-
ately a confirmation of the soliton structure suggested in Fig. 3 must be based
upon such simulations.

5. The Polaron

Landau's original suggestion for self-trapping of electric charge [10] was discussed in detail by PEKAR [37] (who seems to have coined the term "polaron" for a
localized electron plus lattice distortion), by FRÖHLICH [38] and by HOLSTEIN [39].

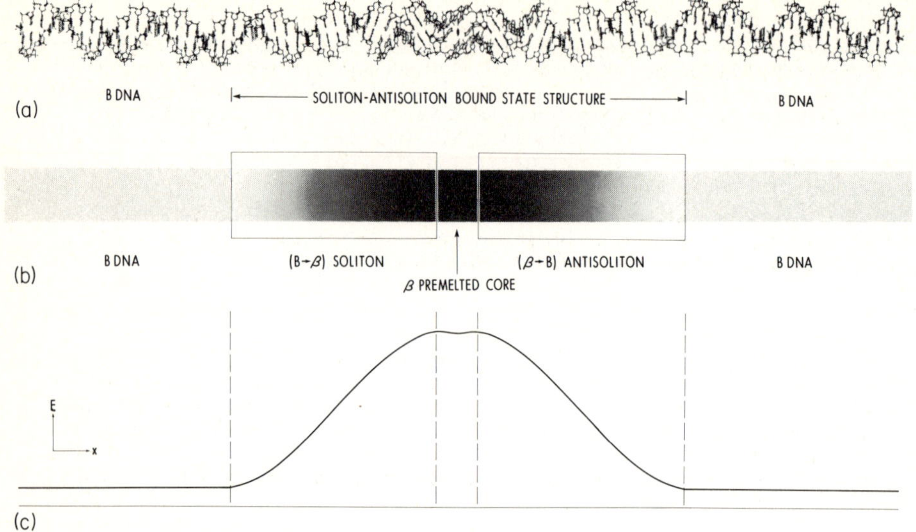

Fig. 3. Sobell's suggestion for a kink-antikink soliton structure in B-conformation DNA. a) Molecular structure in the vicinity of the soliton. b) Schematic
representation of three regions of the soliton. c) Energy density (E) as a function of position (x).

Charge transport is an important function in biological molecules, and, as
DAVYDOV [8] has shown, the polaron may be involved. Analytically the description
of charge transport on an alpha-helix is identical to (4); one merely interprets
a as an electronic wave function and z as longitudinal polarization. Similar
considerations have been used to develop a two-component soliton model for proton
motion in a one-dimensional ice crystal [40-43]. Recently YOMOSA [44] has
developed a theory for proton transport in purple membrane that is based upon a
topological soliton.

6. Solitons of Electrical Polarization

FRÖHLICH [45,46] has suggested that polarization effects might play an important
role in determining the conformational states of biological molecules. He
considered, as a simple example, a sphere which could be elastically deformed
into an ellipsoid of eccentricity η. The elastic energy of deformation would be
$\frac{1}{2}a\eta^2$ where a > 0. The self energy of polarization would be $\frac{1}{2}b(1-c\eta)P^2$ where P is
the polarization vector and c may be positive. Thus the sum of these energies

$$U = \tfrac{1}{2}a\eta^2 + \tfrac{1}{2}b(1-c\eta)P^2 \tag{13}$$

is a minimum at the eccentricity

$$\eta = \tfrac{1}{2}\frac{cb}{a}P^2 . \tag{14}$$

Near this minimum

$$U = \tfrac{1}{2}b\ P^2(1-\tfrac{1}{4}\frac{bc^2}{a}P^2) . \tag{15}$$

Minimizing again with respect to P implies that the original sphere might deform
itself into an ellipsoid with polarization $P = \frac{2}{c}\sqrt{a/b}$ and eccentricity $\eta = 2/c$.
Such a conformational change could be induced, for example, in intrinsic membrane
proteins through the action of the transmembrane potential [47].

The above discussion describes an "electret" which is the electrical analog
of a magnet. BILZ, BÜTTNER and FRÖHLICH [48] have noted that 90% of the materials

230

that display this property are oxides, and oxygen is a common constituent of
organic materials. They have proposed nonlinear wave equations involving interactions of polarization waves with phonons and leading to solitons of both the
topological and the nontopological varieties. The biological significance of
these solitons has been briefly discussed.

Recently TAKENO [19,20] reformulated the quantum mechanical basis for
Davydov's soliton theory. His results reduce to (4) when $J << E_p$, but his dynamic
equations describe polarization of the alpha-helix rather than the probability
amplitude for finding a quantum of amide-I vibrational energy. Thus this picture
nicely complements the ideas developed by Fröhlich.

7. Conformons

In 1972 the term "conformon" was coined independently by VOL'KENSTEIN [49] and
by GREEN and JI [50] to describe a common mechanism for both enzymic catalysis
and biological energy coupling. Vol'kenstein's conformon was defined as a nonlinear state composed of an electron plus local deformation of a macromolecule;
thus it is quite similar to a polaron. The conformon of Green and Ji was defined
as "the free energy associated with a localized conformational strain in biological
macromolecules" and characterized as follows.

"i) The conformon is mobile. The migration of the conformon requires a
relatively rigid protein framework such as the α-helical structure."

"ii) The conformon differs from the generalized electromechanochemical free
energy of protein conformational strains in the sense that the conformon has the
property of a packet of energy associated with conformational strains localized
within a relatively small volume compared with the size of the supermolecule."

"iii) The path of conformon migration need not be rectilinear but will be
dependent on the 3-dimensional arrangement of the linkage system."

"iv) The properties of the conformon are believed to be intimately tied in
with the vibrational coupling between adjacent bonds in polypeptide chains."

From this characterization, the conformon of Green and Ji is seen to be
rather closely related to the Davydov soliton. This relationship has recently
been discussed in a paper by JI [51] which emphasized the ability of the conformon
theory to explain both membrane-associated and membrane-independent coupled
processes whereas the chemiosmotic theory requires a coupling membrane to generate
a transmembrane electrochemical gradient of protons [52]. A general quantum
mechanical formalism for both the Vol'kenstein and the Green-Ji conformons has
also been developed [53].

8. Fröhlich's Theory

In 1968 FRÖHLICH [54,55] introduced a biophysical concept that has stimulated a
number of experimental investigations. He assumed a collection of z oscillatory
modes with frequencies $\omega_1 < \omega_2 < \cdots < \omega_j < \cdots < \omega_z$ which do not interact
directly with each other but can exchange quanta with a heat bath. Nonlinear
interactions between the modes arise from simultaneous absorption and emission of
quanta with different energies. Metabolic energy input to the system was represented by supposing each mode to receive input power represented by the parameter
s. Under these conditions the steady-state number of quanta in each mode is
given by

$$n_j = \frac{A}{\exp[(\hbar\omega_j - \mu)/kT] - 1} \tag{16}$$

where $A > 1$ and $\mu > 0$ but as $s \to 0$, $A \to 1$ and $\mu \to 0$.

Thus for s = 0 (no metabolic pumping of the modes), (16) reduces to the
expected Bose-Einstein expression for the thermal equilibrium number of quanta in

an harmonic oscillator. As s is increased from zero,

$$\mu \to \hbar\omega_1 \,, \tag{17}$$

and (16) implies that the number of quanta in the mode of lowest frequency becomes very large. This effect is similar to Bose-Einstein condensation in superfluids and in superconductors except that the order arises when the metabolic drive (s) is made sufficiently large rather than by lowering the temperature.

This is a generic idea, but the details are important. As a rough estimate of the set of frequencies $\{\omega_i\}$ that might be involved in a real biological system, FRÖHLICH 47 supposed a biochemical molecule with a linear dimension of 100 A to support long wave elastic vibrations leading to frequencies of the order of 10^{11} Hz. Action spectra of microwave induced biological effects provide support for this estimate [56-58].

But Fröhlich has emphasized that both higher and lower frequencies may be involved. "Thus larger units such as DNA-protein complexes might well possess lower frequencies. Higher ones, on the other hand may be based on a combination of the various rotational and vibrational subgroups of relevant molecules," he comments [47].

It is interesting to observe that the amide-I vibration in protein that Davydov proposed as the basis for his soliton model is precisely a "vibrational subgroup of a relevant molecule." Upon adding dissipation and a source of input metabolic energy to (4), one arrives at a system that is very close to Fröhlich's original concept [59]. From this perspective the Davydov and Fröhlich theories appear as complementary explanations for the mystery of energy storage and transport in biological molecules.

9. Resonant Transfer and Lifetime

The fundamental energy-producing reaction in biological systems is generally accepted to be the hydrolysis of adenosine triphosphate (ATP) according to

$$ATP^{-4} + H_2O \to ADP^{-3} + HPO_4^{-2} + H^+ \,. \tag{18}$$

Under normal physiological condition, this reaction produces about 10 kcal/mole (.43 eV) of free energy. McCLARE [61-64] has argued that there are only two ways for this free energy to do useful work: i) By a "constrained equilibrium" mechanism as in a conventional chemical engine (e.g. battery, internal combustion motor, etc.) in which free energy does useful work on a time scale that is long compared with that of thermal relaxation, and ii) A "molecular energy" mechanism in which free energy does useful work on a time scale that is short compared with the thermal relaxation time. Since constrained equilibrium mechanisms are necessarily macroscopic, they cannot be used for biological processes such as muscular contraction and active transport. Thus one is forced to consider a molecular energy mechanism for using the free energy released in (18).

A molecular energy mechanism must act in a time short enough to avoid thermal relaxation, a requirement that suggests resonant energy transfer [60-64]. McCLARE [64] has proposed that the free energy released in (18) may reside in the H^+ ion which could radiate into the amide-A (NH stretch) band of protein from the OH stretch vibration of water. In a similar way one could expect radiation into the first overtone of amide-I (N = 2 in (11)). Since this overtone is in Fermi resonance with amide-I, one would expect direct soliton absorption to be enhanced.

But if energy can radiate into a soliton why can't it just as easily radiate out? To answer this question McCLARE [64] suggested that the energy absorbing state within a protein (he called it an "eximer") might have two symmetry conformations: one symmetric with large dipole moment for easy absorption and the other antisymmetric with small dipole moment to prevent radiation. It is shown in [65] that symmetric and antisymmetric Davydov solitons satisfy McClare's requirements.

If one considers the 1650 cm^{-1} line in Fig. 2 as evidence of soliton absorption, the soliton is not long-lived. It has been suggested [66] that this soliton involves self-trapping by an optical mode (so the Franck-Condon factor (12) remains sufficiently large) which decays in about a picosecond into a true Davydov soliton with self-trapping by acoustic modes.

10. Final Comments

I have two final comments: one specific and the other general.

The soliton theories discussed above tend to be developed in the context of regular secondary structure such as the alpha-helix, but real globular proteins have much more complicated shapes. LOMDAHL [67] has reported initial attempts to extend Davydov's theory to realistic protein geometry. This question is currently under intensive study in the Center for Nonlinear Studies.

My general comment concerns the shockingly poor state of scientific communications in bioenergetics in the recent past. Consider the situation in 1974. The works of FRÖHLICH [45,46,54,55], VOL'KENSTEIN [49], GREEN and JI [50], DAVYDOV [3,4], CARERI [22], and McCLARE [60-64] had all appeared and are, from our present perspective, very closely related. Yet the authors seemed to be almost totally unaware of each other's work. Although Ji knew of Vol'kenstein's paper he only recently learned of Davydov's soliton [51]. Without a specific mechanism, he had little defense against biochemists who (quite correctly) insisted that he explain how this conformon could be long-lived. McClare was in a similar situation. The connection between the work of Fröhlich and that of Davydov has gone unnoticed until the present time [59]. Careri became aware of Davydov's results only in 1982. And so on.

It is hoped that this "introductory phase" of bioenergetic research is coming to a close and that in the future scientists with diverse backgrounds and points of view will be collaborating to find answers to the fundamental biological question posed in the Introduction.

Acknowledgments

It is a pleasure to thank E. Gratton for Fig. 2 and H. M. Sobell for Fig. 3.

References

1. D. E. Green, Science **181**, 583 (1973)

2. D. E. Green, Ann. N.Y. Acad. Sci. **227**, 6 (1974)

3. A. S. Davydov and N. I. Kislukha, Phys. Stat. Sol. (b) **59**, 465 (1973)

4. A. S. Davydov, J. Theor. Biol. **38**, 559 (1973)

5. V. A. Kuprievich and Z. G. Kudritskaya, preprints #ITP 82-62E, 82-63E and 82-64E, Institute for Theoretical Physics, Kiev (1982)

6. G. Careri, U. Buontempo, F. Galluzzi, A. C. Scott, E. Gratton and E. Shyamsunder, Phys. Rev. B (to appear)

7. A. S. Davydov, Physica Scripta **20**, 387 (1979)

8. A. S. Davydov, _Biology and Quantum Mechanics_ (Pergamon, New York 1982)

9. A. S. Davydov, Sov. Phys. Usp. **25**, 899 (1982)

10. L. D. Landau, Phys. Zeit. Sowjetunion **3**, 664 (1933)

11. J. M. Hyman, D. W. McLaughlin and A. C. Scott, Physica **3D**, 23 (1981)

12. A. C. Scott, Phys. Rev. A $\underline{26}$, 578 (1982); ibid. $\underline{27}$, 2767 (1983)

13. A. C. Scott, Physica Scripta $\underline{25}$, 651 (1982)

14. A. C. Scott, Physica Scripta $\underline{29}$, 279 (1984)

15. L. MacNeil and A. C. Scott, Physica Scripta $\underline{29}$, 284 (1984)

16. J. R. Pierce, J. Appl. Phys. $\underline{25}$, 179 (1954)

17. V. K. Fedyanin, V. G. Makhankov and L. V. Yakushevich, Phys. Lett. $\underline{61A}$, 256 (1977)

18. V. K. Fedyanin and V. G. Makhankov, Physica Scripta $\underline{20}$, 552 (1979)

19. S. Takeno, Prog. Theor. Phys. $\underline{69}$, 1798 (1983)

20. S. Takeno, Prog. Theor. Phys. $\underline{71}$, 395 (1984)

21. M. A. Collins, Adv. Chem. Phys. $\underline{53}$, 225 (1983)

22. G. Careri, in Cooperative Phenomena, H. Haken and M. Wagner, eds. (Springer-Verlag, Berlin 1973) p. 391

23. G. Herzberg, Molecular Spectra and Molecular Structure II. Infrared and Raman Spectra of Polyatomic Molecules (Van Nostrand Reinhold, New York 1945)

24. G. Maret, R. Oldenbourg, G. Winterling, K. Dransfeld and A. Rupprecht, Colloid. and Polym. Sci. $\underline{257}$, 1017 (1979)

25. M. B. Hakim, S. M. Lindsay and J. Powell, Biopolymers $\underline{23}$, 1185 (1984)

26. W. N. Mei, M. Kohli, E. W. Prohofsky, and L. L. Van Zandt, Biopolymers $\underline{20}$, 833 (1981)

27. L. L. Van Zandt, K. C. Lu, and E. W. Prohofsky, Biopolymers $\underline{16}$, 2481 (1977)

28. J. M. Eyster and E. W. Prohofsky, Biopolymers $\underline{13}$, 2505 (1974)

29. S. W. Englander, N. R. Kallenbach, A. J. Heeger, J. A. Krumhansl, and S. Litwin, Proc. Nat. Acad. Sci. (USA) $\underline{77}$, 7222 (1980)

30. S. Yomosa, Phys. Rev. A $\underline{27}$, 2120 (1983)

31. S. Takeno and S. Homma, Prog. Theor. Phys. $\underline{70}$, 308 (1983)

32. J. A. Krumhansl and J. R. Schrieffer, Phys. Rev. B $\underline{11}$, 3535 (1975)

33. T. R. Koehler, A. R. Bishop, J. A. Krumhansl, and J. R. Schrieffer, Solid State Comm. $\underline{17}$, 1515 (1975)

34. J. A. Krumhansl and D. M. Alexander, in Structure and Dynamics: Nucleic Acids and Proteins Eds. E. Clementi and R. H. Sarma (Adenine Press, 1983) p. 61

35. H. M. Sobell, in Structure of Biological Macromolecules and Assemblies. Vol. II Eds. F. Jurnak and A. McPherson (Wiley, New York 1984)

36. M. Levitt, Cold Spring Harb. Symp. Quant. Biol. $\underline{47}$, 251 (1983)

37. S. Pekar, Jour. Phys. U.S.S.R. $\underline{10}$, 341 (1946); ibid, 347

38. H. Fröhlich, Adv. in Phys. $\underline{3}$, 325 (1954)

39. T. Holstein, Ann. Phys. $\underline{8}$, 325 (1959); ibid 343

40. V. Ya. Antonchenko, A. S. Davydov, and A. V. Zolotariuk, Phys. Stat. Sol.
    (b) $\underline{115}$, 631 (1983)

41. A. S. Davydov, V. Z. Enolskii, Sov. Phys. JETP $\underline{52}$, 954 (1980)

42. Y. Kashimori, T. Kikuchi, and K. Nishimoto, J. Chem. Phys. $\underline{77}$ 1904 (1982)

43. E. W. Laedke, K. H. Spatschek, M. Wilkins, Jr., and A. V. Zolotariuk, Phys.
    Rev. A (to appear)

44. S. Yomosa, J. Phys. Soc. Japan $\underline{52}$, 1866 (1983)

45. H. Fröhlich, Nature $\underline{228}$, 1093 (1970)

46. H. Fröhlich, Jour. Coll. Phen. $\underline{1}$, 101 (1973)

47. H. Fröhlich, Riv. del Nuovo Cim. $\underline{7}$, 399 (1977)

48. H. Bilz, H. Büttner and H. Fröhlich, Z. Naturforsch. $\underline{36b}$, 208 (1981)

49. M. V. Vol'kenstein, J. Theor. Biol. $\underline{34}$, 193 (1972)

50. D. E. Green and S. Ji, Proc. Nat. Acad. Sci. (USA) $\underline{69}$, 726 (1972)

51. S. Ji, Proceedings of the Second International Seminar on the Living State (R.
    K. Mishra editor) held in Bhopal, India November 1983 (in press).

52. P. Mitchell, Eur. J. Biochem. $\underline{95}$, 1 (1979)

53. G. Kemeny, J. Theor. Biol. $\underline{48}$, 231 (1974)

54. H. Fröhlich, Phys. Lett. $\underline{26A}$, 402 (1968)

55. H. Fröhlich, Int. J. Quant. Chem. $\underline{2}$, 641 (1968)

56. S. J. Webb and A. D. Booth, Nature $\underline{222}$, 1199 (1969)

57. W. Grundler, F. Keilmann and H. Fröhlich, Phys. Lett. $\underline{62A}$, 463 (1977)

58. W. Grundler and F. Keilmann, Phys. Rev. Lett. $\underline{51}$, 1214 (1983)

59. J. A. Tuszyński, R. Paul, R. Chatterjee and S. R. Sreenivasan, Phys. Rev. A
    (to appear)

60. C. W. F. McClare, J. Theor. Biol. $\underline{35}$, 233 (1972)

61. C. W. F. McClare, J. Theor. Biol. $\underline{30}$, 1 (1971)

62. C. W. F. McClare, Nature $\underline{240}$, 88 (1972)

63. C. W. F. McClare, J. Theor. Biol. $\underline{35}$, 569 (1972)

64. C. W. F. McClare, Ann. N. Y. Acad. Sci. $\underline{227}$, 74 (1974)

65. A. C. Scott, Phys. Letts. $\underline{94A}$, 193 (1983)

66. A. C. Scott, "Solitons and Bioenergetics" in Synergetics of the Brain, eds. E.
    Basar, H. Flohr, H. Haken and A. J. Mandell (Springer-Verlag, Berlin 1983)

67. P. S. Lomdahl, "Nonlinear dynamics of globular proteins" in Nonlinear Electro-
    dynamics in Biological Systems, eds. W. R. Adey and A. F. Lawrence
    (Plenum, New York 1984)

Experiments for the Detection of Solitons in Biopolymers

Irving J. Bigio, Clifford T. Johnston, and Scott P. Layne
Los Alamos National Laboratory, Los Alamos, NM 87545, USA

Introduction

Unfortunately it is not possible to sneak inside a biological system with a camera and snap a picture of an object labeled "soliton". And yet that is not unlike what we are trying to do when we ask solitons in biological systems to reveal themselves to experimental scrutiny. The sign must say "soliton" and not "exciton" or "conformational change" or any other plausible identification, because the theory of nonlinear interactions is relatively new to biophysics and it has not yet been firmly established experimentally. Moreover, such confirmation is not likely to come from any single definitive experiment. Rather, we expect to build up a large body of evidence, from a variety of experiments that approach the problem from different angles, which when taken all together would constitute a convincing case.

In planning the experimental approach to be taken, we have been partially guided by what can and cannot be calculated (with a reasonable level of confidence). One asks, "What experimental manifestations of solitons can actually be calculated and predicted by the numerical model?", since a proper theory should have predictive capabilities. As discussed in the chapter by A.C. Scott, the primary motivation is to understand the fundamental mechanisms of energy transport in hydrogen-bonded biopolymers. The potential significance of solitons in biopolymers has been discussed by Davydov[1] and others during the intervening years since he first introduced the concept in 1973.

Experimental Methods

As described by Scott, the self-trapped states that we are concerned with invoke an interaction, via a nonlinear coupling, between the amide-I vibration (mainly $C=O$ stretch) of the peptide group and some other molecular motion. Scott [2] has considered two different applications of the theory. In the case of the alpha-helix that molecular motion is due to longitudinal sound waves which invoke the motion of whole peptide groups relative to each other, thus causing changes in the hydrogen bond lengths. Whereas in the case of acetanilide (ACN) it is only the motion (out of plane) of the proton associated with the hydrogen bond.

The two main concerns of this experimental program (and most experiments aimed at demonstrating the existence of some effect) are 1) to be able to turn the effect on and off at will, and 2) to determine what might be the most easily addressed experimental manifestations of the existence of the effect.

<u>Controlled production</u>: In the case of ACN, which is grown in clean single crystals, it may be possible for solitons to be produced by

direct electromagnetic excitation since the out-of-plane motion of the proton (which couples with the C=O vibration) can occur fast enough for such an interaction. As a source of such direct optical excitation we have constructed a tunable carbon-monoxide (what else?!) laser which can be tuned to emit at precisely the desired frequency.

On the other hand, the longer time required for the motion of whole peptide groups in the alpha-helix would be expected to prevent the **direct** optical excitation of Davydov solitons. Thus, controlling the presence of solitons in biological systems obviously entails some understanding of the natural processes by which the solitons are generated in living organisms. (If they don't exist in natural circumstances then the problem suddenly becomes less interesting.) This sequence of events is not well established. While it appears to be fairly well accepted that the initial energy-generating mechanism for most protein activities is the hydrolysis of adenosine triphosphate (ATP), the steps that lead from there to the formation of a vibrational soliton are less well understood. Davydov [1] has postulated that the entry point for the energy in the molecule is the amide-I vibration of the peptide group, and Scott[3] has noted the resonance between vibrationally excited water (at about 1646 cm^{-1}) and the Davydov soliton itself (at about 1647 cm^{1}). Again because of the time constants involved, one might wish to invoke collisional transfer mechanisms rather than radiative coupling. Nonetheless, since vibrationally excited water may be formed by the recombination of the proton released in the hydrolysis of ATP, this speculation is quite appealing.

Experimentally, the above speculations mean that by controlling the biochemical processes that affect the hydrolysis of ATP, or perhaps by directly exciting the water molecules (again, with the CO laser) we can vary the number of solitons present in a given sample. One example of such control is to synchronize the growth cycles of all or most of the members of a colony of yeast or bacteria. It is assumed since the organisms are metabolizing more (hence processing more ATP) during certain parts of their growth cycles, that the largest density of solitons is likely to be present in the sample at that time.

<u>Manifestations</u>: The theoretical model yields a number of physical attributes of the biological solitons which distinguish them from other modes of the system. These include the following:
    -the characteristic energy of the soliton state itself and also the presence of other resonances that may appear because of the nonlinear interactions.
    -the characteristic lifetime of the soliton which is expected to be longer than that of normal modes.
    -the characteristic velocity for cases (such as the alpha-helix) in which the soliton propagates.
    -the fact that if we understand the nature of the soliton we can externally control the conditions for its presence.
The most flexible approach to search for these types of manifestations is the use of optical **spectroscopies**: more specifically, infrared-absorption spectroscopy, Raman spectroscopy and Brillouin spectroscopy.

<u>Spectroscopies</u>: Infrared spectroscopy has the advantage of being very sensitive, but it is limited to dipole-allowed transitions. Also it encounters the experimental difficulty for fluid samples that the medium in which the sample is dissolved (or suspended) as well as the vessel containing the sample must both be transparent at the infrared wavelengths of interest.

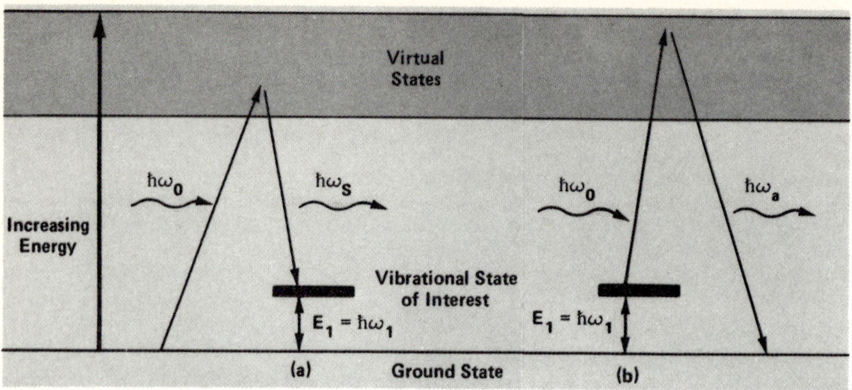

Fig. 1. The two cases of Raman scattering: (a) Stokes scattering and (b) anti-Stokes scattering

Raman spectroscopy, on the other hand, is much less sensitive (i.e., the scattering cross sections are typically orders of magnitude smaller than the infrared-absorption cross sections for the same transition when both are allowed). However, it can address a larger variety of symmetries; and since it is performed with visible-wavelength lasers it does not suffer from problems of transparency of the medium or vessel. Figure 1 depicts the two cases of Raman scattering: (a) Stokes scattering, and (b) anti-Stokes scattering.

In general there are experimental difficulties often encountered in Raman spectroscopy of real biological samples. One problem is due to the fact that they are typically not homogeneous, but are collections of cells (or other structures) suspended in a fluid. Moreover, those cells are typically dynamic: changing in size and shape with time. This results in a time-varying elastic scattering (either Mie or Rayleigh) of the laser light, consequently yielding a time variation of the background noise level. Thus when a Raman spectrum is recorded under such conditions with a commonly-used scanning monochromator (which is scanned in time) and photodetector, false "spectra" can appear unless specific discrimination methods are employed. The apparati used in our experiments totally avoid this problem by recording entire portions of the spectrum simultaneously. The system shown in Fig.2 includes a polychromator (a monochromator without an exit slit) which disperses a large portion of the Raman spectrum onto a photodiode-array detector that is part of an "optical multichannel analyser" (OMA). Thus, temporal variations in the background noise are not localized spectrally but

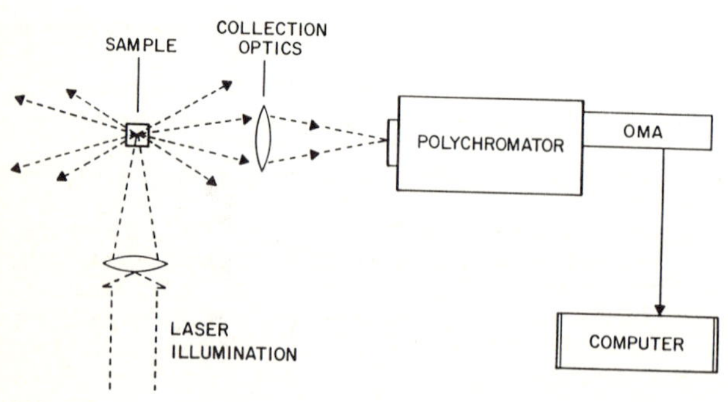

Fig. 2. Schematic of the experimental setup for Raman spectroscopy

238

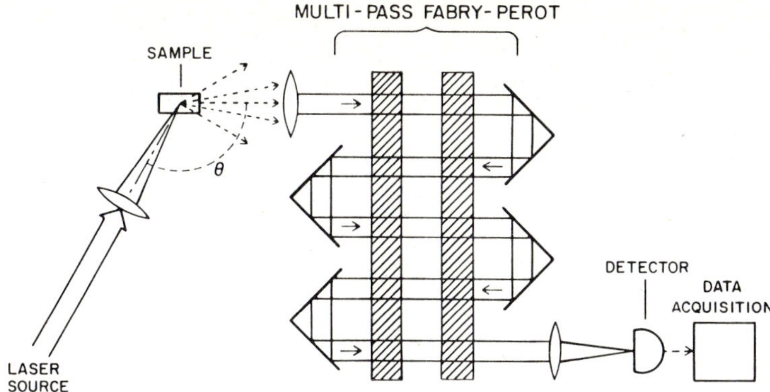

Fig. 3. Experimental arrangement for Brillouin spectroscopy

properly spread out over the spectrum. Signal-averaging techniques are also more effective under these conditions.

Another, and more difficult to resolve, problem with Raman spectroscopy of real biological samples is their very low density. (Yeast, for example, stop growing at densities of 10^8-10^9 per cm^3.) This results in Raman signals which are generally too weak to be distinguished from noise. If there is a fortuitously located (spectrally) electronic transition in the appropriate molecule, then resonance Raman spectroscopy can be employed. Enhancements of the scattering cross section can range from 10^2 to 10^4. Thus, this problem can sometimes be sufficiently ameliorated with the use of a tunable dye laser to provide the illumination at the appropriate frequency.

Brillouin spectroscopy mainly addresses resonances of very low frequency (from <1 cm^{-1} down to <10^{-3} cm^{-1}) which are generally associated with acoustic and other long-range modes of the system. Since Brillouin scattering involves frequency shifts which are exceedingly small when compared with the laser frequency itself (often less than one part in 10^6) the discrimination against Rayleigh and diffusely scattered light must be excellent. Although experimentally difficult to implement and for practical purposes limited to the above frequency range, a device that can provide such a level of discrimination is the multi-pass Fabry-Perot interferometer, which is shown schematically in Fig.3. Contrast ratios exceeding 10^{20} can be achieved in an instrument of this sort. Experiments are planned in the near future which make use of a system of this type.

Experimental Measurements

To date most of the actual experimental measurements have been done on ACN. In 1973 Careri and coworkers [4] first noted the unusual spectral feature in the region of the amide-I resonance that grew in strength as the temperature was lowered, and were unable to ascribe it to any conventional interpretation. Ten years later [5], in collaboration with Scott, they determined that this feature coincides with the characteristic energy calculated for a vibrational soliton in that system and that the temperature dependence might also be in harmony with the self-trapped state model.

Our highly detailed spectroscopic studies of ACN have provided a variety of new information, some of it corroborative of the soliton interpretation and some of it as yet unexplained and requiring further analysis. For example, if the "soliton" line at 1650 cm^{-1} derives from the "normal" amide-I line at 1665 cm^{-1}, then both lines should have the same symmetry (polarization). Figure 4 shows some data which confirm that expectation.

On the other hand Fig.5 shows data which are more problematic. The spectrum in Fig.5 was taken from an ACN crystal in which we substituted C^{13} for the C^{12} in the peptide group but not elsewhere. From a simple reduced-mass calculation one would expect to see the amide-I resonance lines simply shifted down by ~38 cm^{-1}. However,

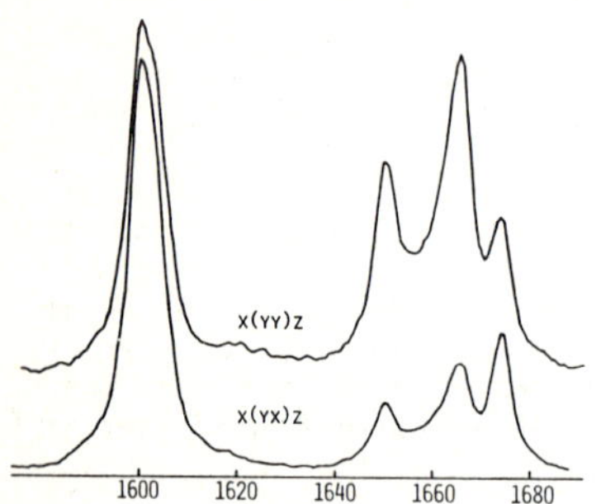

Fig.4. Raman spectra of C^{12}-ACN at 100 K for two different sets of polarization/orientation. The amplitude ratios for the lines at 1650 and 1665 are equal.

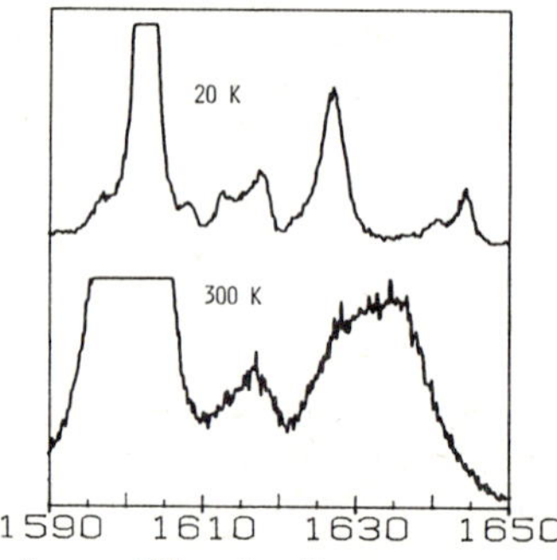

Fig.5. Raman spectra of C^{13}-substituted ACN at two temperatures.

the line that appears at 1612 cm^{-1}, where the soliton line would be expected is considerably weaker than the soliton line in the C^{12} data (but still stronger than a small phenyl-ring feature also at 1612 cm^{-1} in the C^{12} data. Moreover, a new broad feature appears on the higher-frequency side of the shifted amide-I resonance. One can speculate about a possible interference with an infrared-active phenyl-group line at 1620 cm^{-1} preventing the soliton from forming, or about some combination of Fermi resonance effects, but at the time of writing a proper analysis of the data has not yet been carried out. We have also generated considerable data with N^{15}-substituted ACN and deuterated ACN, and those data should also be reconciled.

The characteristic lifetime of the soliton is another issue that can be addressed with Raman spectroscopy. Under thermal equilibrium one normally expects the ratio of populations of two states to be given by the Boltzman distribution function: $N_u/N_l = exp(-E/kT)$, where the subscripts u and l represent the upper and lower states respectively, and E is the energy separating them. Since the intensity of Raman scattered light is directly proportional to the population of the given state from which the scattering occurs, it is also true that the ratio of anti-Stokes to stokes intensities, $I_a/I_s = N_u/N_g$ (the lower state is usually the ground state). Thus, by measuring I_a/I_s for a given transition (and knowing the temperature) we can determine whether there is a larger population in the excited state than would be expected from thermal equilibrium considerations. Such an occurance might be expected for states whose lifetime (due to self-trapping which is a non-equilibrium

condition) is longer than normal. Very preliminary observations of
this sort with ACN reveal some features in the phonon region which
are worthy of further investigation.   But at the time of writing
the results are too ambiguous to draw any conclusions.

We have also recently begun taking data in a concerted and
rigorous effort to observe Raman spectral features from samples of
living bacteria. Webb [6] has reported seeing a variety of lines
which were strongest when the cells were synchronized in their
growth cycles. Working with the same bacteria (E. coli) we have
carefully demonstrated synchronization using a currently accepted
state-of-the-art method. We are now in the process of taking
spectral data with sample concentrations similar to those used by
Webb (10^7-10^8 per cm^3) and expect to report the results in a future
publication.   Since our apparati allow us to acquire data quickly
compared with the life cycles of the bacteria, we can also quickly
concentrate the sample to much higher densities ($10^{11}/cm^3$) and take
a spectrum before the metabolic conditions change. This experiment
will also soon be carried out.

We have not limited ourselves to optical spectroscopies. In the
case of ACN which can be grown in large single crystals, we have
availed ourselves of x-ray diffraction analysis (to eliminate the
possibility that structural changes with temperature could be
responsible for the effect) and NMR spectroscopy (to verify isotopic
substitutions). Neutron scattering studies are also a possibility
for future investigation.

<u>Conclusions</u>

In conclusion it can be stated that there is generally a burden of
proof, somewhat more severe than usual, that is placed on a theory
that purports a new explanation of phenomena in a field to which the
theory itself is new. That burden is the challenge of this
experimental effort, which nonetheless must remain objective in its
conclusions. Measurements on both ACN and bacteria continue, and
experiments on other biological systems are planned. Hopefully at
the next convening of a meeting of this sort, there will be more
definitive results to report.

REFERENCES

1. A.S. Davydov, "The theory of contraction of proteins under their
excitation", J. Theor. Biol. 38, p.559 (1973).
A variety of topics can also be found in: A.S. Davydov, <u>Biology and
Quantum Mechanics</u>, Oxford: Pergamon Press, 1982.

2. A.C. Scott, "Solitons in biological molecules," in <u>Comments on
Molecular and Cellular Biophysics,</u> in press (1984).

3. A.C. Scott, private communication.

4. G. Careri, "Search for cooperative phenomena in hydrogen-bonded
amide structures," in <u>Cooperative</u> Phenomena, H. Haken and M. Wagner,
eds. (Springer-Verlag, 1973).

5. G.Careri et al.,"Infrared absorption in acetanilide by solitons,"
Phys. Rev. Lett. 51, p.304 (1983).

6. S.J. Webb, Phys. Reports 60, p.201 (1980).

Solitary Excitations in DNA Double Helices

S. Yomosa

Sugiyama Jogakuen University, Nagoya 464, Japan

1. Introduction

Biopolymers which carry wide varieties of correlations and couplings
can be regarded as higher order nonlinear systems in which biological
functions proper to them are performed. Therefore it is expected that
the theory of nonlinear waves will provide adequate tools to clarify the
molecular mechanisms of biological functions.
   The earliest suggestion for a polymer soliton was made by DAVYDOV
[1] in an attempt to construct a molecular theory of muscle contraction.
Davydov's soliton is characterized by the intramolecular vibrational
exciton of amide I coupled with the lattice deformation of an H-bonded
peptide-chain in α-helical protein. Recently SCOTT and his coworkers
[2] have attempted to generalize Davydov's theory by taking into account
long range nature of exciton transfer, and have determined the threshold
level of the coupling parameter value to form a soliton by doing extensive numerical calculations. On the other hand , TAKENO [3] has
studied the vibrational soliton by employing a coupled oscillator-
lattice model and shown that the classical picture of the coupled
vibrational soliton appears to be more appropriate to describe vibrational energy transfer in α-helical proteins. Recently we presented
a theory of nonlinear lattice soliton in α-helical protein to elucidate
the molecular mechanism of muscle contraction[4]. We regarded the one-
dimensional chain of peptide groups joined together by H-bonds in
α-helical proteins as a Toda-lattice where the potential of H-bonding
interaction between peptide groups has a remarkable nonlinearity.
We showed that, if we assume that the energy released by the hydrolysis
of ATP molecules is used for the formation of a soliton, the nonlinear
lattice soliton gives a surprisingly reasonable value of maximum
amplitude of lattice contraction or lattice tension and the energy
of this soliton can be transformed with a high efficiency to a mecha-
nial work.
   Recently the existence of an open state in deoxyribonucleic acid
(DNA) and synthetic polynucleotide double helices has been demonstrated
by hydrogen-deuterium exchange measurements[5-7]. Assuming a mobile
open unit [8] diffusing along the double helix, ENGLANDER et al.[9]
suggested that the open state in DNA may be described as a solitary
excitation. Previously, we proposed a soliton theory [10] to give
a theoretical explanation of the open states in DNA duplexes. Since
our theory had a defect in the expression of H-bonding energy between
complementary base pairs, we presented a revised analysis in which the
H-bonding energy has a more general and reasonable form [11]. Here we
adopted   the   simplified model of the previous paper:
that each nucleotia. base can rotate around an axis parallel to the
helical axis of the duplex accompanying a rotation of the sugar and
the phosphate of the nucleotide to which the base belongs. Our
Hamiltonian gives two types of sine-Gordon equations, and four modes
of solitary excitations are found. The average soliton number
density of our system is given, using earlier results [12,13]
on the statistical mechanics of nonlinear fields.

TAKENO and HOMMA [14] have developed dynamics of DNA double helices
as the <u>plane base-rotator model</u> where chaotic behavior has been
investigated together with solitary excitations.

In this talk we wish to report briefly our revised theory of solitary
excitations in DNA double helices.

2. Hamiltonian and Soliton Solutions

The B form of DNA double helices is schematically represented in Fig.
1(a), where each arrow shows the direction of the base attached to the
strand, and the conjugated base pairs are indicated by the pairs of
arrows arranged in horizontal parallel planes separated by a distance
of a=3.4 A. The z axis is a tenfold screw axis. Fig. 1(b) shows a
projection of the nth complementary base pair in the x-y plane, where
B_n and B_n' denote the complementary bases. The directions of B_n and B_n'
in the horizontal plane are specified by the rotational angles of the
directional vectors $P_n B_n$ and $P_n' B_n'$ of the bases around the axes P_n and
P_n', respectively, which are denoted by χ_n and χ_n' , where the rotational
axes of nth and n'th nucleotides, P_n and P_n', respectively, are parallel
to the z axis.

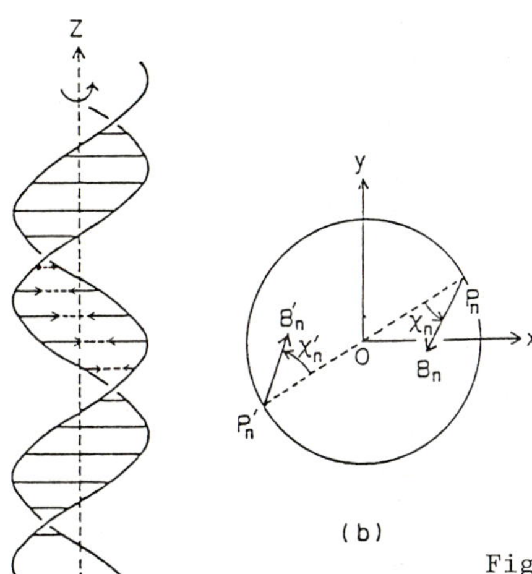

Fig. 1.(a) Schematic representation
of the Watson- Crick model,
(b) horizontal projections of the
complementary base pair.

Considering that the H-bonding energy between the nth base pair
$V_n(\chi_n,\chi_n')$ is a periodic function of the rotational angles χ_n and χ_n'
with the periods 2π, and assuming that V_n is an even function with
respect to χ_n and χ_n', we expand V_n in the form of a double Fourier
series,

$$V_n(\chi_n,\chi_n') = \sum_{p,q=0}^{\infty} B_{pq}^{nn'} \cos(p\chi_n)\cos(q\chi_n') \tag{1}$$

Approximating it by the first four terms, and adopting mean parameter
values for double and triple H-bonds in A-T (adenine-thymine) and G-C
(guanine-cytosine) base pair $\bar{B}_{pq}^{nn'}=B_{pq}$, and considering the symmetry
relations $B_{01}=B_{10}$, the total energy of H-bonds between complementary

base pairs can be represented by the following form

$$\sum_n [A(1-\cos\chi_n)+A(1-\cos\chi_n')+B(1-\cos\chi_n\cos\chi_n')], \tag{2}$$

where A and B are the constants.

By assuming that the stacking energy between intrastrand adjacent bases and the torsional energy of the nucleotide strand are both functions of the relative torsional angles between adjacent bases and that these energies are given in the same functional form, the sum of the stacking and torsional energies may be written as

$$\sum_n \{S[1-\cos(\chi_n-\chi_{n-1})]+S[1-\cos(\chi_n'-\chi_{n-1}')]\}, \tag{3}$$

where S is the constant. In the above two expressions the B form is taken as the zero level of the energy.

Introducing the fields of rotational angles in the continuum approximation, $\chi_n(t)\rightarrow\chi_n(z,t)$, $\chi_n'(t)\rightarrow\chi_n'(z,t)$, the Hamiltonian of DNA and synthetic polynucleotide double helices can be written alternatively as follows:

$$H = \int\frac{dz}{a}[\ \frac{1}{2}I(\dot\chi^2+\dot\chi'^2)+A(1-\cos\chi)+A(1-\cos\chi')+B(1-\cos\chi\cos\chi')$$

$$+ \frac{1}{2}Sa^2(\chi_z^2+\chi_z'^2)] \tag{4}$$

$$= \int\frac{dz}{a}\{\ \frac{1}{2}I(\dot\psi^2+\dot\phi^2)+2A(1-\cos\frac{\psi}{2}\cos\frac{\phi}{2})+B[1-\frac{1}{2}(\cos\psi+\cos\phi)]$$

$$+ \frac{1}{4}Sa^2(\psi_z^2+\phi_z^2)\} \ , \quad \text{where} \tag{5}$$

$$\psi=\chi+\chi', \quad \phi=\chi-\chi'. \tag{6}$$

The first term represents the kinetic energy of the rotational motion of the nucleotide base accompanied by the nucleotide sugar and phosphate around the axis P in which I is the mean value of the moment of inertia of the nucleotide around axis P. From the Lagrangians corresponding to (4) and (5), we obtain the following Euler-Lagrange equations, respectively:

$$I\ddot\chi+A\sin\chi+B\sin\chi\cos\chi'-Sa^2\chi_{zz}=0, \tag{7a}$$

$$I\ddot\chi'+A\sin\chi'+B\sin\chi'\cos\chi-Sa^2\chi_{zz}'=0, \tag{7b}$$

$$I\ddot\psi+2A\sin\frac{\psi}{2}\cos\frac{\phi}{2}+B\sin\psi-Sa^2\psi_{zz}=0, \tag{8a}$$

$$I\ddot\phi+2A\sin\frac{\phi}{2}\cos\frac{\psi}{2}+B\sin\phi-Sa^2\phi_{zz}=0. \tag{8b}$$

Assuming that, by taking an adequate acidity and salt concentration, we can realize a solvent condition $A\cong0$ in which the H-bonding energy between the base pair is almost equal to that between the bases and water molecules of the solvent, we study the dynamics of our system including the case $A=0$.

244

A. The case A=0

In this case we obtain the following eight sets of solutions from (7) and (8);

$$\chi=2n\pi, \quad \chi'=2m\pi, \tag{9a}$$

$$\chi=(2n+1)\pi, \quad \chi'=(2m+1)\pi; \tag{9b}$$

$$\chi=\phi_s(\zeta-\zeta_0), \quad \chi'=2m\pi, \tag{10a}$$

$$\chi=2n\pi, \quad \chi'=\phi_s(\zeta-\zeta_0), \tag{10b}$$

$$\chi=\phi_s(\zeta-\zeta_0)+\pi, \quad \chi'=(2m+1)\pi, \tag{10c}$$

$$\chi=(2n+1)\pi, \quad \chi'=\phi_s(\zeta-\zeta_0)+\pi, \tag{10d}$$

$$\chi=m'\pi+\tfrac{1}{2}\phi_s(\zeta-\zeta_0), \quad \chi'=-m'\pi+\tfrac{1}{2}\phi_s(\zeta-\zeta_0), \tag{11a}$$

$$\chi=n'\pi+\tfrac{1}{2}\phi_s(\zeta-\zeta_0'), \quad \chi'=n'\pi-\tfrac{1}{2}\phi_s(\zeta-\zeta_0'), \tag{11b}$$

where n,m,n' and m' are integers and $\phi_s(\zeta-\zeta_0)$ is the soliton solution of the following sine-Gordon equation,

$$I\ddot{\chi}+B\sin\chi-Sa^2\chi_{zz}=0, \tag{12}$$

where $\zeta=z-vt$. The first two of these solutions correspond to the ground-state solutions, the following four are the uncoupled 2π soliton solutions where the motions of χ and χ' are not coupled and the last two are the in-phase and out-of-phase coupled π soliton solutions, respectively, where the motions χ and χ' are coupled. These solutions (9a),(9b),(10a)-(10d),(11a),(11b) are shown in Figs. 2(a)-2(h).

We wish to propose here the role which may be played by a soliton in DNA. It may be supposed that if the condition $A\simeq0$ may be effectively realized by changing the acidity and salt concentration of the solvent adequately, then the coupled π soliton is the solitary excitation with the lowest energy. As seen in Figs. 2(g) and 2(h), this soliton is very interesting physiologically, because an opening or a closing of the structure of DNA can easily be realized by motion of this soliton. Thus a movement of a coupled π soliton alone is sufficient for an opening of the structure of DNA in the duplication of DNA and

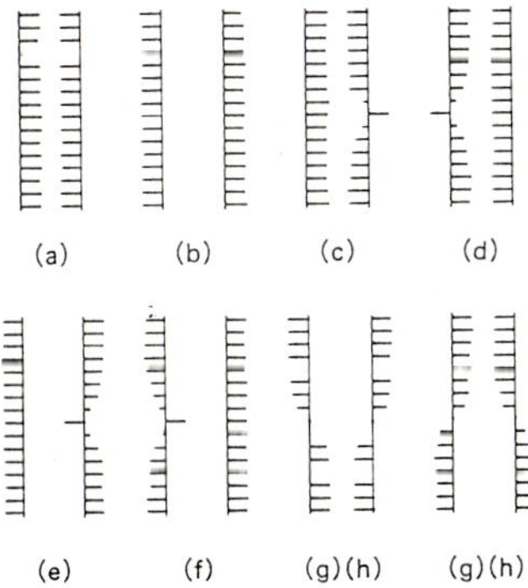

(a)   (b)   (c)   (d)

(e)   (f)   (g)(h)   (g)(h)

Fig. 2 Ground states and solitary excitations in the case A=0 are schematically shown.

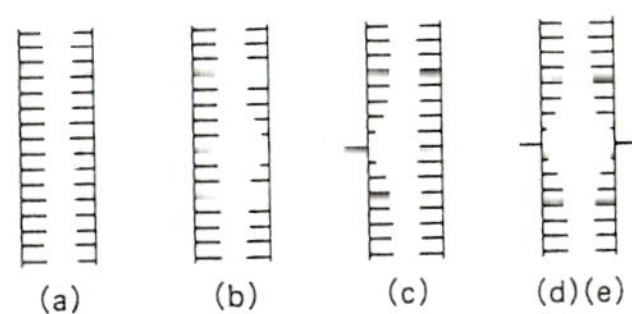

(a)   (b)   (c)   (d)(e)

Fig. 3 Ground state and soliton solutions in the case A≠0 are schematically shown.

the transcription of mRNA; there is no need to consider such an unlikely
motion as a continuous rewinding of the double strands of DNA, as
has been supposed so far.

B.   The case $A \neq 0$

In this case we obtain the following sets of solutions from (7) and
(8);

$$\chi = 2n\pi, \quad \chi' = 2m\pi, \tag{13}$$

$$\chi = \tilde{\phi}_s(\zeta - \zeta_0), \quad \chi' = 2m\pi, \tag{14a}$$

$$\chi = 2n\pi, \quad \chi' = \tilde{\phi}_s(\zeta - \zeta_0), \tag{14b}$$

$$\chi = \chi' + 2m\pi = \bar{\phi}_s(\zeta - \zeta_0), \tag{15a}$$

$$\chi = -\chi' + 2n\pi = \bar{\phi}_s(\zeta - \zeta_0), \tag{15b}$$

where n and m are integers and $\tilde{\phi}_s(\zeta - \zeta_0)$ is a soliton solution of the
sine-Gordon equation,

$$I\ddot{\chi} + (A+B)\sin\chi - Sa^2\chi_{zz} = 0 \tag{16}$$

and $\bar{\phi}_s(\zeta - \zeta_0)$ is a soliton solution of a double sine-Gordon equation:

$$I\ddot{\chi} + A\sin\chi + \frac{B}{2}\sin2\chi - Sa^2\chi_{zz} = 0. \tag{17}$$

The first one of these solutions corresponds to the ground state,and
the following two are the uncoupled 2π soliton solutions and the last
two are the in-phase and the out-of-phase coupled 2π soliton solutions.
These solutions, (13),(14a),(14b),(15a),(15b),are shown schematically
in Figs. 3(a)-3(e),respectively. From Figs. 2 and 3 it can be seen
that the lengths of open configurations are about 5a-10a.
   The breather solutions corresponding to these soliton solutions can
also exist as the kink-antikink bound states. These solitons and
breathers in DNA may serve as a nucleation center for RNA polymerase-
promoter recognition, but direct experimental evidences have not been
found.

3.   Estimations of the Average Soliton Number Density
     and the Parameter Values

Using the results of statistical mechanics of one-dimensional scalar
fields governed nonlinear wave equations, the average soliton number
density in the polynucleotide duplex is obtained as

$$n = (2\pi)^{-1/2}(8/\tilde{l})(\beta E_s^{(0)})^{1/2}e^{-\beta E_s^{(0)}}, \tag{18}$$

where $\tilde{l}$ and $E_s^{(0)}$ are given as

$$\tilde{l} = (S/(A+B))^{1/2}a, \quad E_s^{(0)} = 8[(A+B)S]^{1/2} \tag{19}$$

In the above we assume that the uncoupled 2π solitons which have the
lowest energy in the case $A \neq 0$ are only excited in the solvent condition where the H-D exchange measurements [5-7] were performed.
   By dividing the double helix with total length L into N segments
having a length of open configuration Δz, and denoting the number of
segments in open configuration and closed configuration by N_{op} and
N_{cl}, respectively, the equilibrium constant can be written as $K = N_{op}/N_{cl}$
$\cong N_{op}/N = nL/(L/\Delta z) = n\Delta z$. Then logK is given as

$$\log_{10}K=\log_{10}[(2\pi)^{-1/2}(8\Delta z/\tilde{l})]+ \frac{1}{2}\log_{10}(\beta E_s^{(0)})-\beta E_s^{(0)}\log_{10}e. \qquad (20)$$

NAKANISHI and TSUBOI[7] and TEITELBAUM and ENGLANDER[8] calculated the equilibrium constant K from the observed H-D exchange rate constant. The experimental data obtained by them in the temperature region lower than $47°$C where double helical structure is kept as a whole were plotted on a logarismic scale of K against reciprocal absolute temperature. By comparison between the theoretical result and their experimental data, we can estimate the rest energy of soliton, length of open configuration as $E_s^{(0)}$=0.35 and Δz=12.51. The continuum approximation used by us is justified by the condition $\tilde{l}/a>1$ (S>A+B). If we assume $A+B\cong S$, we can estimate the H-bonding energy and the sum of the stacking andtorsional energies by using the relation (19) as $2(A+B)\cong 0.1$eV, $4S \cong 0.2$eV. Then from (19) we obtain $\tilde{l}\cong a$, and we can estimate the length of open configuration and the upper limit of soliton velocity as Δz 12.5a, $v_0\cong 1.1\times10^4$cm/sec, assuming an approximate mean value of the moment of inertia of nucleotide $I=MR^2$, $M=308m_p$, $R\cong 4$A. Here v_0 means the velocity of the transverse shearing wave in DNA. From the experimental data, we can see that $K\cong 2\times10^{-4}$ at 300K, then we can estimate soliton number density and the mean number density of bases in open state as about $n\cong K/\Delta z\cong 2\times10^{-5}$/a, $5n\cong 10^{-4}$/a, respectively, at 300K.

References

1 A.S.Davydov: J. Theor. Biol. 38,559 (1973),Studia Biophsica 47, 221 (1974), and Biology and Quntum Mechanics (Pergamon,New York, (1982)
2 A.C. Scott: Phys. Rev. A 26 , 578 (1982)
3 S.Takeno:Prog. Theor. Phys. 71 (1984)395
4 S. Yomosa: J. Phys. Soc Jpn 53, No.10 (1984)
5 C.Mandal, N.R.Kallenbach, and S. W. Englander: J. Mol. Biol.135, 391 (1979)
6 M.Nakanishi,M.Tsuboi,Y.Saijo,andT.Nagamura: FEBS Lett.81,61 (1977)
7 M.Nakanishi and M.Tsuboi: J. Mol. Biol. 124, 61 (1978)
8 H.Teitelbaum and S. W. Englander: J. Mol.Biol.93,55(1975); 92,79 (1975)
9 S. W. Englander,N.R.Kallenbach,A.J.Heeger,J.A.Krumhansl, and S.Litwin: Proc. Natl. Acad. Sci. USA 77,7222 (1980)
10 S. Yomosa: Phys. Rev. A 27, 2120(1983)
11 S.Yomosa: Phys. Rev. A 30,474 (1984)
12 J.A.Krumhansl and J. R. Schrieffer: Phys. Rev. B 11,3535 (1975)
13 J. F. Currie,J. A. Krumhansl,A.R.Bishop,and S.E.Trullinger: Phys. Rev. B 22, 477(1980)
14 S. Takeno and S. Homma: Prog. Theor. Phys. 70, 308(1983)

Part VI

Solitons and Chaos

Pattern Selection and Low-Dimensional Chaos in Systems of Coupled Nonlinear Oscillators

Alan Bishop

Center for Nonlinear Studies and Theoretical Division, Los Alamos National Lab.
Los Alamos, NM 87545, USA

The longtime behavior of a number of one- and two-dimensional driven, dissipative, dispersive, many-degree-of-freedom systems is studied. It is shown numerically that the attractors are characterized by strong mode-locking into a small number of (nonlinear) modes. On the basis of the observed profiles, estimates of chaotic attractor dimensions, and projectons into nonlinear mode bases, it is argued that the same few modes may (in these extended systems) give a unified picture of spatial pattern selection, low-dimensional chaos, and coexisting coherence and chaos. Analytic approaches to this class of problem are summarized.

2. Introduction

In recent years considerable attention has been given to the properties of low-dimensional maps as models for complicated dynamics in higher-dimensional dynamical systems [1]. This attention has been merited by the proof of "universal" properties in classes of one-dimensional maps [1]. However, with few exceptions, the low-dimensionality has been introduced explicitly by restricting consideration to models with a very small number of degrees-of-freedom. On the other hand, equally active research has focused on the subject of spatial pattern selection in non-equilibrium nonlinear systems with many degrees-of-freedom (e.g. convection cells [2], reaction-diffusion systems [3]). In these cases mode-locking is very strong and a small number of modes dominate the spatial structure and temporal evolution in a nonlinear partial differential equation (p.d.e.) or large system of coupled ordinary differential equations (o.d.e.'s).

The perspective we will emphasize here is that the phenomena of pattern formation, low-dimensional chaos and coexisting coherence and chaos can be intimately connected in perturbed, dissipative dynamical systems with many degrees-of-freedom. More specifically, chaotic dynamics may develop by chaotic motions of the collective coordinates identifying the dominant (determining) patterns in the quiescent regimes. In this way only a small loss of mode-locking is responsible for the temporal chaos coexisting with spatial coherence -- only a small number of modes become unstable and the many remaining ones retain strong coherence. There are many examples of this scenario [4], which gives the problem of identifying and testing (nonlinear [5]) mode reduction schemes a general mathematical and physical importance [5]. Physically, it is essential in experiments to gain information on both time and space correlations; chaotic diffusion may be dominated by motion of the coherent pattern; possible consequences for fractal structure of the patterns need to be investigated.

In this brief summary we report six examples in the form of driven, damped nonlinear p.d.e.'s (and their discrete analogs) in one and two spatial dimensions. Especially in the 1-d cases, solitons [7] (or near-solitons) are fundamental excitations of the underlying unperturbed Hamiltonian systems, and motivate nonlinear mode reductions. Four of the examples are based on perturbed sine-Gordon (SG) equations

$$\phi_{tt} - \phi_{xx} + \sin\phi = F(x,t) - \varepsilon\phi_t \quad . \tag{1}$$

Here ϕ is a scalar field and x, t are space and time, respectively. Subscripts denote derivates. $F(x,t)$ is a forcing field and ε a damping constant. Note that neither pattern selection nor chaos occurs in (1) without dispersion [6], i.e. in the

overdamped limit. In the four examples below we have studied eqn. (1) numerically
on a finite line (length L) and with a high density spatial mesh (i.e.,
approximating the p.d.e.):

Case 1 Here L = 24 with 120 grid points and periodic boundary conditions,
$\varepsilon = 0.2$ and $F(x,t) = \gamma \sin \omega_d t$ with $\omega_d = 0.6$. The initial data is a static "pulse"
profile (the actual shape of the pulse is not very important -- the number of co-
existing attractors and hysteresis are small [8,9]). As reported elswhere [8], this
system undergoes a spontaneous spatial period doubling for $0.6 \lesssim \gamma \lesssim 0.9$ but remains
simply periodic in time. Typical spatial profiles are shown in Fig. 1a, and should
be thought of as "breather-soliton" wavetrains [7-10]. As $\gamma \to 0.9$, the duration
of a "chaotic" initial transient diverges, resulting in temporal chaos for $0.9 \lesssim \gamma$
$\lesssim 1.4$. Accompanying this chaos are large ($\gg 2\pi$) variations in the spatial average
of $\phi(x,t)$ [denoted by $\bar{\phi}(t)$] and large amplitude ($\sim 2\pi$) spatial variations in ϕ relative to $\bar{\phi}$. Instantaneous spatial correlation functions suggest [8] strong structural disorder. However, following the evolution in detail through a period of the
driving field (Fig. 1b) shows transparently that the basic coherent structures in
the quiescent pre-chaotic regime are preserved, but that their mode-locking relative
to each other has been (chaotically) broken so that the structure fails to repeat by
a small amount after each driver period. Consequently, we can decompose the field
at each instant of time into either two "breather solitons" or two "kink-solitons"
and two "antikink-solitions" (at instants of kink-antikink collision the field ϕ
may appear to be flat -- see Fig. 1b). Furthermore, chaotic evolution of $\bar{\phi}(t)$
through multiples of 2π does not take place via single particle dynamics but rather
through the slow diffusion of the kinks (antikinks) -- as with thermally assisted
transport in such systems [11]. We have confirmed the identification of a small
number of coherent "soliton" modes in the chaotic regime by projecting the field
at successive times onto a true soliton basis [9]. An example is shown in Fig. 2.
This small number of dominating collective modes suggests that the chaos will be
governed by a low-dimensional strange attractor [12]. We have checked the dimension (ν) using the algorithm of Grassberger and Procaccia [13]. We estimate [13]
that $\nu(\gamma = 1.0) = 2.5 \pm 0.3$, and in fact the dimension is found to lie within this
range throughout the chaotic regime. Thus ν is indeed low [14]. [For $\gamma \lesssim 0.9$, ν
is 1.0, as expected. ν also approaches 1.0 for large $\gamma \gtrsim 10$ where the nonlinear
potential is a small perturbation on the dynamics.]

Case 2 Here conditions are the same as case 1, except that the initial data is
a single static kink with periodic boundary conditions mod (2π). The attractor is

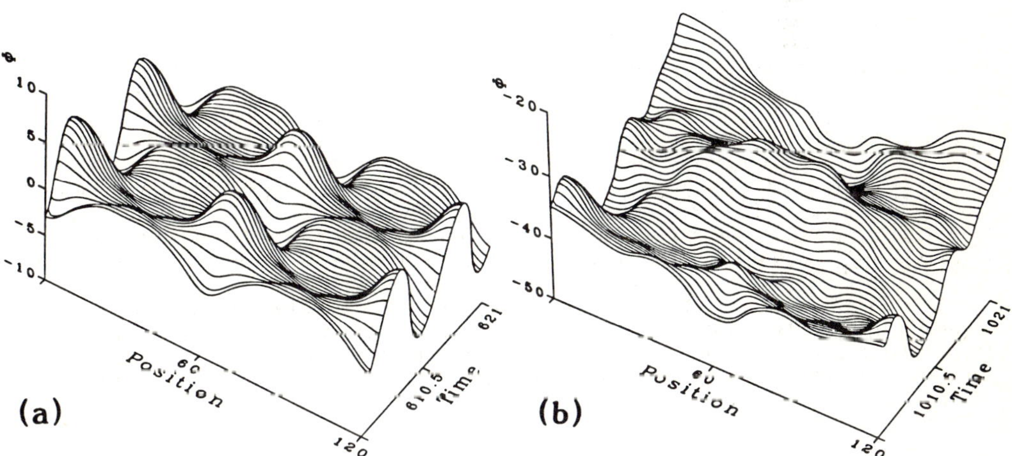

Figure 1. Space-time evolutions of $\phi(x,t)$ for the SG system (Case 1) through two
driving periods for $\varepsilon = 0.2$, $\omega_d = 0.6$, with periodic boundary conditions, and driving
strengths (a) $\gamma = 0.8$, which results in periodic time evolution, (b) $\gamma = 1.0$, which
results in chaotic kink-antikink motion (nearly repeating every driving period)

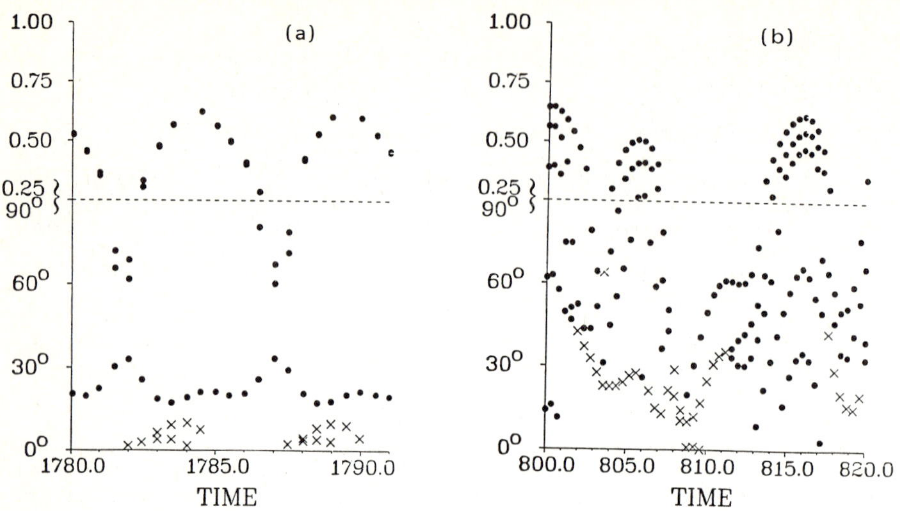

Figure 2.   Nonlinear spectral decompositions for Case 1, using the "soliton" transform on a periodic line [9]. The time evolution is shown of zeroes of the discriminant in the complex eigenvalue (λ) plane for the associated linear scattering
operator [9]. Zeroes occur either : (i) on the imaginary axis, symmetrically about
1/4. Dots denoting such zeroes above 1/4 are shown in the top portions of each
plot; or (ii) in pairs symmetric about the imaginary axis. Only the zeroes in the
positive real quadrant are shown. If these lie on the circle of radius 1/4 (non-
translating states) they are denoted by dots and their angle relative to the real
axis is shown in the lower portion of each plot. Otherwise the zeroes are denoted
by crosses and only their angle is shown, not their distance from the origin.
Figure 2(a) is for Case 1 with $\Gamma = 0.89$ where the attractor is periodic in time and
period-1/2 in space. Note the periodic evolution of the zeroes (the period of the
driving field is $2\pi/0.6 \simeq 10.5$). The crosses here correspond to wavelength $k = 2$
and 4 radiation modes. Figure 2(b) shows the "chaotic" evolution at $\Gamma = 1.0$. Note
that only a few collective modes dominate even this situation. Complete details
and further examples are given in Ref. [9]

found [8] to be a kink plus a "breather" and, for $\gamma \geq 0.9$, these coherent structures
move randomly with respect to each other producing temporal chaos. The dimension
of the chaos is again correspondingly low: e.g., for $\gamma = 1.0$, we estimate [13]
$\nu = 2.6 \pm 0.4$.

   Case 3 Here [15] we adopted boundary conditions appropriate to a finite Josephson
junction oscillator in zero magnetic field [16], viz. $\phi_x(x = 0,t) = \phi_x(x = L,t) = 0$.
The number of grid points = 120, $L = 6$, $\varepsilon = 0.1$, and $F(x,t) = \gamma_0 + \gamma \sin \omega_d t$ with
$\gamma_0 = 0.35$. Even with single kink initial conditions i.e. on the first "zero field
step" (ZFS) [16] a great variety of dynamical behaviors (including jumping between
ZFS's) are observed [15]. For our present purposes we report an example which
illustrates one prevalent source of chaos: for $\omega_d = 1.25$ and $\gamma = 1.2$, the kink
initial data is attracted to a nonchaotic "symmetric" state [16] on the third ZFS
as shown in Fig. 3a. Increasing γ to 1.5, we find a chaotic long-time evolution
in which the strongly mode-locked coherent structures of the third ZFS are still
dominant but their relative mode-locking has been (chaotically) broken -- see
Fig. 3b. Consistently, we estimate [13] the attractor dimension to be low:
$\nu = 2.5 \pm 0.4$.

   Case 4 In this case [17] we used outflow boundary conditions, $L = 40$, 800
grid points, $\varepsilon = 0.05$, and completely flat initial data ($\phi(x,t = 0) = \phi_t(x,t = 0)$
= 0. Chaotic evolution is induced by choosing a spatially inhomogeneous driving
field $F(x,t) = \gamma$, x [15,25], $F(x,t) = 0$ elsewhere. Our estimates [13] of ν for

252

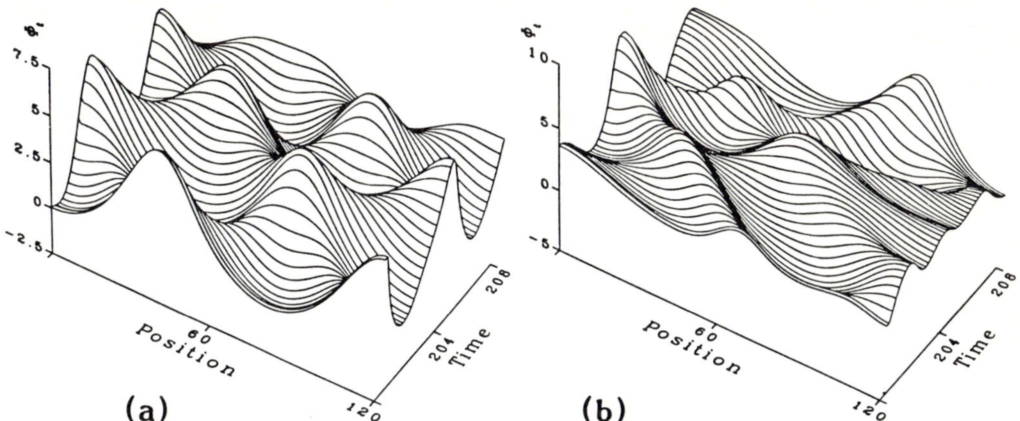

Figure 3. Space-time evolutions of $\phi_t(x,t)$ for the SG system (Case 3) through 1.6 driving periods for $\varepsilon = 0.1$, $\omega_d = 1.25$, with Neumann boundary conditions, dc driving $\gamma_0 = 0.35$, and ac driving strengths (a) $\gamma = 1.2$, which produces a nonchaotic standing wave state on the third ZFS, (b) $\gamma = 1.5$, which produces a chaotic pair of driven pulses

$0 \leq \gamma < 5$ show that ν increases from 1.0 at $\gamma = 0$ to 2.8 ± 0.4 for $\gamma = 2$ (with a rapid increase at the chaotic threshold $\gamma \sim 0.4$) and then rapidly decreases to 1.0 at larger values of γ (as in Case 1). Again the low dimension is entirely consistent with an examination of spatial profiles: once more ϕ evolves by the separation of coherent kink-antikink pairs which are periodically nucleated at the center of the line [17].

Case 5 Here we studied [18] a strongly perturbed magnetic chain of classical spins $\vec{S}_n = (S_n^x, S_n^y, S_n^z) = (\cos\theta_n \cos\phi_n, \cos\theta_n \sin\phi_n, \sin\theta_n)$, $n = 1, 2, 3 \ldots N$, governed by the Hamiltonian

$$H = - \sum_{n=1}^{N} \vec{S}_n \cdot \vec{S}_{n+1} + \alpha \sum_{n=1}^{N} \left(S_n^z\right)^2 - \vec{B} \cdot \sum_{n=1}^{N} \vec{S}_n \quad , \tag{2}$$

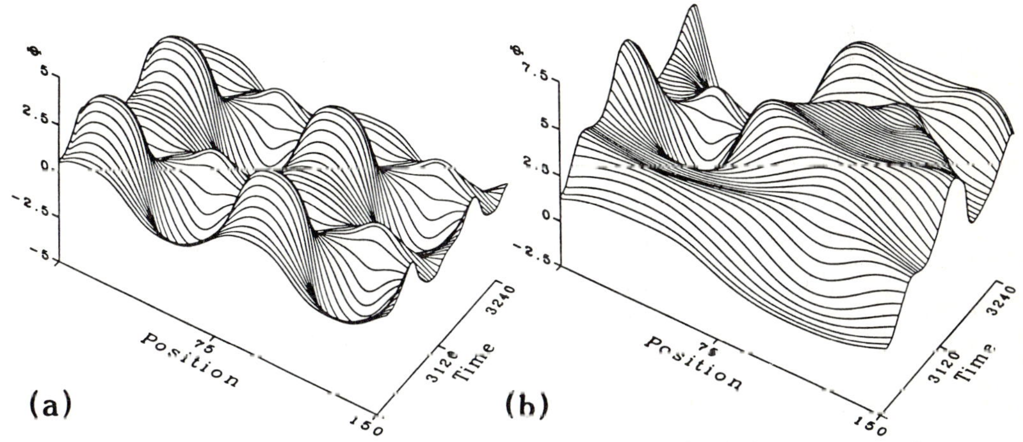

Figure 4. Space-time evolutions of the in-plane angle $\phi(x,t)$ for the easy-plane ferromagnet (Case 5) through two driving periods for $\varepsilon = 0.1$, $\omega_d = 0.05144$, with periodic boundary conditions, and ac driving strengths (a) $B_0^x = 0.03429$, resulting in a standing wave pattern, (b) $B_0^x = 0.02743$, resulting in intermittency between smooth standing wave structures and chaotic kink-antikink dynamics

where $\alpha > 0$ is an easy-plane anisotropy (ϕ_n and θ_n are the in-plane and out-of-plane angles) and $\vec{B}$ is an applied magnetic field. A Gilbert-Landau dissipation term [18] of strength ε was added to the equations of motion following from (2). Again a great variety of chaotic and non-chaotic evolutions are observed [18], depending on magnetic field configurations and parameter values. However, spontaneous pattern formation, consequent low-dimensional chaos, and coexisting coherence and chaos are once again <u>prevalent</u> phenomena. An example is illustrated in Fig. 4, where we have used periodic boundary conditions, $N = 150$, $\varepsilon = 0.1$, $\alpha = 0.1907$ (so as to relate to the material $CsNiF_3$ [18]), $\vec{B} = (B^x, B^y, B^z) = (B_0^x \sin(\omega_d t), 0, 0)$, $\omega_d = 0.05144$, and <u>random</u> initial data $\phi_n(t = 0)$, $\theta_n(t = 0)$. In Fig. 4a, $B_0^x = 0.03429$. The period-$\frac{1}{2}$ spatial structure seen in Fig. 4a forms spontaneously as the long time attractor with simply-periodic entrained motion. Decreasing B_0^x to 0.02743 we have entered a chaotic regime characterized (Fig. 4b) by chaotic motions of the coherent nonlinear mode components of the precursor period-$\frac{1}{2}$ spatial pattern. The Grassberger-Procaccia estimate [13] of the attractor dimension is correspondingly low --, e.g., for $B_0^x = 0.02743$, we find $\nu = 1.9 \pm 0.4$ for S^z.

Our final example, <u>Case 6</u>, corresponds to the SG Eq. (1) extended to <u>two</u> spatial dimensions. A 10x10 square was used [19] with 161x161 particles and <u>various</u> boundary conditions and initial data. The general scenario of space-time competitions found in one-dimension <u>is</u> confirmed in these cases. Detailed results are given elsewhere, [19] but Fig. 5 shows a typical non-chaotic, higher spatial symmetry precursor attractor (periodic boundary conditions were used) which eventually (for $1.0 \lesssim \Gamma \lesssim 1.2$) gives way to low-dimensional ($\nu \sim 2.4 \pm 0.3$) chaos and coherent lowest-symmetry spatial patterns. The higher spatial dimension also admits new phenomena, including [19] transition from a metastable, simply periodic, higher symmetry <u>two</u>-dimensional pattern to a final attractor which is higher symmetry and simply periodic but with variations in only <u>one</u> spatial direction: an example is illustrated in Fig. 6.

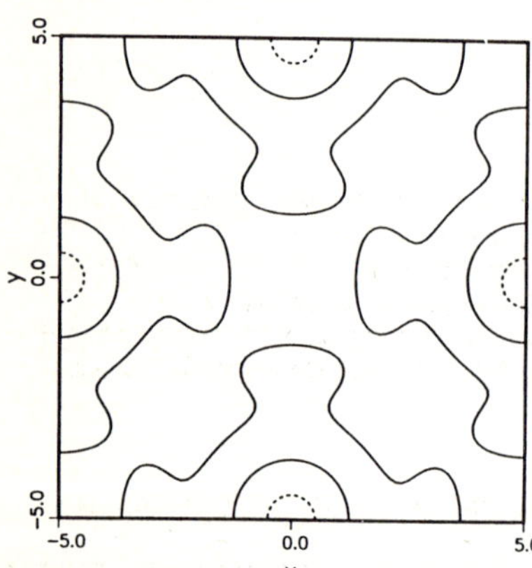

Figure 5. Homogeneously driven, damped SG system in <u>two</u> space dimensions (a 10 x 10 square) with periodic boundary conditions -- case 6. Parameters are as for case 1 ($\varepsilon = 0.2$, $\omega_d = 0.06$) and a random pulse profile was used as initial data [19]. Here $\Gamma = 0.9$. The attractor is simply periodic in time but with a spontaneous higher spatial symmetry -- period - 1/2 on a $\pi/4$-rotated $\sqrt{2} \times \sqrt{2}$ lattice. [Lines are ϕ-intensity contours: solid and dashed lines signify above and below spatial average ϕ in units of 0.04π]

In conclusion, we have illustrated with a diverse collection of driven, dissipative, dispersive p.d.e.'s that a small number of determining modes may be <u>typically</u> responsible for both spontaneous pattern formation in quiescent regimes and for low-dimensional chaos in subsequent chaotic regimes. We emphasize that complex temporal behaviors <u>are</u> certainly observed in general parameter regimes for our systems (subharmonics, quasi-periodicity, etc.), especially if strong symmetries are not imposed through boundary conditions or initial data [8,15,18]. Here, however, we have deliberately concentrated on relatively simple circumstances where the <u>competition</u> between pattern selection and complex time-dependence (which we consider of most fundamental importance) is isolated.

The symbiotic relationship between analysis and numerical simulation has a rich history in studies of strongly nonlinear systems (e.g. the Fermi-Pasta-Ulam problem, the recent history of exact soliton equations). The problems considered here are

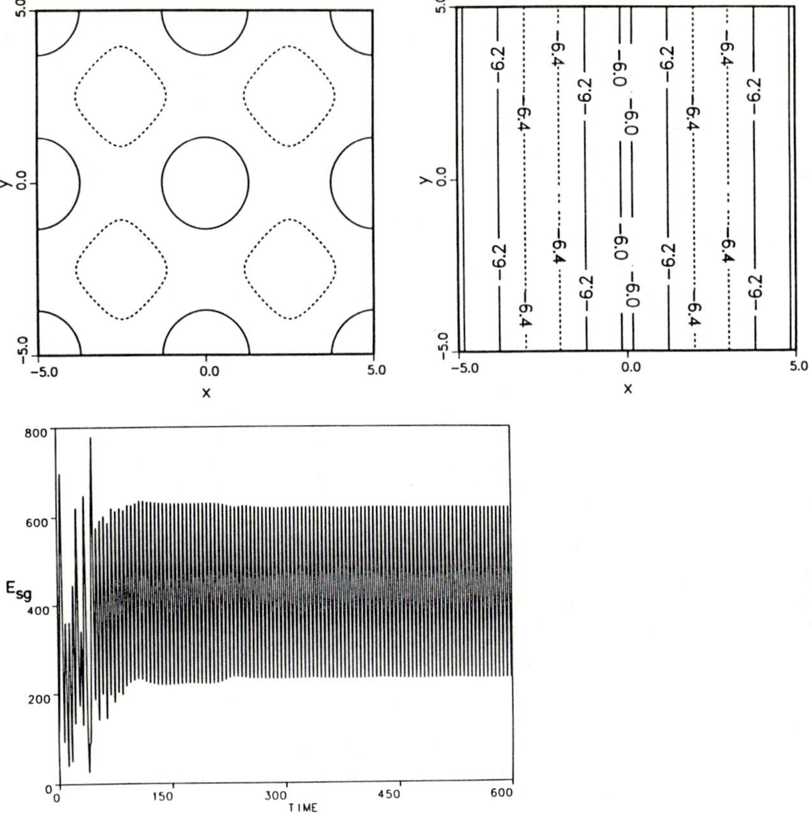

Figure 6. Same as Fig. 5 but with Γ = 1.6. After a "chaotic" initial transient, the time evolution is simply periodic and with a spontaneous higher spatial symmetry. However, there is also pattern conversion from a period-1/2 pattern on a √2 x √2 lattice (Fig. 6(a)) to a purely <u>one</u>-dimensional period-1/2 pattern (Fig. 6(b)). Figure 6(c) exhibits this pattern discrimination in terms of the total sine-Gordon energy as a function of time -- note that the one-dimensional attractor has a slightly lower energy

further fine examples, our results having stimulated new questions and directions of analytic attack on the part of mathematicians and physicists. The kinds of questions being addressed include: (i) Adequately identifying collective coordinates to suggest optimal mode reductions (and low-dimension map constructions) which can be compared with those of the full equation. An important issue here is to clarify the role of the extended modes. Are they or the collective modes the drivers or the slaves? Clearly, in general this is a grey area of self-consistent mutual renormalization with extended modes "dressing" the coherent structures. There is certainly evidence [8,20] for selected extended modes being strongly affected and playing a dominating role. Projections onto unperturbed soliton bases [9,21,35] will allow us to address this issue with more control; (ii) Understanding relationships between the coherent patterns we find and bounds [5] on the number of determining modes, fractal structures, Liapunov exponents [21,22], diffusion coefficients, and chaotic attractor dimensions; (iii) Elucidating criteria for pattern selection in nonlinear dispersive [6,23] systems such as that studied here.

Analytic approaches currently focus on: (a) Existence and (local) [24] stability analyses, which are particularly appropriate if the patterns are low-amplitude (e.g.

for $\varepsilon \ll 1$, $\gamma \ll 1$ and $\omega_d \lesssim 1$, for which a low-amplitude breather may stabilize).
Techniques include analyses for a uniform state [25,26], one-breather state [25]
(generalizing the approach of Kaup and Newell [27]), and multi-breather wave-trains
(using, e.g., algebraic-geometric soliton representations for a periodic SG line
[10]).   Applications of the Melnikov test for "horseshoe" chaos [28] are also being
attempted analytically and numerically; (b) Explicit truncations to a small number
of coupled modes, e.g. a breather wave-train [25,33] (plus long-wavelength radiation
if "background" or spatial-average motion is important), or small number of most
unstable extended modes [20], or modes extracted numerically by projections onto
unperturbed solitons [9,21,35]; (c) Adding thermal noise to the equation of motion
to yield a Langevin description and allowing construction of a Fokker-Planck equation
for the field's probability distribution.   This last approach has the synergetic
attraction that, after averaging over the noise distribution, non equilibrium statistical mechanics systems can [29] be cast in the form of a finite, anisotropic equilibrium Hamiltonian model in one higher spatial dimension.   The Hamiltonians produced
in this way[30] have the feature of "competing interactions" which is currently
popular [31] in equilibrium statistical mechanics and responsible for (experimentally
observed) inhomogeneous mean-field states.   These synergetic parallels are evident
in the attractors of 1-D studies [8,9,15,18] and even more strikingly in 2-D [19]
-- as for example in 2-D $\rightarrow$ 1-D pattern transitions (c.f. Fig. 6), which should be
compared with, e.g., hexagonal $\rightarrow$ striped discommensuration array phase transitions
in 2-D incommensurate systems [31].

Studying chaotic dynamics of p.d.e.'s and other many-degree-of-freedom systems
is a most exciting and topical area in dynamical systems theory.   Consequently,
literature and understanding are rapidly changing, but we would be remiss not to
conclude with a representative list of relevant publications.   Classifying loosely
by the type of equation, these include: SG-like systems with periodic [8,9,21],
Neumann [15] or absorbing [17] boundary conditions, including 2-D [19] cases and
discrete generalizations [32]; the cubic Schrödinger equation [33], including close
analogs in plasma models [22] and bistable optical ring oscillators [20], and complex generalizations [34]; the Toda lattice and generalizations [35]; and classical
spin chains [18].   We also recommend the somewhat different but complementary studies
of coupled map lattices [36] and of cellular automata [37].

It is a pleasure to acknowledge the invaluable support of my colleagues, especially J. C. Eilbeck, K. Fesser, P. S. Lomdahl, D. W. McLaughlin, O. H. Olsen,
E. A. Overman, I. Satija and G. Wysin.   This work was supported by the USDOE.

References

1. See Physica 7D (1983).
2. e.g. M. Cross, Phys. Rev. A 25, 1065 (1982).
3. e.g. M. Meinhardt, "Models of Biological Pattern Formation" (Academic Press 1982).
4. Examples of coesisting coherence and chaos include: clumps and cavitons in
    turbulent plasmas; filamentation in lasing mediums; large scale structures in
    turbulent fluids (e.g. modon "blocking" patterns controlling atmospheric flow
    and weather, or gulf stream "rings" in oceanography); and perhaps even the red
    spot of Jupiter! In all cases the coherent structures are long-lived and with
    slower dynamics than the single-particle turbulence -- a unifying practical con-
    cern is their effect on transport and predictability (in space and time). There
    are also increasing numbers of controlled laboratory scale observations (e.g.
    in convection cells, water wave surface solitons) as well as probable
    applications in biological contexts.
5. In some cases rigorous bounds on the number of determining modes have recently
    been established (e.g. C. Foias, et al., Phys. Rev. Lett. 50, 1031 (1983)), and
    it has been possible to bound the attractor (fractal) dimension by the number of
    determining modes (e.g. O. Manley, et al., preprint (1984), B. Nicolaenko and
    B. Scheurer, preprint (1984)). In certain cases (e.g. for some reaction-diffu-
    sions problems) even a truncated set of linear modes can be accurate. (See J. C.
    Eilbeck, J. Math. Biol. 16, 233 (1983), B. Nicolaenko et al., Proc. Acad. Sc.
    Paris 298, 23 (1984)).
6. A. R. Bishop, E. Domany and P. S. Lomdahl (unpublished results).

7. e.g., R. K. Dodd, et al. "Solitons and Nonlinear Wave Equations" (Academic Press 1982).

8. D. Bennett, A. R. Bishop, S. E. Trullinger, Z. Physik B $\underline{47}$, 265 (1982); A. R. Bishop, et al., Phys. Rev. Lett. $\underline{50}$, 1095 (1983); and Physica $\underline{7D}$, 259 (1983).

9. E. A. Overman, D. W. McLaughlin, A. R. Bishop, preprint (1984).

10. N. Ercolani, et al., preprint (1984).

11. See, e.g., M. Büttiker and R. Landauer, in "Physics in One Dimension", eds. J. Bernasconi and T. Schneider (Springer 1981).

12. See, e.g. J. D. Farmer et al., Physica $\underline{7D}$, 153 (1983).

13. P. Grassberger and I. Procaccia, Phys. Rev. Lett. $\underline{50}$, 346 (1983). The calculation of various attractor "dimensions" remains in an early state of development (see Ref. 12). In particular, important questions remain such as sensitivity to the scale of structures and the spatial patterns (similar questions apply to Liapunov exponents -- see refs. 21, 22). We emphasize that our error estimates quoted here are realistically conservative. Typically we used an embedding dimension of 5-10 and 80,000 data points.

14. Assuming only 2 breathers (or 4 kinks) as a truncated modal set, the maximum dimension of the space containing the attractor is 8. Our initial data symmetry reduces this to 4. The presence of dissipation will typically further reduce the "active" dimension. Our estimates of ν are generally in the range 2 - 2.5. This is entirely reasonalbe in view of our estimates (unpublished) of ν for a chaotic single particle with similar damping and driving strengths: there the maximum dimension is 2 but we generally find $\nu = 1.1 - 1.3$.

15. A. R. Bishop, J. C. Eilbeck, G. Wysin, APS March Meeting Bulletin (1984), and preprint; See also M. P. Soerensen, et al., Phys. Rev. Lett. $\underline{51}$, 1919 (1983).

16. P. S. Lomdahl, et al., Phys. Rev. B $\underline{25}$, 5737 (1982).

17. J. C. Eilbeck, P. S. Lomdahl, A. C. Newell, Phys. Lett. $\underline{87A}$, 1 (1981).

18. G. Wysin and A. R. Bishop, APS March Meeting Bulletin (1984), and preprint.

19. O. H. Olsen, P. S. Lomdahl, A. R. Bishop, J. C. Eilbeck, preprints (1984).

20. D. W. McLaughlin, J. V. Moloney, A. C. Newell, Phys. Rev. Lett. $\underline{51}$, 75 (1983); and preprint (1984).

21. M. Imada, J. Phys. Soc. Jpn. $\underline{52}$, 1946 (1983).

22. G. D. Doolen, et al., Phys. Rev. Lett. $\underline{51}$, 335 (1983).

23. Of course there are many different mechanisms for pattern selection and self-organization depending on the context -- see, e.g., "Fronts, Interfaces and Patterns," eds. A. R. Bishop, L. J. Campbell, P. J. Channell, Physica D (1984). In particular, our cases should be contrasted with those where diffusion or reaction - diffusion dominates, even though physical questions (such as mode reduction) can be quite similar.

24. Some degree of hysteresis and coexistence of attractors persists in these many-degree-of-freedom systems, although it is typically much less pronounced than for a single oscillator.

25. N. Ercolani and D. W. McLaughlin, unpublished.

26. S. E. Trullinger et al., unpublished.

27. D. J. Kaup and A. C. Newell, Proc. Roy. Soc. (London) $\underline{A361}$, 413 (1978).

28. See, e.g., P. J. Holmes and J. E. Marsden, Archive Rational Mechanics and Analysis, $\underline{76}$, 135 (1981).

29. See, e.g., E. Domany, Phys. Rev. Lett. $\underline{52}$, 871 (1984).

30. A. R. Bishop, and E. Domany, unpublished.

31. See, e.g., P. Bak, Rep. Prog. Phys. $\underline{45}$, 587 (1982).

32. J. Oitmaa and A. R. Bishop, preprint (1984). The effect of a discrete lattice is to produce many metastable states becuase of the Peierls-Nabarro pinning forces. The situation is similar to that of large scale dynamics in a discrete discommensurate model (c.f. Refs. 1,31), and all combinations of "order" and "chaos" in both space and time and possible.

33. e.g. N. Bekki and K. Nozaki, these proceedings, and references therein.

34. e.g. H. T. Moon et al., Physica $\underline{7D}$, 135 (1983); K. Nozaki and N. Bekki, Phys. Rev. Lett. $\underline{51}$, 2171 (1983).

35. K. Fesser, et al., preprint (1984); B. Paulus, et al., Phys. Lett. $\underline{102A}$, 89 (1984), and preprint.

36. K. Kaneko, these proceedings.

37. See, e.g., "Cellular Automata," eds. J. D. Farmer, T. Toffoli, S. Wolfram (North-Holland Amsterdam 1984).

Solitons and Chaos in the Sine-Gordon System

Peter L. Christiansen

Laboratory of Applied Mathematical Physics, The Technical University of Denmark
DK-2800 Lyngby, Denmark

The perturbed sine-Gordon system models long Josephson tunnel junction oscillators.
Soliton dynamic states and chaotic intermittency between the states have been found
computationally.

1 Modelling the Josephson Oscillator

The Josephson tunnel junction oscillator with overlap geometry illustrated in Fig. 1
is modelled by the perturbed sine-Gordon equation [1]

$$\phi_{xx} - \phi_{tt} - \sin\phi = \alpha\phi_t - \beta\phi_{xxt} - \gamma \, . \tag{1}$$

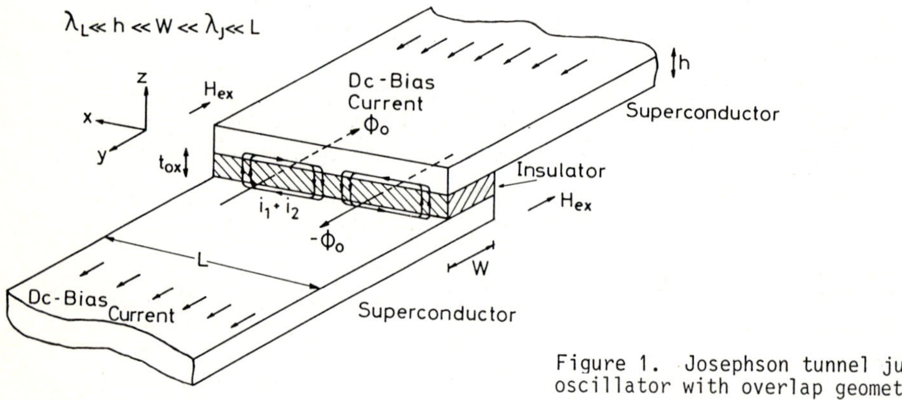

Figure 1. Josephson tunnel junction
oscillator with overlap geometry [2]

Here x is a normalized space coordinate measured along the oscillator and t is
normalized time. The difference between the phases of the order parameter of the
two superconductors is denoted $\phi = \phi(x,t)$. The right hand side of the equation
contains the terms of the classical sine-Gordon equation while the terms on the
left hand side model losses ($\alpha\phi_t$ and $\beta\phi_{xxt}$) and constant bias current (γ).

At the ends of the junction the homogeneous open-end boundary conditions

$$\phi_x(0,t) = \phi_x(\ell,t) = 0 \tag{2a}$$

corresponding to zero current (since the current on the junction is proportional to
ϕ_x) or the inhomogeneous boundary conditions

$$\phi_x(0,t) = \phi_x(\ell,t) = \eta \tag{2b}$$

corresponding to external normalized magnetic field, η, are applied. The normalized
length of the oscillator is denoted by ℓ.

For the computational modelling of the oscillator initial conditions of the form

$$\phi(x,0) = F(x) \quad \text{and} \quad \phi_t(x,0) = G(x) \tag{3}$$

are used. Here the functions F and G are chosen such that stationary states of
$\phi(x,t)$ are obtained without spending too much computer time on transient evolutions
of the solution. In practice, the final ϕ and ϕ_t distributions from a nearby stationary state in parameter space (α, β, γ, ℓ, η) are often used as the functions F
and G respectively.

2 Soliton Dynamic States

The classical sine-Gordon equation has 2π-kinks, anti-kinks as well as breathers
as soliton solutions [3]. Similarly, the perturbed sine-Gordon equation (1) has
soliton solutions in the looser sense of the word. Figure 2 thus shows a soliton
(the x-derivative of a 2π-kink) travelling in the negative x-direction, being reflected into an anti-soliton at $x = 0$ with boundary condition (2a), travelling in
the positive x-direction and being reflected into a soliton at $x = \ell$ with boundary
condition (2a).

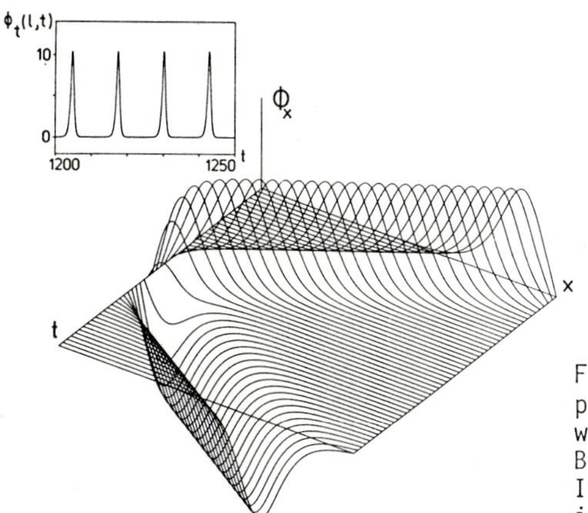

Figure 2. Computer solution of the
perturbed sine-Gordon equation (1)
with $\alpha = 0.05$, $\beta = 0.02$, $\gamma = 0.35$.
Boundary conditions (2a) with $\ell = 6$.
Initial conditions (3) with one soliton. The inset shows $\phi_t(\ell,t)$ [2]

The energy input in the system due to the γ-term in (1) balances the dissipation
due to the α and β-terms in (1). We designate such a stationary state a soliton dynamic state.

The periodic motion of the soliton (and anti-soliton) is responsible for the
emission of electromagnetic radiation, typically in the GHz-range, from the oscillator. The DC-component of the voltage which is proportional to the time-derivative
ϕ_t has been computed for different values of the applied bias current γ in (1). The
resulting curve shows agreement with experimentally measured IV-curves for the junction [4].

Each soliton dynamic state corresponds to a branch of the IV-characteristic called
a zero field step in case of zero external magnetic field. The soliton dynamic state
shown in Fig. 2 gives rise to the first zero field step.

Fourier analysis of ϕ_t provides the computational power spectrum for the radiation from the oscillator which shows agreement with experimental measurements of
the power spectrum [4], [5].

Recently, we have investigated a number of soliton dynamic states occurring for
different boundary conditions [6], [7], [8]. Some of the results are summarized in
Fig. 3, which illustrates the corresponding soliton trajectories in the xt-plane.
Thus the diagram ZFS 1 represents Fig. 2.

3 Chaotic Intermittency between Soliton Dynamic States

In order to tune the electromagnetic radiation from the oscillator a constant external magnetic field may be applied. This situation is modelled by boundary condition (2b) for the overlap geometry. Figure 4 shows the resulting soliton dynamic

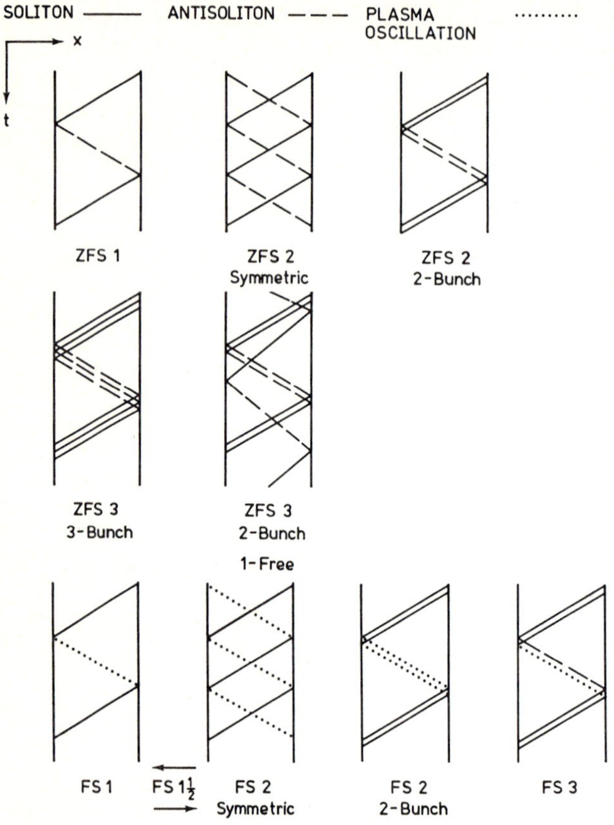

Figure 3. Soliton dynamic states. ZFS = zero field step corresponding to boundary conditions (2a). FS = Fiske step corresponding to boundary conditions (2b)

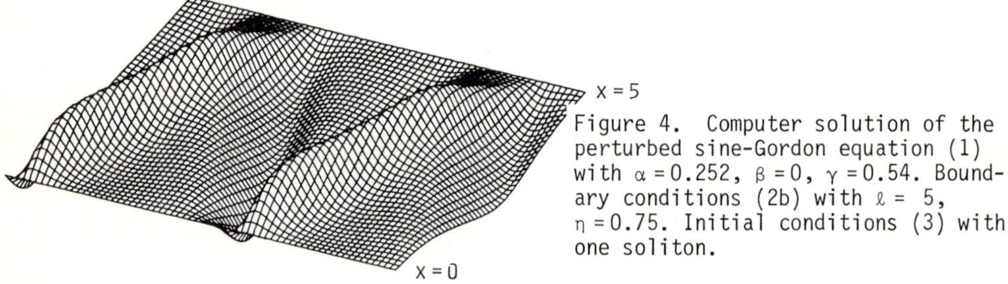

Figure 4. Computer solution of the perturbed sine-Gordon equation (1) with $\alpha = 0.252$, $\beta = 0$, $\gamma = 0.54$. Boundary conditions (2b) with $\ell = 5$, $\eta = 0.75$. Initial conditions (3) with one soliton.

state. At x = 0, energy is then absorbed due to the boundary condition such that the incident soliton is reflected not into anti-soliton (which requires the minimal energy, 8 in normalized units for the sine-Gordon soliton) but into plasma oscilla-tions. At x = ℓ, these oscillations regenerate a soliton due to the energy input produced by the boundary condition. The state gives rise to a branch of the IV-char-acteristic for the oscillator called a Fiske step. The dynamic state shown in Fig. 4 yields the first Fiske step. The diagram FS 1 in Fig. 3 represents Fig. 4. With two solitons present on the oscillator we get the second Fiske step (FS 2).

In a narrow portion of parameter space (α, β, γ, ℓ, η) the oscillator switches back and forth between FS 1 and FS 2 [7]. This chaotic intermittent switching gives rise to a special branch ("FS $1\frac{1}{2}$") in the computed IV-characteristic between the two branches, FS 1 and FS 2. Recent experimental measurements have perhaps also revealed such structures [9].

The soliton formation effectively occupies infinitely many degrees of freedom and thus projects the sine-Gordon system down to a non-linear low-dimensional system. This system exhibits chaotic intermittency between two critical points (corresponding to the two soliton dynamic states).

Acknowledgement
The financial support of the European Research Office of the United States Army through contract No. DAJA 37-82-C 0057 is gratefully acknowledged.

References
1 D.W. McLaughlin and A.C. Scott: Phys. Rev. A 18, 1652 (1978)
2 P.S. Lomdahl, O.H. Soerensen, and P.L. Christiansen: Phys. Rev. B 25, 5337 (1982)
3 A.C. Scott, F.Y.F. Chu, and D.W. McLaughlin: Proc. IEEE 61, 1443 (1973)
4 P.L. Christiansen, P.S. Lomdahl, A.C. Scott, O.H. Soerensen, and J.C. Eilbeck:
   Appl. Phys. Lett. 39, 108 (1981)
5 P.S. Lomdahl, O.H. Soerensen, P.L. Christiansen, A.C. Scott, and J.C. Eilbeck:
   Phys. Rev. B 24, 7460 (1981)
6 M.P. Soerensen, P.L. Christiansen, R.D. Parmentier, and O. Skovgaard: Appl. Phys.
   Lett. 42, 739 (1982)
7 M.P. Soerensen, N. Arley, P.L. Christiansen, R.D. Parmentier, and O. Skovgaard:
   Phys. Rev. Lett. 51, 1919 (1983)
8 M.P. Soerensen, R.D. Parmentier, P.L. Christiansen, O. Skovgaard, B. Dueholm,
   E. Joergensen, V.P. Koshelets, O.A. Levring, R. Monaco, J. Mygind, N.F. Pedersen,
   and M.R. Samuelsen: Phys. Rev. B (in press)
9 M. Cirillo, G. Costabile, S. Pace, R.D. Parmentier, and B. Savo (to appear)

Soliton Propagation Properties in a Josephson Transmission Line

J. Nitta, A. Matsuda, and T. Kawakami

Musashino Electrical Communication Laboratory, Nippon Telegraph and Telephone
Public Corporation, Musashino-shi, Tokyo 180, Japan

1. Introduction

Recently, the soliton concept in nonlinear dispersive waves is applied in solid
state physics[1],[2].   A sine-Gordon soliton in particular is widely used as a
model for dislocations in crystals, domain walls in ferromagnetics and excitations
in the charge-density-wave(CDW) state. A quantized magnetic flux also exhibits
the sine-Gordon soliton properties in a Josephson transmission
line(JTL)[3],[4],[5].   However, an actually fabricated JTL obeys the following
modified sine-Gordon equation, taking perturbational losses into account:

$$\frac{\partial^2\phi}{\partial x^2} - \frac{\partial^2\phi}{\partial t^2} = \sin\phi + \alpha\frac{\partial\phi}{\partial t} - \beta\frac{\partial^3\phi}{\partial^2 x\partial t} - \gamma , \qquad (1)$$

where α , β and γ are associated with quasiparticle   tunneling   loss,
superconducting rf loss and an artificially provided bias current, respectively.

In the following sections, soliton propagation properties in such a perturbed
sine-Gordon system will be described for the purpose of clarifying the effects of
the perturbational terms.   In addition to a standard nondamped JTL, a well damped
JTL has been investigated experimentally with the use of a single quantized
magnetic flux observation system[6] and numerically with the use of computer
simulations.   Threshold properties for a resistively coupled JTL have been also
investigated for the purpose of soliton propagation control.

2. Standard Nondamped Josephson Transmission Line[7]

The fabricated JTL is a Nb/Nb-oxide/Pb tunnel junction in an overlap geometry as
shown in Fig.1. The size is 50 μm wide and 1 cm long. Both ends of the JTL are
terminated by Au thin film resistors.   Observed soliton waveform examples are
shown in Fig.2, as a function of input pulse height with a fixed pulse width. The
input pulse waveform is shown in the inset of Fig.2. The effective applied input
voltage is reduced to a mismatch factor of $2Z_0/50$, where Z_0 is the JTL
characteristic impedance. The single-input-pulse is dissociated into soliton
trains during propagation, as expected from the soliton propagation theory[8].
The number of solitons varies according to changes in input pulse height.

To compare with experimental results, Eq.(1) is numerically integrated with the
real   input   waveform and the appropriate boundary condition. The damping
parameter, α , and bias current, γ , are experimentally determined. The
experimentally obtained interval times between each respective soliton and

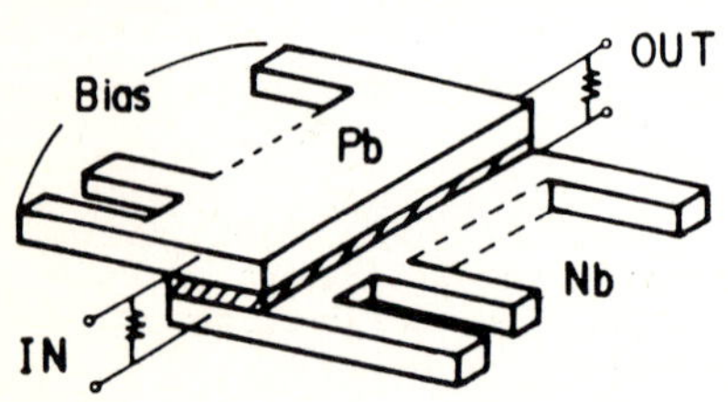

<u>Fig.1</u>   Fabricated JTL structure

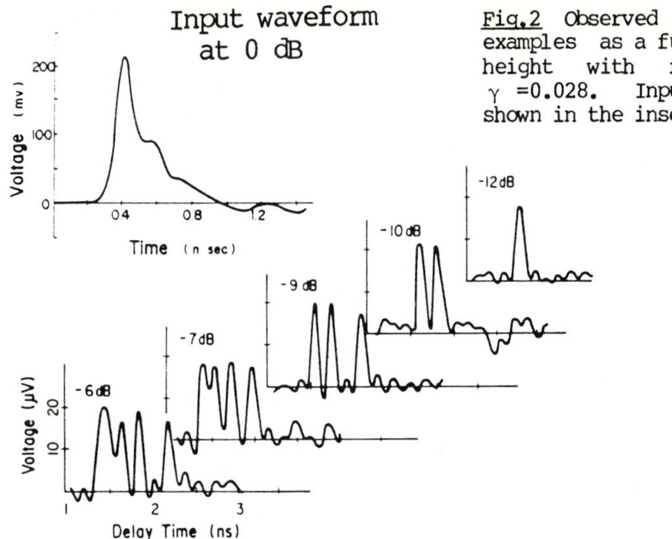

Fig.2 Observed soliton waveform examples as a function of input pulse height with fixed bias current γ =0.028. Input pulse waveform is shown in the inset.

propagation delay times agree with the numerical simulations, if large β is assumed.

As one of the features in a sine–Gordon soliton, it is expected that the waveform will show velocity dependence. Bias current γ dependence of soliton waveforms is examined for a fixed input pulse height as shown in Fig.3. With an increase in γ , the soliton waveform steepens with an associated increase in propagation velocity, especially for γ <0.04. For 0.04< γ <0.4, the soliton width is almost constant, because the measurable minimum pulse width is limited to about 120 ps in our system. When 0.5< γ <0.69, two or more solitons propagate in the JTL, but stable propagation is still maintained. However, when γ exceeds 0.71, the JTL jumps to the uniform voltage state. Then, the sin Φ term is averaged out and the JTL shows linear responses. The waveform at γ =0.71 actually agrees with the input pulse waveform.

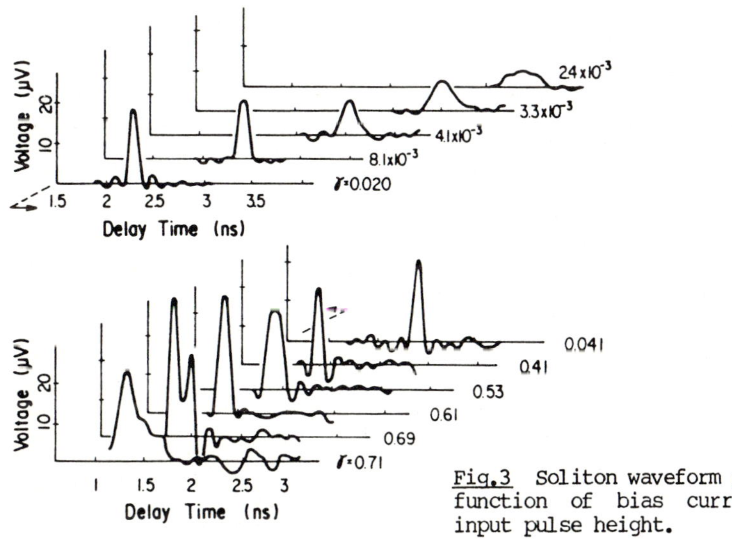

Fig.3 Soliton waveform profiles as a function of bias current with fixed input pulse height.

<u>3. Well Damped Josephson Transmission Line[9]</u>

For the purpose of clarifying the effects of damping and bias terms, a well damped(large α) JTL was fabricated by adding an external shunt resistance parallel to the junction. For a large α and γ case, where the perturbation approach can not be justified, it was thought to be instructive to investigate the soliton propagation properties experimentally. The fabricated JTL is shown in Fig.4, where a Cr layer is used as a damping resistor interconnected to the Pb upper electrode by an Au layer. Observed soliton waveform examples with a fixed bias current, γ =0.545, are shown in Fig.5, as a function of input pulse height. In the present well damped JTL, stabler and better separated multisoliton propagations were observed even in the high bias level($\gamma > 0.9$), than in the nondamped JTL. With a decrease in input pulse height, the number of output solitons decreases, as in the nondamped JTL case. These experimentally obtained results verify the numerical simulations[10],[11].

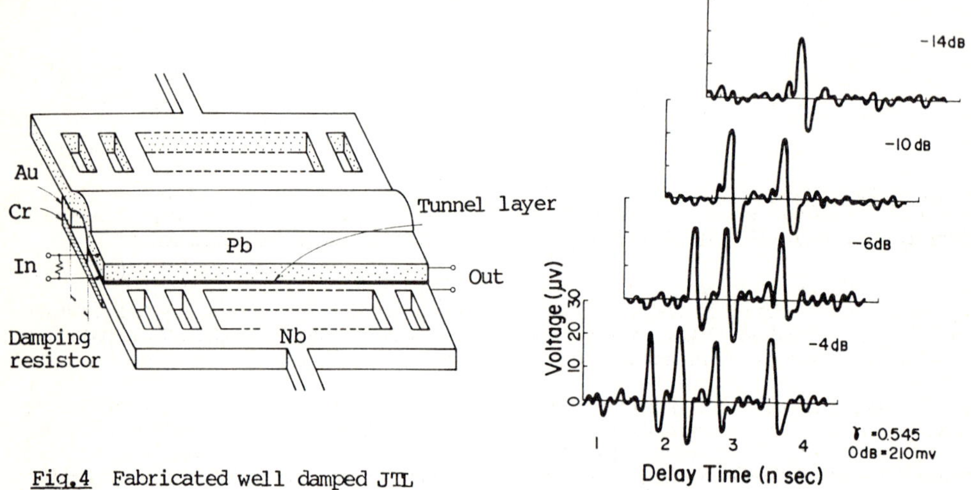

<u>Fig.4</u> Fabricated well damped JTL

<u>Fig.5</u> Observed soliton waveform examples as a function of input pulse height with fixed bias current γ =0.545.

Figure 6 shows a comparison of velocities between expermentally obtained average velocities and theoretical final velocities. The experimental average velocities can be directly calculated from the soliton propagation delay times and fabricated JTL length. Theoretical final velocities were estimated with the use of the MCLAUGHLIN-SCOTT equation[12],[13]:

$$u = u_0 \left\{1+\left(\frac{4\alpha}{\pi\gamma}\right)^2\right\}^{-1/2} . \qquad (2)$$

Here, the maximum propagation velocity is $u_0=8.3\times10^6$ m/s for the present JTL. The damping parameter was experimentally determined to be α =0.66 using the normal resistance component of the JTL's I-V characteristics. This value was much larger than the nondamped JTL case(α =0.0041). The experimental average velocities, including velocity changes after soliton formation, seem to agree approximately with the final velocities. Although eq.(2) was obtained by the perturbation approach with an assumed 2π -kink solution for the unperturbed sine-Gordon equation, it can be applied even in the large α and γ case. This agreement can be regarded as a feature of a well damped JTL, where the

264

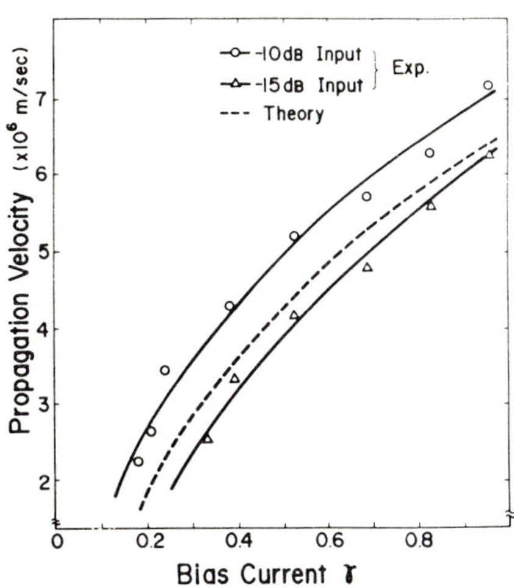

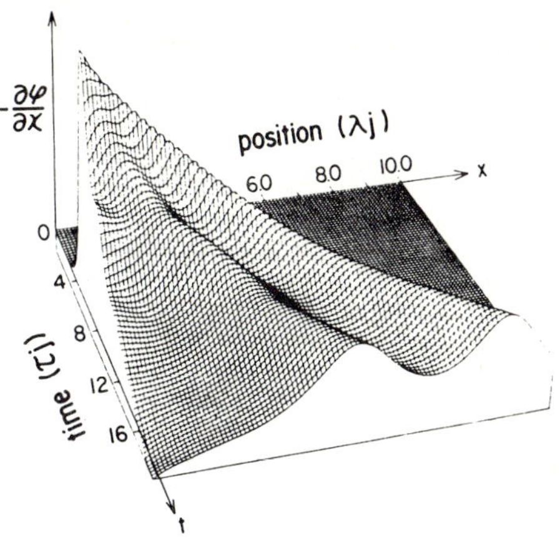

Fig.7 Soliton formation process for α =0.66, γ =0.545 and input of -10 dB. Space and time coordinates are normalized by Josephson penetration depth(552 μ m) and Josephson plasma period(64 ps) of the present JTL, respectively. Position x=0 corresponds to input termination.

Fig.6 Comparison of velocities between experimentally obtained results and MCLAUGHLIN-SCOTT theory.

soliton rapidly reaches the final velocity. The difference in velocities between the -10 dB and -15 dB input cases shows that the higher input pulse heights provide solitons with faster initial velocities.

The multisoliton formation process, where interaction among solitons is important, is also simulated for α =0.66, γ =0.545 and input pulse -10 dB as shown in Fig.7. The current distribution curve(- ∂Φ/∂x) corresponds to the local magnetic field. The current distribution at the JTL input end(x=0) indicates the effective applied current. As time passes, the current distribution peak at the input end splits into two peaks having different initial velocities. The respective peaks, which correspond to two solitons, repulsively interact even for large separation and approach the final velocity. Finally the stable and well separated solitons propagate in the well damped JTL.

4. Resistively Coupled Josephson Transmission Line[14]

For the purpose of soliton propagation control, soliton transmission threshold properties have been investigated experimentally with regard to a resistively coupled Josephson transmission line(RCJ). The fabricated RCJ consists of two JTLs(JTL1 and JTL2) interconnected by an Au series resistor, R_C , where a flux quantization condition does not hold , as shown in Fig.8. Input voltage is fed to the termination resistor, R_t , of the JTL1. Both bias currents, $γ_1$ and $γ_2$

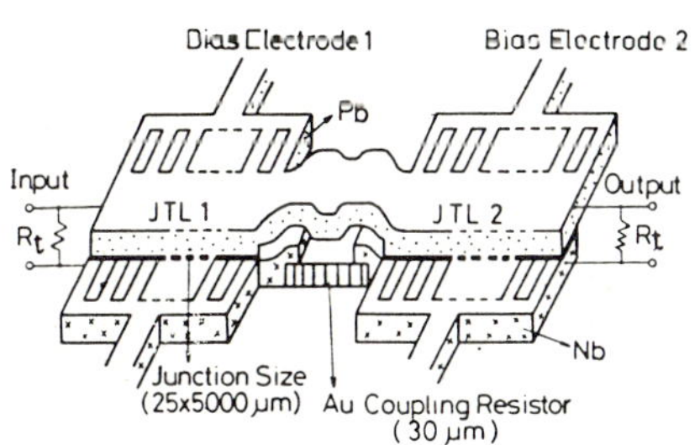

Fig.8 Fabricated structure of resistively coupled JTL.

are independently and homogeneously supplied through the action of comb-like electrodes.

It has been previously reported that the incident soliton ·, passes more easily through the coupling resistors[11],[14],[15] in the higher bias levels, because they provide a soliton with Lorentz force. However, unexpected threshold properties have been observed in the multisoliton input case. Figure 9 shows observed soliton waveform examples at the output end as a function of γ_2 with fixed input -20 dB(0 dB $= 560$ m V) and $\gamma_1 = 0.22$. An intermediate nontransmission bias region exist between high and low bias transmission regions.

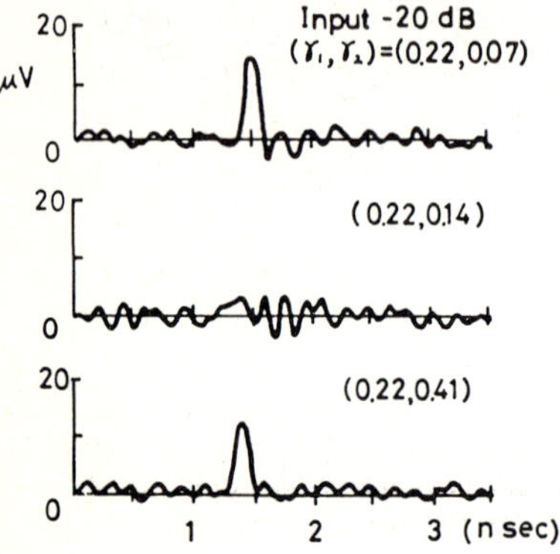

Fig.9 Observed soliton waveform examples as a function of γ_2 with fixed bias $\gamma_1 = 0.22$ and input of -20 dB.

Figure 10 shows the soliton threshold properties between the transmission and nontransmission through the coupling resistor R_C as a function of the two bias levels(γ_1 , γ_2). Open circles and triangles indicate the transmissions at -20 dB and -23 dB inputs, respectively, while filled symbols indicate the nontransmissions. To compare these experimentally obtained results with numerical simulations, the modified sine-Gordon equation(1) is numerically integrated with a real input waveform and experimentally obtained JTL parameters. In numerical simulations of the present RCJ, the termination resistor, R_t , is treated as a fitting parameter. The boundary condition at the coupling resistor can be described by

$$\frac{\partial \Phi_1}{\partial x} + \beta \frac{\partial^2 \Phi_1}{\partial x \partial t} = \frac{\partial \Phi_2}{\partial x} + \beta \frac{\partial^2 \Phi_2}{\partial x \partial t} \quad ,$$

$$\frac{\partial \Phi_2}{\partial t} = \frac{\partial \Phi_1}{\partial t} + \left\{ R_c + L_s \frac{\partial}{\partial t} \right\}\left\{ \frac{\partial \Phi_1}{\partial x} + \beta \frac{\partial^2 \Phi_1}{\partial x \partial t} \right\} \quad , \qquad (3)$$

where Φ_1 and Φ_2 indicate the phase differences on both sides of the coupling resistor, R_c , and L_s represents the stray inductance component at R_c . The rf loss coefficient is assumed to be 0.028, according to Ref.7. In Fig.10, the shadowed regions represent the numerically obtained transmission regions where R_t is assumed to be $0.67 Z_0$. The dependences on bias levels agree with the experimental results. Transmission or nontransmission can be controlled by manipulation of the bias current levels.

Figure 11 shows the numerically obtained propagation process, at $\gamma_1 = 0.11$, $\gamma_2 = 0.25$ and input -20 dB, near the coupling resistor. Two incident solitons change to damping oscillations in the JTL2 due to low bias levels of both lines. However, under a special condition where the respective damping oscillations are in phase, a new soliton can be created in the JTL2. It has been clarified that the island-like propagation region in Fig.10 corresponds to such a inphase condition.

266

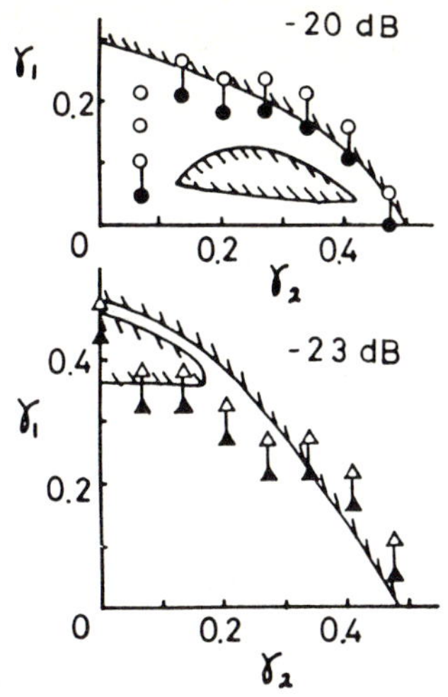

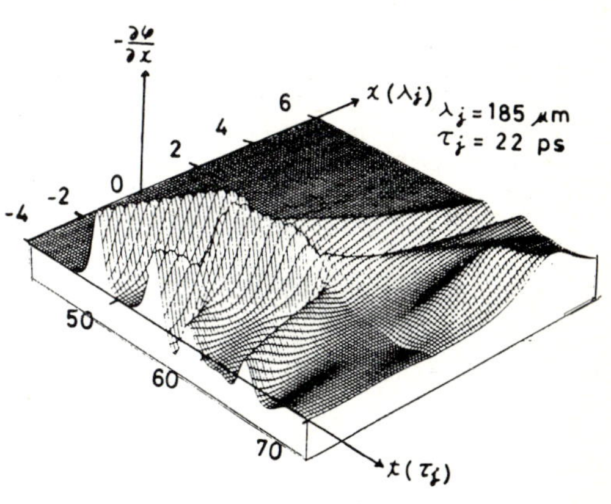

Fig.10 Transmission threshold as a function of γ_1 and γ_2. Open circles and triangles indicate transmissions. Filled symbols indicate nontransmissions.

Fig.11 Numerically obtained propagation process near R_C at γ_1=0.11, γ_2=0.25 and input of -20 dB. Position x=0 corresponds to coupling resistor R_C position.

5. Conclusions

Soliton propagation properties in the JTLs have been investigated experimentally and numerically. In the nondamped JTL, it has been observed that a single-input-pulse is dissociated into multisoliton, and propagation waveform is dependent on soliton propagation velocity. In the well damped JTL, stable and well separated multisolitons have been observed even in the high bias current level. The dependence of soliton propagation velocity on the bias current level can be interpreted in terms of the MCLAUGHLIN-SCOTT theory. In the resistively coupled JTL, whether the incident soliton transmit or not depends on the input pulse height and the bias current levels.

Acknowledgements

The authors would like to express their thanks to H. Takayanagi for his kind advice, and to Y. Kato, T. Kimura and T. Yamada for their continuous support.

Reference

[1] A.C. Scott, F.Y.F. Chu and D.W. McLaughlin: Proc. IEEE 61 1443 (1973)
[2] A.R. Bishop and T. Schneider: Soliton and Condensed Matter, Springer Series of Solid State Science 8, Springer-Verlag (1978)
[3] A.C. Scott: Nuovo Ciment 69B 241 (1970)
[4] T.A. Fulton, R.C. Dynes and P.W. Anderson: Proc. IEEE 61 28 (1973)
[5] K. Nakajima, Y. Onodera, T. Nakamura and R. Sato: J.Appl.Phys. 45 4095 (1974)
[6] A. Matsuda and S. Uehara: Appl.Phys.lett. 41 770 (1982)
[7] A. Matsuda and T. Kawakami: Phys.Rev.Lett. 51 694 (1983)
[8] A.C. Scott, F.Y.F. Chu and S.A. Reible: J.Appl.Phys. 47 3272 (1976)
[9] J. Nitta, A. Matsuda and T. Kawakami: J.Appl.Phys. 55 2758 (1984)
[10] K. Nakajima, Y. Onodera and Y. Ogawa: J.Appl.Phys. 47 1620 (1976)
[11] A. Matsuda and H. Yoshikiyo: J.Appl.Phys. 52 5727 (1981)
[12] D.W. McLaughlin and A.C. Scott: Appl.Phys.Lett. 30 545 (1977)
[13] D.W. McLaughlin and A.C. Scott: Phys.Rev. A18 1652 (1978)
[14] J. Nitta, A. Matsuda and T. Kawakami: Appl.Phys.Lett. 44 808 (1984)
[15] H.S. Newman and K.L. Davis: J.Appl.Phys. 53 7026 (1982)

A Soliton as an Attractor of a Driven Damped Nonlinear Schrödinger Equation

N. Bekki and K. Nozaki

Department of Physics, Nagoya University, Nagoya 464, Japan

1. Introduction

Recent studies have shown that some driven damped soliton systems
have a chaotic soliton exhibiting coherent spatial structure and
temporal chaos [1]. These studies indicate that a low dimensional
chaotic attractor corresponding to a chaotic soliton is embedded
in such driven damped soliton systems. When forcing and damping
effects are small enough and a forcing field consists of two oscillating components, a soliton of the nonlinear Schrödinger equation
has been shown to become a chaotic attractor described by a couple
of ordinary differential equations (obtained by means of the first
order perturbation theory) [2].

In this paper we investigate attracting solitons of a driven
damped nonlinear Schrödinger equation for the case that a forcing
term consists of a <u>single frequency oscillating field</u>. For a single
frequency oscillating force, the first order perturbation theory
gives only a phase-locked soliton corresponding to a fixed point
attractor in the space of a soliton parameter [3]. However, as the
forcing term becomes larger, it is anticipated that the first order
perturbation theory breaks down and the fixed point attractor bifurcates to a limit cycle corresponding to a soliton which has an
oscillating amplitude. Here, we show by means of a numerical method
that as the strength of the forcing field increases, the phase-locked
soliton bifurcates to an amplitude-oscillating soliton and subsequent
bifurcations lead to a chaotic solitary wave as a chaotic attractor.
The bifurcation process from an amplitude-oscillating soliton to
a chaotic solitary wave is approximately described by means of a
one-dimensional map similar to the quadratic map.

2. Bifurcations of a Soliton

We focus our attention on a bifurcation process from a coherent
soliton to a chaotic solitary wave in the following driven damped
nonlinear Schrödinger equation.

$$iq_t + q_{xx} + 2|q|^2 q = -i\gamma q - i\varepsilon e^{i\omega t} \quad , \tag{1}$$

where $\gamma(>0)$, $\omega(>0)$ and ε are real constants. If γ and ε are sufficiently small, the following phase-locked soliton is shown to be a
fixed point attractor of (1) by means of numerical and perturbation
methods [2,3].

$$q(x,t) = 2\eta \text{ sech } 2\eta x \cdot \exp\{-i(\psi+\pi/2)\} \quad , \tag{2}$$

$$4\eta^2 = \omega \ , \qquad \psi = \psi_0 - \omega t \ , \qquad \sin \psi_0 = 2\gamma\sqrt{\omega}/\pi\varepsilon \quad , \quad \text{where} \tag{3}$$

$$|\pi\varepsilon/2\gamma\sqrt{\omega}| > 1 \quad . \tag{4}$$

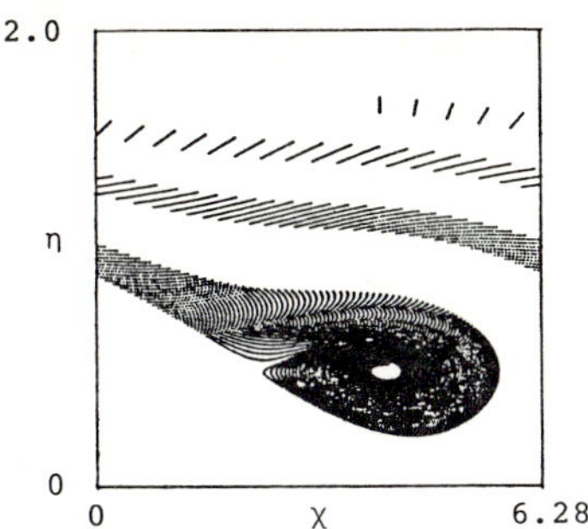
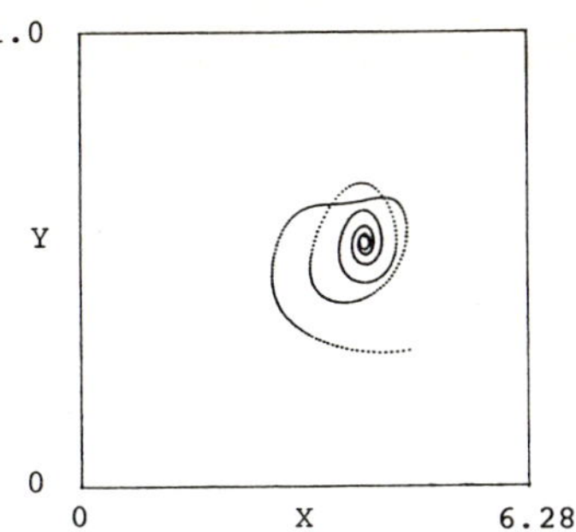

Fig.1 A basin of a fixed point attractor (χ=4.02, η=0.5), where χ=3π/2-ψ-ωt, for ε=0.1, γ=0.1 and ω=1.

Fig.2 A trajectory of a soliton attracted to the fixed point for ε=0.1, γ=0.1 and ω=1.

Figure 1 shows a basin of the attractor in the parameter space of soliton (3π/2-ψ-ωt, η) for γ=0.1, ε=0.1 and ω=1.

Our question is what happens to this fixed point attractor when ε becomes larger. In order to answer this question, (1) is solved numerically by means of the pseudo-spectral method [4]. In our numerical simulations, γ and ω are fixed (γ=0.1 and ω=1) while ε is varied. As an initial condition, we take a single soliton (2) with η=0.3 and ψ=0, which falls in the basin of the fixed point attractor (Fig.1). Of course, we have confirmed that attractors obtained here do not depend on the choice of initial conditions as long as initial profiles belong to the basin of attractors. From (3), ε must be greater than 0.064 in order that the fixed point attractor exists. For 0.064 $<\varepsilon\lesssim$ 0.105, our numerical simulations confirm the existence of the fixed point attractor (Fig.2). When ε becomes greater than 0.105,the fixed point bifurcates to a periodic cycle (representing an amplitude-oscillating soliton) in the phase space (X,Y) (Fig.3(a)), where

$$X = \arg\{q(0,t)\} - \omega t \quad ,$$
$$Y = |q(0,t)|/2 \quad . \tag{5}$$

The center position of the periodic cycle (X_C , Y_C) is approximately determined by the phase-lock condition (3) as

$$\sin(3\pi/2-X_C) \approx 2\gamma\sqrt{\omega}/\pi\varepsilon \quad , \qquad Y_C \approx \sqrt{\omega}/2 = 0.5 \quad .$$

Further increase of ε causes a sequence of period-doubling type bifurcations: a period 2 cycle exists in 0.141 $\lesssim\varepsilon\lesssim$ 0.145 (Fig.3(b)) and a period 4 cycle appears at $\varepsilon\approx$0.146 (Fig.3(c)). For $\varepsilon\gtrsim$0.148, a chaotic solitary wave emerges as shown in Fig.3(d). Figure 4 shows a profile of the amplitude of a chaotic solitary wave in the real space (x,t). As ε increases further, a period 3 cycle is found to be embedded in the chaotic region near $\varepsilon\approx$0.156 (Fig.5).

Properties of the bifurcation sequence are similar to those of the quadratic map[5]. In fact, an approximate one-dimensional map is found to be embedded in our system. Let us take successive minima of the phase defined in (5), X_1 , X_2 , X_3 , $\cdots$ and plot X_{n+1} versus X_n , then we obtain a one-dimensional map except that X_{n+1} sometimes happens to take the nearly same value as X_n (Fig.6). If we interpret two successive same X_{n+1} and X_n as one X_n , we have

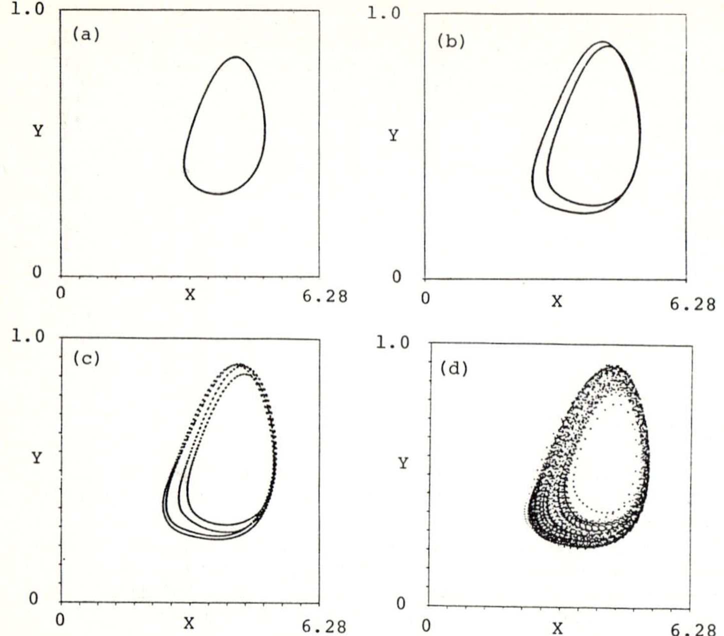

Fig.3   (a) A trajectory of a period 1 limit cycle corresponding to an amplitude-oscillating soliton for ε=0.13   (b) A period 2 limit cycle for ε=0.144   (c) A period 4 limit cycle for ε=0.1468 (d) A trajectory of a chaotic attractor for ε=0.1495.

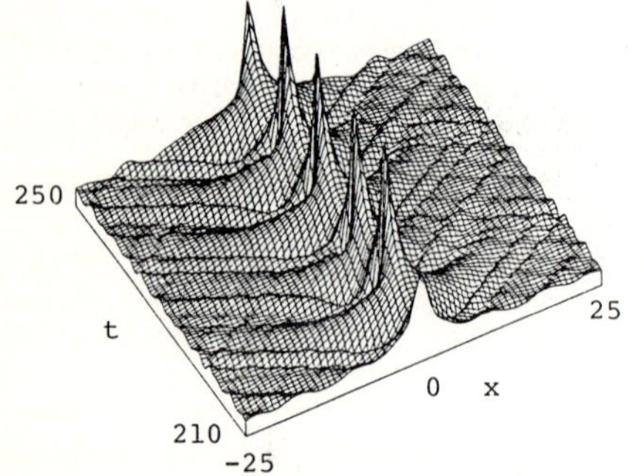

Fig.4   A profile of the amplitude of a chaotic soliton for ε=0.1495 and initial conditions η=0.3 and ψ=0 in (2).

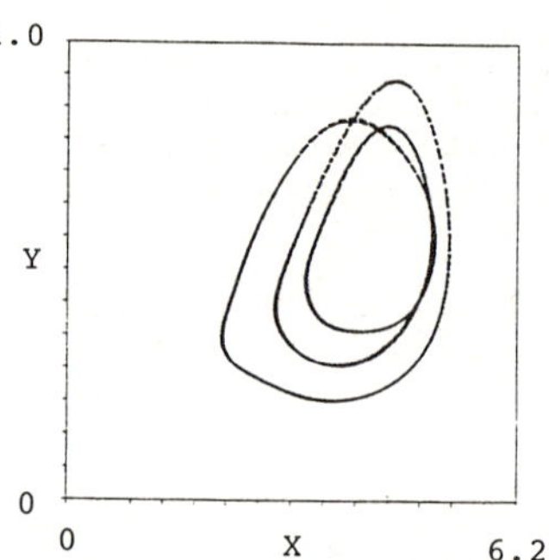

Fig.5   A period 3 limit cycle embedded in a chaotic region for ε=0.155985.

a one-dimensional map similar to the quadratic map. In such a map, the sequence of period-doubling bifurcations terminates at infinite period but at a finite value of an external parameter. Beyond this value lie regions of chaos, in which stable periodic cycles such as a period 3 obtained here are embedded. Similarities between our map and the quadratic map give us a reasonable conjucture that there exists a limiting value of ε, beyond which a chaotic soliton appears.

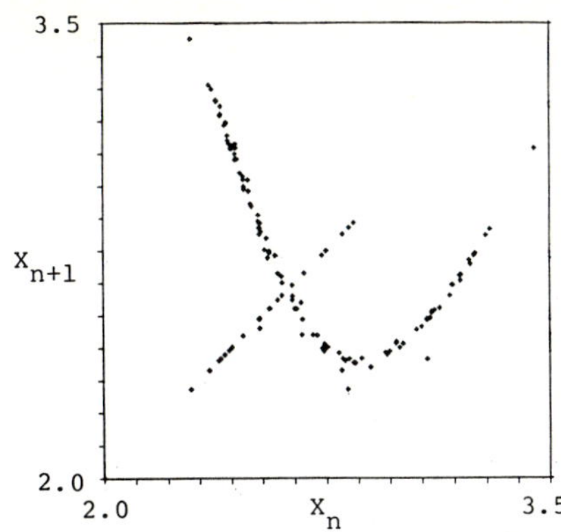

Fig.6 A plot of successive minima of phase (X_n , X_{n+1}) for a chaotic solitary wave $(\varepsilon=0.1495)$.

3. Concluding Remarks

Our numerical simulations of the driven damped nonlinear Schrödinger equation show a sequence of bifurcations from a phase-locked soliton to a chaotically oscillating solitary wave. This sequence of bifurcation can not be described by the first order soliton perturbation theory, but it is expected that a higher order soliton perturbation theory including effects of radiations may give some insight into bifurcation phenomena of a soliton studied here. For example, if contributions from higher order perturbations can be replaced by another oscillating force with a frequency different from ω, we can show by means of a soliton perturbation theory that the fixed point attractor turns into a limit cycle similar to that obtained here and subsequent period-doubling type bifurcations lead to a chaotic attractor.

Although the whole picture of a basin of attractors, as obtained here, is not yet known, both a single soliton and two solitons are attracted to the chaotic and periodic attractors for some initial values of soliton parameters.

The authors thank Dr. Y. Kaneda for help in making our numerical code.

References

1. K. Nozaki and N. Bekki: Phys. Rev. Lett. <u>50</u>, 1226 (1983);
   A.R. Bishop, K. Fesser, P.S. Lomdahl, W.C. Kerr, M.B. Williams
   and S.E. Trullinger: Phys. Rev. Lett. <u>50</u>, 1095 (1983);
   D.W. McLaughlin, J.V. Moloney and A.C. Newell: Phys. Rev. Lett.
   <u>51</u>, 75 (1983).
2. K. Nozaki and N. Bekki: Phys. Lett. <u>102A</u>, 383 (1984).
3. D.J. Kaup and A.C. Newell: Phys. Rev. <u>B18</u>, 5162 (1978).
4. D.G. Fox and S.A. Orszag: J. Comp. Phys. <u>11</u>, 612 (1973);
   B. Fornberg and G.B. Whitham: Phil. Trans. Roy. Soc. <u>289</u>, 373 (1978).
5. For example, A.J. Lichtenberg and M.A. Lieberman: Regular and
   Stochastic Motion (Springer-Verlag, Berlin, 1903), P.397.

Kinks and Spatially Complex Behavior in One-Dimensional Coupled Map Lattices

Kunihiko Kaneko

Department of Physics, Faculty of Science, University of Tokyo, Hongo, Bunkyo-ku
Tokyo 113, Japan

1. Introduction and Models

Recent studies on low-dimensional dynamical systems have had great success
in elucidating the onset of turbulence and various aspects of chaos [1].   The
success, however, is limited to systems with a few number of excited modes. Then,
what happens in a system with a large number of excited modes and with a spatial
complexity? Can the low-dimensional chaos be an elementary process for the
turbulence? Furthermore, spatial patterns are important for the understanding of
turbulence. How are the patterns in nonlinear systems characterized? What is the
effect of spatial patterns on the onset of chaos?

On the other hand, solitons have played an important role in the nonlinear
field theory. Integrable systems, however, are very rare (perhaps nongeneric) and
it will be important   to extend the notion of "soliton" as a topological
excitation to nonintegrable systems. One well-known excitation in one-
dimensional systems is a "kink", which is important not only in integrable systems
but also in nonintegrable ones such as a ϕ^4-system. Here, we extend the notion of
kinks to dissipative systems and study the effect of chaos on the stability of
kink-antikink states and the bifurcation of such states to chaos.[7]

Recently there have been some interesting works on "soliton and chaos"[2-5].
From the very start our study is quite different from them.   We use the
"coupled map lattice", not a partial differential equation, as a model [7-9]. The
coupled map lattice is given by

$$(\text{I}) \quad x_{n+1}(i)=f(x_n(i))+D((x_n(i+1)+x_n(i-1))/2-x_n(i))$$

$$(\text{II}) \quad x_{n+1}(i)=f(x_n(i))+D((f(x_n(i+1))+f(x_n(i-1)))/2-f(x_n(i)))$$

where i=1,2,..,N denote one-dimensional lattice sites with a periodic boundary
condition ($x_n(N+1)=x_n(1)$). The function f(x) is chosen to be $1-Ax^2$ (i.e., logistic
map). Since the bifurcation sequences and properties of attractors have been
extensively investigated for the logistic map, it is useful to take the coupled
logistic lattice (CLL) for the study of spatial structures of coupled chaos. See
ref.[6] for the coupled logistic map with N=2. Here we note that the model (I)
corresponds to the "previous coupling" (i.e., the coupling is given by the values
$x_n(i)$'s at the previous step n), while the model (II) corresponds to the "future
coupling" (i.e., it is given by the values $f(x_n(i))$'s which are the values of
$x_{n+1}(i)$'s (at the future step) in the absence of the coupling D).

2. Period-doublings of kink-antikink structures

One typical pattern of CLL ((I) and (II) with $f(x)=1-Ax^2$) appears for the
parameter region A where the period-doubling bifurcations proceed for the logistic
map [10]. The pattern consists of flat regions and domain boundaries, which may be
called kinks or antikinks. The attractor is a cycle with a period 2^n (note that
the values x(i)'s for the kink positions also take the same values after the 2^n
iterations). In a domain, x(i)'s take almost the same values as the values for
the stable cycle with the period 2^n for a single logistic map. The phase of the

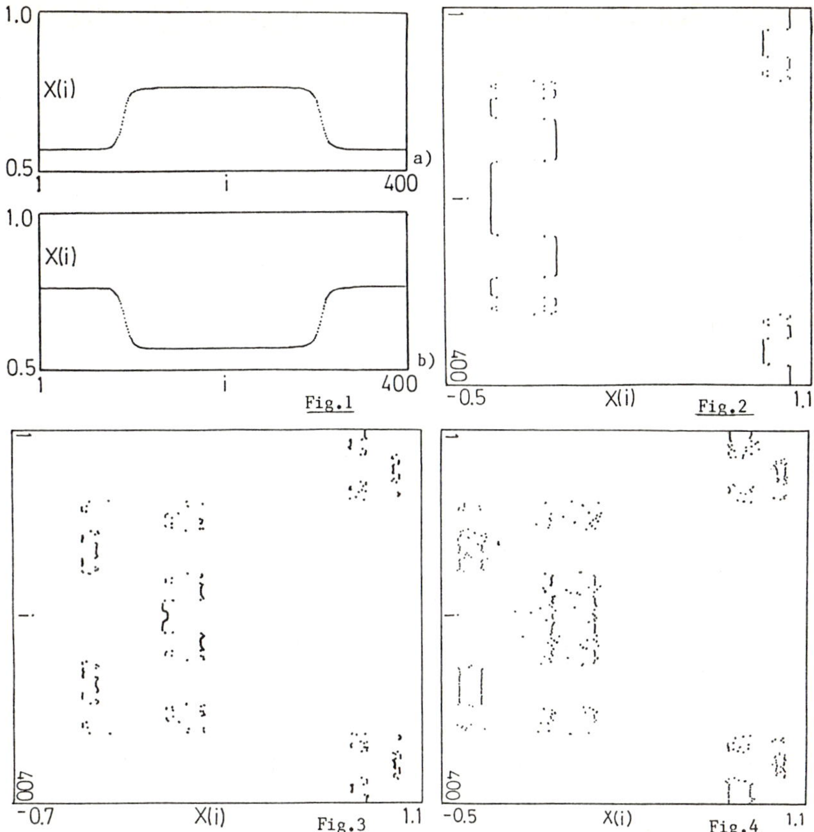

Figs. 1-4: Snapshot of the CLL (II) with the initial condition $x_0(i)=\sin(\pi i/N)$ with N=400 and D=0.05, after the transients are discarded. The value A is chosen to be 0.755 (Fig.1), 1.3 (Fig.2), 1.395(Fig.3), and 1.43(Fig.4). The period of the attractor is 2 for Fig.1 (the pattern repeats a and b alternatively), 4(Fig.2), 16(Fig.3), while the attractor is chaos for Fig.4.

2^n-cycle, however, differs by domains. An example of the kink pattern is shown in Figs. 1a)b), where the period is 2. As A is increased, period-doubling bifurcations occur (see Fig.2 for a 4-cycle). The following points should be noted:

I) The width of a kink decreases as A is increased. For example, the width is 42 (for A=0.752), 27 (A=0.755), 18 (A=0.76), 8 (A=0.8), and 4 (A=0.9) for the model (II) with N=200 with the initial condition $x_0(i)=\sin(\pi i/N)$. (Here we regard that a site belongs to a kink if the value x(i) differs by more than 1/50 from a flat region; we note that the kink structure with a period two appears at A~0.75.) The slope of the map $f^2(x)$ between x_- and x_+ (x_- and x_+ are the two stable points with a period two for the logistic map ; the flat regions take these values) becomes steeper with the increase of A-0.75, which destabilizes the kinks with a long width. (The slope of the map corresponds to U'(x) for the Hamilton systems with the potential U(x).) The width of a kink increases as the coupling is increased for the model (II).

II) Period-doubling occurs not only in "time" but also in "space", if the kink-antikink patterns exist. Thus, the period-doubling brings about the spatial structure with more complexity. Small kink-antikink structures disappear by the decrease of N or by the increase of the coupling. (See Fig. 3 for the structure with a period 16, where small kink-antikink structures cannot be seen.)

III) Many attractors coexist for the CLL with large N. They are obtained by the change of the initial condition $x_0(i)$. If the coupling vanishes, k^N attractors coexist for the parameter region where a stable cycle with a period k exists. It seems that the number of the attractors in the presence of the coupling is at least of the same order. Some of the attractors are characterized by the position of kinks and the width of the domains. The increase of the number of the attractors by the period-doubling may imply that our dynamical system can be an information source by the change of the parameter.

IV) As A is increased further after the period-doubling bifurcations, the successive band mergings occur for the single logistic map, which appear as the "domain mergings" in our model. See Fig. 4 for the 4-band structures. The domain regions still exist stably. In a domain, the motion is chaotic, but the domain boundaries do not move by the iterations.

3. Transition from torus to chaos in alternating structures

Another interesting pattern in CLL is an "antiferro-like" structure, which has already appeared in Fig.4. The structure is an alternating one, i.e., the structure with a wavelength two and appears by the "zigzag" instability. An example is given in Fig. 5, where the even (odd) sites are denoted by $x(\bullet)$. The attractor for the figure is a cycle with a period two. We note that there can be kinks in the antiferro-like structure as can be seen in Fig. 5, which is rather analogous to solitons in polyacetylene [11]. As the nonlinearity A is increased, a Hopf bifurcation occurs and a torus appears (see Fig. 6 for the pattern) and the transition from torus to chaos proceeds [12]. We note that kinks continue to exist even after the transition from torus to chaos. By taking a suitable initial condition, we can change the number of kinks. Kinks cause the modulation of the phase of a torus, which accelerates the onset of chaos.

The zigzag instability appears in a wide range of parameters (which appears more easily in the model (I) than (II)) and the alternating structure is frequently seen, though the spatial structure is chaotically modulated in many cases (see e.g., Fig. 4).

4. Spatio-temporal Intermittency

Another interesting pattern in CLL is spatio-temporal intermittency. As an example, we choose the parameter A slightly larger than 1.75, where a stable

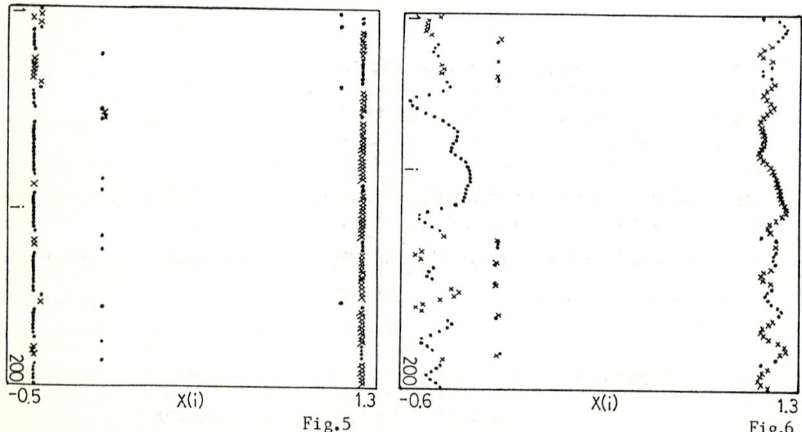

Figs. 5-6: Snapshot of the CLL (I) with the initial condition $x_0(i)=\sin(2\pi i/N)$ with N=200, after the transients are discarded. D=0.2. A=0.7 for Fig. 5 (cycle with period two) and A=0.8 for Fig. 6 (torus). Even sites are denoted by x, while odd ones are shown by $\bullet$.

274

three-cycle appears from chaos through intermittency [13]. The three-cycle is given by $x_1^* \sim 1$, $x_2^* = f(x_1^*) \sim -0.75$, and $x_3^* = f(x_2^*) \sim 0.01$. The dynamics of CLL consists of the burst regions and the laminar regions (i.e., $x \sim x_1^*, x_2^*,$ or x_3^*) where the neighboring sites take almost the same values (i.e., $|x(i+1)-x(i)|$ is small). Since the three-cycle is stable for D=0, the kink-antikink pattern in §2 might be expected. The stability of the pattern, however, is very small. We have not observed a kink with a finite width for $A \sim 1.752$. This is explained as follows: In the case in §2, topological chaos does not exist for $A < 1.4011\cdots$ for the map $x_{n+1}=f(x_n)$, while it exists in the present case. Thus an orbit of the map $x_{n+1}=f(x_n)$ with x far from x_1^*, x_2^*, and x_3^* shows a chaotic transient. Therefore, x(i)'s which deviate largely from x_1^*, x_2^*, and x_3^* are hard to remain stably at their positions and they bring about the bursts, which propagate to other sites. We also note that $|x(i+1)-x(i)|$ cannot be small in the burst, since the sensitive dependence to initial condition in the dynamics $x_{n+1}=f(x_n)$ due to the existence of the topological chaos makes the distance $|x(i+1)-x(i)|$ large even if it is initially small. To sum up; **kinks are unstable if the topological chaos exists in the "one-body" mapping $x_{n+1}=f(x_n)$ and they bring about the propagating bursts.**

To study the bursts from kinks in more detail, we choose the initial condition $x_0(i)=0.0013(\sim x_3^*)$ for $1 < i \leq N/2$ and $-0.75(\sim x_2^*)$ for $N/2 < i \leq N$ and study how the bursts at i=N/2 and N propagate. Here, the model (II) is chosen, though the similar phenomena have been observed also for the model (I). (Quantitative details are different between the two models.) We regard x(i) as a "burst" if $|x(i+1)-x(i)| > 0.1$, while it is regarded as laminar otherwise. In Fig. 7, the propagation of the bursts is shown, where the dot is plotted for the sites with a burst. Roughly speaking, the speed of the propagation decays by $(D-D_c)^a$ with $a=0.75\pm0.1$ as the coupling D is decreased ($D_c \sim 0.00105$ for A=1.752). For $D' < D < D_c$, the two bursts collide without spreading and a homogeneous state with period three appears after the iterations $n > n_c$ (n_c increases as $D \rightarrow D_c-0$ or $D \rightarrow D'+0$). For $D < D'$ the initial steps at i=N/2 and N do not move and the initial pattern does not change. That is, the coupling can be regarded as zero for $D < D'$ and a direct product state of logistic maps remains.

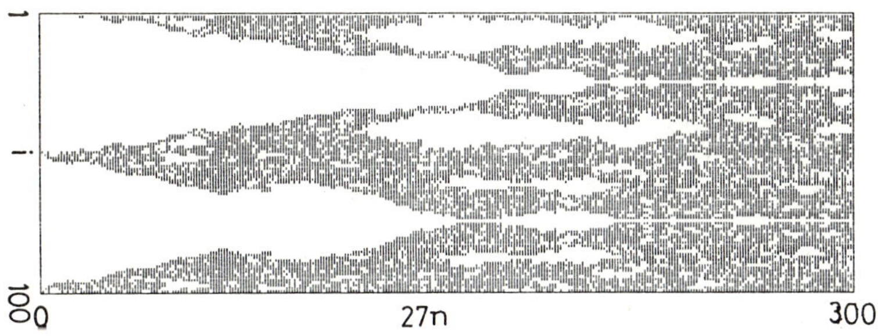

Fig. 7: Bursts in CLL (II) with N=100 , A=1.752, and D=0.0012. The dots are plotted if $|x_{27n}(i+1)-x_{27n}(i)| > 0.1$, for $n=1,2, \cdots,300$.

In fact, a unique parameter value D_c cannot be defined which distinguishes the two regions. Near $D \sim D_c$, there is a fine stripe structure in the parameter space D. For example, at $D=D_1$, a homogeneous state is attained, while at $D=D_2 < D_1$, bursts exist, and at $D=D_3 < D_2$, a homogeneous state is again attained, and so on. Since the basin structure is fractal in general if the topological chaos exists [6,14], the above structure is not so surprising. The parameter region where the structure is observed, however, is rather small and we can roughly define D_c.

The distribution P(j) about the size of the laminar region (j denotes the length of a sequence of a laminar state in time or in space) decays exponentially about j. At $D \sim D_c$, the decay rate of the exponential is rather small and the pattern of laminar and burst states seems to show a self-similar structure (see

275

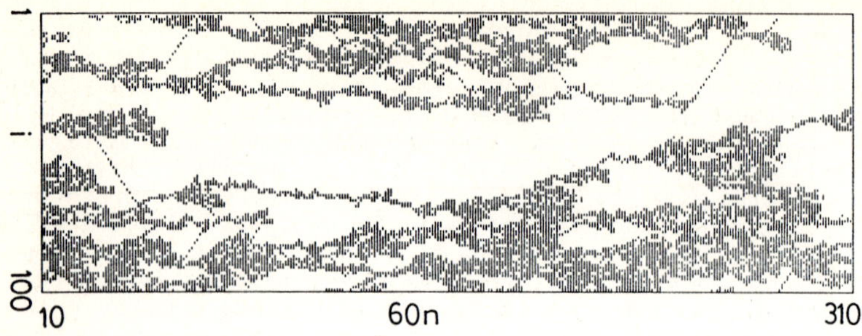

Fig. 8: Bursts in CLL (II) with N=100 , A=1.752, and D=0.00105 with the initial condition $x_0(i)=0.2\sin(2\pi i/N)$. The dots are plotted if $|x_{60n}(i+1)-x_{60n}(i)|>0.1$, for n=10,11,···,310.

Fig. 8) The beautiful self-similar patterns in cellular automata have recently been observed by S. Wolfram [15].

In the present section we have used a reduction to {0,1} with the choice of 0=laminar and 1=burst. We may construct a probabilistic cellular automaton (PCA) as a model for the spatio-temporal intermittency. Here, the probability is introduced through the chaotic motion of $x_{n+1}=f(x_n)$ and the coarse graining to the two values. We studied a PCA with a rule which may model the above phenomena. The pattern in the PCA is rather analogous to Figs. 7 and 8. As a parameter in the rule of the PCA is changed, the speed of the propagation of the burst is decreased, till the bursts die out in the process of steps. The coarse graining from a coupled map lattice to a PCA is an interesting problem of statistical mechanics.

5.Summary and discussions

In the present paper we have studied the three typical patterns of CLL with the emphasis on the kink-antikink patterns. We have not yet observed the diffusion or the propagation of a kink in CLL if it stably exists, but it will be of importance to study such phenomena in general nonlinear systems. If the chaos is localized in a domain as is the case in our model, the dynamics of the system may be reduced to the motion of a kink and the local chaos. Such reduction will be a relevant method of statistical mechanics.

Another method of a reduction is given in §4, where the reduction to "burst" and "laminar" is performed. The Figures 7 and 8 look quite similar to the evolution of a perturbation in a turbulent flow [16]. Will the spatio-temporal intermittency in coupled map lattices be relevant to understand the intermittency in turbulence?

It may be also useful to introduce the "pattern entropy" to characterize quantitatively the complexity of the various patterns in our model. Using a reduction to symbolic dynamics with a finite state, an entropy in a one-dimensional space can be defined in a manner similar to the usual entropy for a time series [17]. Detailed studies will be left to the future.

Another interesting coupled map lattice is "coupled circle lattice" (CCL), i.e., the model (II) with $f(x)=x+A\sin(2\pi x)+D$. The case with N=2 has recently been studied in connection with the stability of a three-torus [18,19]. The CCL will be relevant to study the entrainment and turbulence of the phase motions of coupled oscillators and to check the stability of a high-dimensional torus.

Acknowledgement

The author would like to thank the Japan Society for the Promotion of Science for financial support. He would also like to thank Prof. P. Grassberger for useful correspondence and to thank LICEPP for the facility of FACOM M190.

References

1. See for recent advances, **Chaos and Statistical Methods**(ed. Y. Kuramoto, Springer, 1984; Proceeding of the previous Kyoto Summer Institute)
2. J. C. Eilbeck, P. S. Lomdahl and A. C. Newell, Phys Lett.**87A** (1981) 1
3. D. Bennet, A. R. Bishop and S. E. Trullinger, Z. Phys. **B47** (1982) 265
4. K. Nozaki, Phys Rev. Lett. **49** (1982) 1883; K. Nozaki and N. Bekki, Phys. Lett. **102A** (1984) 383
5. M. Imada, J. Phys. Soc. Jpn. **52** (1983) 1946
6. K. Kaneko, Prog. Theor. Phys. **69**(1983) 1427; see also J. M. Yuan, M. Tung, D.H. Feng, L.M. Naruducci, Phys. Rev. **A28** (1983) 1662;T. Hogg and B.A. Huberman, Phys. Rev. **29A** (1984) 275; see also for the case with $N \lesssim 6$, I. Waller and R. Kapral, preprint (1984) submitted to Phys Rev. A
7. K. Kaneko, Prog. Theor. Phys. **72** (1984) No. 3
8. T. Yamada and H. Fujisaka, Prog. Theor. Phys. **70** (1983) 1240 and preprint (1984) to appear in Prog. Theor. Phys.
9. Y. Aizawa, preprint (1984) Kyoto Univ.
10. M.J. Feigenbaum, J. Stat. Phys. **19** (1978) 25, **21** (1979) 669
11. W.P. Su, J.R. Schrieffer and A.J. Heeger, Phys. Rev. Lett. **42**, (1979) 1698
12. A summary of the recent studies on the transition from torus to chaos can be seen in K. Kaneko, Ph. D. Thesis, Univ. of Tokyo, 1983 (unpublished)
13. Y. Pomeau and P. Manneville, Comm. Math. Phys. **74** (1980) 189; see also J.E. Hirsch, B.A. Huberman and D.J. Scalapino, Phys. Rev. **A25** (1982) 519
14. S. Takesue and K. Kaneko, Prog. Theor. Phys.**72** (1984) 35
15. S. Wolfram, Rev. Mod. Phys. **55** (1983) 601; P. Grassberger, F. Krause, and T. Twer, J. Phys. **A17** (1984) L105
16. See also R. J. Deissler, Phys. Lett. **100A** (1984) 451
17. V.M. Alekseev and M.V. Yakobson, Phys. Rep. **75** (1981) 287; see also S. Wolfram, Physica **10D** (1984) 1
18. C. Grebogi, E. Ott and J. A. Yorke, Phys. Rev. Lett. **51** (1983) 339 and preprint (1984)
19. K. Kaneko, Prog. Theor. Phys. **71** (1984) 282 and to appear in **Theory of Dynamical Systems and Its Applications to Nonlinear Problems** (ed. H. Kawakami, World Sci. Co. Pub.)

Photograph of the Participants of the Seminar

1. Wada, Y.
2. Ono, Y.
3. Kawasaki, T.
4. Ito, H.
5. Ogata, M.
6. Inoue, K.
7. Tu, G.-Z.
8. Toda, M.
9. Lewis, Z.V.
10. Nagamachi, S.
11. Inawashiro, S.
12. Katsura, S.
13. Suzuki, M.
14. Tagami, Y.
15. Sakuma, T.
16. Scott, A.C.
17. Elgin, J.N.
18. Mertens, F.G.
19. Takeno, S.
20. Yu, L.
21. Kawabata, C.
22. Nakamura, A.
23. Murase, M.
24. Ichikawa, Y.-H.
25. Yokota, M.
26. Tsuzuki, T.
27. Matsukawa, H.
28. Yoshida, H.
29. Watanabe, S.
30. Terai, A.
31. Suzuki, Y.Y.
32. de Veça, H.J.
33. Jackson, E.A.
34. Hirota, R.
35. Wadati, M.
36. Ohkuma, K.
37. Takayama, H.
38. Sasaki, K.
39. Kawahara, T.
40. Satsuma, J.
41. Nakayama, T.
42. Ishikawa, M.
43. Nozaki, K.
44. Yoshida, H.
45. Kawamoto, S.
46. Ozawa, K.
47. Sasai, M.
48. Ohfuti, Y.
49. Sogo, K.
50. Kaup, D.J.
51. Christiansen, P.L.
52. Campbell, D.K.
53. Kimura, Y.
54. Gibbon, J.D.
55. Yomosa, S.
56. Gibbons, J.
57. Theodorakopoulos, N
58. Nakano, H.
59. Homma, S.
60. Martinez, L.
61. Olmedilla, E.
62. Bishop, A.R.
63. Ishimori, Y.
64. Kawata, T.
65. Hirayama, M.
66. Nishiyama, S.
67. Sasaki, R.
68. Lakshmanan, M.
69. Tajiri, M.
70. Kuniba, A.
71. Nakamura, K.
72. Nakajima, K.
73. Nitta, J.
74. Ozaki, M.
75. Mizobuchi, Y.
76. Honda, K.
77. Yamashita, M.
78. Toyoki, H.
79. Horibe, M.
80. Takano, K.
81. Sabata, H.
82. Matsumoto, K.
83. Oishi, S.
84. Harada, H.
85. Konno, K.
86. Nishikawa, I.
87. Ikeda, R.
88. Matsumoto, S.
89. Uchinami, M.
90. Ito, M.
91. Fukutome, H.
92. Kako, F.
93. Kaneko, K.
94. Kawamura, H.
95. Konishi, T.
96. Kato, H.
97. Miyashita, T.
98. Onodera, Y.
99. Adachi, S.
100. Kato, M.
101. Sumide, T.
102. Takizawa, E.I.
103. Honda, K.
104. Saitoh, N.
105. Yoneyama, T.
106. Hioki, H.
107. Tsuru, H.

Index of Contributors

Solitons and Condensed Matter Physics

Proceedings of the Symposium on Nonlinear (Soliton) Structure and Dynamics in Condensed Matter, Oxford, England, June 27–29, 1978

Editors: **A. R. Bishop, T. Schneider**
Revised 2nd printing. 1981. 120 figures. XI, 342 pages. (Springer Series in Solid-State Sciences, Volume 8) ISBN 3-540-09138-6

"… It is of an unusually high quality. The contributing authors are clearly authorities in their fields, and try to convey new information and insight in the psychology and sociology of solitons instead of repeating a welltrodden script under the milky gaze of a sleep audience… one has the feeling that the authors tried to put their own specialty in the language or the context of condensed matter physics, and that (with some homework by the reader) genuine transmission of information has been achieved.
This is sufficiently rare in proceedings of meetings as to be commended, and this is a good book to have in one's departmental library." *Contemporary Physics*

Springer-Verlag
Berlin
Heidelberg
New York
Tokyo

Solitons

Editors: **R. K. Bullough, P. F. Caudrey**
With contributions by numerous experts
1980. 20 figures. XVIII, 389 pages. (Topics in Current Physics, Volume 17)
ISBN 3-540-09962-X